SECOND EDITION

WORLDS
Apart

6l⁹⁵

TITLES OF RELATED INTEREST FROM PINE FORGE PRESS

SECOND EDITION

WORLDS
Apart

Social Inequalities in a Global Economy

Scott Sernau

Indiana University South Bend

PINE FORGE PRESS
An Imprint of Sage Publications, Inc.
Thousand Oaks • London • New Delhi

For information:

Pine Forge Press
An imprint of Sage Publications, Inc.
2455 Teller Road
Thousand Oaks, California 91320
E-mail: order@sagepub.com

Sage Publications Ltd.
1 Oliver's Yard
55 City Road
London EC1Y 1SP
United Kingdom

Sage Publications India Pvt. Ltd.
B-42, Panchsheel Enclave
Post Box 4109
New Delhi 110 017 India

Printed in the United States of America

Library of Congress Cataloging-in-Publication Data

Sernau, Scott.
Worlds apart: Social inequalities in a global economy / Scott Sernau.— 2nd ed.
     p. cm.
Includes bibliographical references and index.
ISBN 1-4129-1524-4 (pbk.)
   1. Social stratification. 2. Equality. I. Title.
HM821.S47 2006
305—dc22

                                              2005003218

This book is printed on acid-free paper.

05   06   07   10   9   8   7   6   5   4   3   2   1

Acquiring Editor:	Benjamin Penner
Production Editor:	Diana E. Axelsen
Copy Editor:	Judy Selhorst
Typesetter:	C&M Digitals (P) Ltd.
Indexer:	Mary Mortensen
Cover Designer:	Michelle Lee Kenny

Contents

Preface

Social stratification and inequality have remained at the core of sociological thinking from the classical theorists on through the work of current scholars, who are demonstrating new interest in issues of race, class, and gender. Yet the concept of stratification itself can be a challenging one to teach and to study. Students are often more interested in learning about the particular aspects of inequality that they see affecting themselves than they are in examining the whole structure of social inequality. Students who have never been encouraged to think of their own experiences in terms of social class and social structure may approach the whole topic with apathy. This is not to blame students—the failure to think in terms of class is a problem deeply rooted in our society. Students may also face a course on social stratification with a certain dread: Those who are math-phobic may worry about too many statistics, and those from relatively privileged backgrounds may worry that they will be the subject of finger-pointing by "radical" professors. Although I have always tried to connect the course I teach about inequality to the lived experiences of my students and their communities, I admit that I have probably also assigned readings that have often contributed to both apathy and angst on the part of students.

At the same time that I've been teaching courses on inequality over the past decade, I have also had the privilege of editing the American Sociological Association's syllabus and instructional materials collection for inequality and stratification, and I have organized workshops on teaching courses in this subject matter at the annual meetings of various professional societies. In attending these workshops, I have realized that although instructors are often passionate about the topic, they have their own angst in teaching it. They want students to understand the foundations of classical theory in a way that actually illuminates their current studies; they don't want students to see those foundations as just the work of "old, dead Germans." Instructors want to incorporate exciting new material on race, class, and gender while still giving students a solid grounding in the core concepts. They are often eager to include material on the globalized economy while still helping students to understand changes in their own communities. And above all, they are struggling to find ways to help students see the relevance—even the urgency—of this material to the society we are currently making and remaking. Their plea has been for materials that are organized but not pat, hard-hitting but not preachy; they are looking for ways to help students both care deeply and think deeply about the topic.

This book is an effort to answer that plea. The language and the examples I use here are straight from current headlines and everyday experience—straightforward without oversimplifying difficult issues. The classical theorists get their say, not just in a perfunctory overview at the beginning but throughout the entire book, as their ideas give foundation to current topics. At the same time, discussion of the divides of race and gender is not just appended to the chapters but integrated into the analysis and the narrative so that students can begin to grasp how differing dimensions of inequality interrelate. Likewise, the theme of global change and the globalization of our times is integral to each chapter. Rather than tack some comparative material onto the end of each chapter, I place the U.S. experience in a global context throughout. In my teaching I have found that the way to help students see the relevance and importance of global material is to link it directly to their own lived experience, and I have brought that approach to this book.

This is not a book by committee, and I have not tried to make it sound like one. I occasionally relate personal experiences (they are, as one speaker noted, the only kind I have) and close-to-home examples. My hope is that students in turn will be able to relate the material to their own lives and communities and the changes they are witnessing in both.

The first two chapters explore the background to a sociological study of inequality: the lively debate that has swirled around the topic since the very first civilizations and the emerging global economy that provides the context for understanding a society's struggles with poverty and inequality. Chapter 3 gives expanded attention to the intersection of race, class, and gender—along with the related dimensions of age, sexuality, ethnicity, and religion—as a way to provoke thoughtful reflection on how these are intertwined in our social world. It presents students with a challenge to think systematically, maybe for the first time, about how social inequalities of class, race, and gender have affected who they are and what Max Weber would have called their life chances. Chapters 4, 5, and 6 explore how class, race, and gender divide U.S. and global social structure. These three chapters are followed by two that round out Max Weber's analysis of the dimensions of inequality: Chapter 7 addresses prestige and lifestyle, and Chapter 8 discusses political power. These chapters bring the ideas of Weber, Thorstein Veblen, and C. Wright Mills to life with current examples of changing lifestyles and patterns of consumption as well as debates about such things as campaign finance reform. The chapters in Part III look at the challenges posed by inequality: education and mobility, poverty and place, public policy, and the role of social movements. These chapters examine the classic studies of mobility but also the current debates on educational reform; the realities of urban, suburban, and rural poverty; the challenges of public policy, from the New Deal to welfare reform and beyond; and the struggles of both old and new social movements. The final chapter, on social movements, is both a call to understanding—linking the labor movement, the women's movement, and the civil rights movement just as previous chapters linked class, gender, and race—and a call to action. It describes new movements whose successes show that despite real societal constraints, positive action toward a more just society is possible.

The combination of critical thinking and personal involvement is carried into the "Making Connections" exercises at the end of each chapter. These provide students with links to reliable sources of further information through both the world and the World Wide Web. They also offer students options for exploring the topics discussed in the chapters in more detail, applying concepts to their own experiences, backgrounds, and local communities. These wide-ranging exercises amplify the local-global connections made in the book and give students and instructors the opportunity to deepen and extend the learning process. The message throughout this volume is that although there are no easy answers, we must not assume that there are no answers. Rather, we must accept the challenge to move on to deeper understandings and to new and better questions. My hope is that every reader finds here a challenge to move from apathy and angst to analysis and action.

Acknowledgments

The Sage/Pine Forge people have been a wonderful team and a delight to work with. Jerry Westby, senior editor, helped to craft this project in its early stages, and Ben Penner, acquisitions editor, has provided insights, enthusiasm, and encouragement throughout. It has been a pleasure to work with an editor who not only understands "market forces" but also truly grasps social forces and the important social justice issues of our day. This is a rarity, and I have enjoyed the collaboration. Diana Axelsen, production editor, cheerfully and efficiently organized everything, including last-minute updates, into a coherent whole. Theresa Sexton was a wonderful research assistant who collected and updated data. Judy Selhorst repaired my problematic prose and carefully checked sources. The book is also enriched by a collaboration with two talented young photographers, Catherine Alley and Elena Grupp, who provided photo essays set in rural Honduras, the Navajo reservation of Arizona, and the old industrial corridor of South Bend, Indiana. Their eye for the challenges and harsh realities as well as the beauty and cultural richness of struggling places is a wonderful complement to the message of the text.

Books such as this live and die at the hands of reviewers, and I've been fortunate to have had some of the best. Thanks go to those who read the early drafts of the first chapters and provided the insights to build this into a much stronger book.

For the first edition:

William L. Breedlove, College of Charleston

Jean Davison, American University

John A. Noakes, Franklin and Marshall College

Blaine Stevenson, Central Michigan University

Richard Tardanico, Florida International University

Robert Wendt, Millikin University

For the second edition:

William L. Breedlove, College of Charleston

Jon Pease, University of Maryland

Robert Wood, Rutgers University

Anna Zajicek, University of Arkansas

Joya Misra, University of Massachusetts, Amherst

Alexandra Hrycak, Reed College

Kebba Darboe, Minnesota State University, Mankato

Michaela Simpson, Western New England College

Alan Brown, University of Delaware

Christina Myers, Emporia State University

Michael Bourgeois, University of California, Santa Barbara

Their resounding enthusiasm for the book and its contributions kept me writing, while at the same time their painstaking critiques of the chapters kept me honest and constantly refining the material. The book's final form owes a great deal to their suggestions that I reorganize some of the chapters to present the material with maximum clarity as well as to highlight important issues concerning race and gender, global economic change, and social movements.

Personal thanks go to my wife, Susan, and my family for their support and understanding when deadlines approached and weekends at the beach became weekends at the computer.

PART I

Roots of Inequality

1

The Great Debate

An imbalance between rich and poor is the oldest and most fatal ailment of all republics.

—Plutarch, Greek philosopher (c. 46–120 A.D.)

Inequality, rather than want, is the cause of trouble.

—Ancient Chinese saying

The prince should try to prevent too great an inequality of wealth.

—Erasmus, Dutch scholar (1465–1536)

Consider the following questions for a moment:

Is inequality a good thing? And good for whom? This is a philosophical rather than an empirical question—not is inequality inevitable, but is it good? Some measure of inequality is almost universal; inequalities occur everywhere. Is this because inequality is inevitable, or is it just a universal hindrance (perhaps like prejudice, intolerance, ethnocentrism, and violence)?

Is inequality necessary to motivate people, or can they be motivated by other factors, such as a love of the common good or the intrinsic interest of a particular vocation? Note that not everyone, even among today's supposedly highly materialistic college students, chooses the most lucrative profession. Volunteerism seems to be gaining in importance rather than disappearing among college students and recent graduates. Except for maybe on a few truly awful days, I would not be eager to stop teaching sociology and start emptying wastebaskets at my university, even if the compensation for the two jobs were equal. What is it that motivates human beings?

Inequality by what criteria? If we seek equality, what does that mean? Do we seek equality of opportunities or equality of outcomes? Is the issue one of process? Is inequality acceptable as long as fair competition and equal access exist? In many ways, this might be the American ideal. Would you eliminate inheritance and family advantages for the sake of fairness? What would be valid criteria for equality? Would education be a criterion? Note that this implies that education is a sacrifice to be compensated and not an opportunity and privilege in its own right. Would talent be a criterion? Does it matter how talent is employed? For instance, should talented teachers be compensated as well as talented basketball players, or better? Think about this one carefully, for talent is not a completely benign criterion. Unless they are social Darwinists, most people would not want to see those with severe physical or mental limitations left destitute.

How much inequality is necessary? Should societies seek to magnify or minimize differences among individuals and groups? Is the issue of inequality a matter of degree? In such a view, the problem is not with inequality but with gross inequality. If so, should there be limits on inequality? And at which end of the spectrum? Would you propose a limit on how poor someone can be? Would you propose a limit on how rich someone can be? Rewarding individuals according to talent raises the issue of magnifying versus minimizing human differences. Currently, we tend to magnify differences greatly. It is not uncommon for the CEO of a major firm to garner 100 times the income of a factory worker in that firm. Although the CEO may be very talented and very hardworking, it is hard to image that he (or, rarely, she) is 100 times as clever, intelligent, or insightful as the workers, and he cannot work 100 times as much, as that would far exceed the number of hours in a week. Human differences are smaller than we sometimes imagine. Let's assume that we use IQ, an arguably flawed measure, as our criterion. Normal IQ ranges from about 80 (below this people are considered mentally handicapped and might need special provision) to 160 (this is well into the genius range). If everyone were to receive $500 of annual income per IQ point, then the least mentally adept workers would receive $40,000 and the handful of geniuses would receive $80,000—not much of a spread compared with the realities of modern societies. In compensation, should societies magnify or minimize human differences in ability?

The Historical Debate

The questions posed above are as current as the latest debate in the U.S. Congress and as ancient as the earliest civilization. They have dogged thinkers throughout the entirety of human history—that is, as long as we have been committing thought to writing and as long as we have had sharply stratified societies. Some of the earliest writings that have survived consist of rules of order and justice. Attempts to bring these together—that is, to answer the question of what constitutes a just social order—have been sharply divided from the beginning. In his study of the sweep of inequality across human societies, Gerhard Lenski (1966) divides the responses to this question into the "conservative thesis" and the "radical antithesis." The conservative thesis is the argument that inequality is a part of the natural or divine order of things. It cannot, indeed should not, be changed. Although this view has dominated history, it has almost from the very beginning been challenged by a counterargument, an antithesis. The radical antithesis is that equality is the natural or divine order of things; inequality, in this view, is a usurpation of privilege and should be abolished or at least greatly reduced.

Arguments from the Ancients

Some of the earliest writings that survive consist of laws, codes, and royal inscriptions. It is perhaps not surprising that most of the ancient rulers, sitting at the pinnacles of their stratified societies, were conservative on the issue of inequality. Hammurabi, king of ancient Babylon around 1750 B.C., was one of the very first to set down a code of laws, a "constitution" for his kingdom. In one sense, Hammurabi was very progressive. Rather than ruling by whim and arbitrary fiat, he set down a code of laws that specified the rights and duties of his subjects along with the penalties they faced for infractions. But Hammurabi did not consider all his subjects to be created equal. His laws differed for a "Man," essentially a title of nobility, and for the common man, who apparently did not possess full manhood status. (His laws tended to ignore women altogether, except as the property of their men.) For the same infraction, a common man might have had to pay with his life, whereas a Man would only have had to pay so many pieces of silver. Many modern American judicial reformers have noted that most of the people on prison death rows in the United States are poor, and that the wealthy can secure the best lawyers with their "pieces of silver." Corporate crimes are much more often punished with fines than with prison terms. The idea that laws apply differently to different classes of citizens is very ancient, and in this, Hammurabi and his counselors were "conservatives."

About the time that Hammurabi was formulating his laws, the Aryan invaders of India were establishing a **caste system** that formalized, and in some ways fossilized, a stratified society with fixed social positions. According to the Hindu laws of Manu, the different castes came from different parts of the body of the deity Vishnu. This image of parts of society as parts of a body would reemerge again in medieval Europe as well as in early sociological descriptions. In India, the ruling Brahmin caste was said to have come from the Great Lord Vishnu's head, whereas the lowly outcaste came from his feet. The laws of Manu stated:

But in order to protect this universe, He, the most resplendent one, assigned separate duties and occupations to those who sprang from his mouth, arms, thighs, and feet.

Thus each person is in an appropriate position according to his or her caste's divine origins—teacher, soldier, cattle herder, lowly servant—"for the sake of the prosperity of the worlds." We might note other origins of the castes as well: Those in the upper classes were largely descended from the conquerors, whereas those in the lower classes were mostly descended from the conquered.

The conservative thesis of an unchanging order of rulers and ruled, privileged and common, received one of its first recorded challenges in the writings of the Hebrew prophets. Often coming from outside the established religious system, these rough-edged oracles stood before kings and denounced not only the idolatry the rulers practiced but also their oppression of the poor.

As early as 1000 B.C., the prophet Nathan denounced King David's adultery with Bathsheba not for its sexual immorality (the king had many wives and "concubines," or sexual servants) but because it robbed a poor man of his only wife. The prophet Micah denounced the wealthy of his day in strong language:

> They covet fields and seize them,
> and houses, and take them.
> They defraud a man of his home,
> a fellow man of his inheritance.
> Therefore, the Lord says:
> I am planning disaster against this people,
> from which you cannot save yourselves.
>
> (Micah 2:2–3, New International Version)

Likewise, the book of Isaiah is filled with prophetic challenges to religious hypocrisy amid the poverty of the times:

> Yet on the day of your fasting, you do as you please
> And exploit all your workers . . .
> Is not this the kind of fasting I [the Lord] have chosen:
> to loose the chains of injustice and untie the cords of the yoke,
> to set the oppressed free and break every yoke?
> Is it not to share your food with the hungry
> and to provide the poor wanderer with shelter?
>
> (Isaiah 58:3, 6–7, New International Version)

At times the prophets were heeded, although more often they were scorned or killed. Yet their writings offer striking examples of the antiquity of the radical antithesis.

A radical contemporary of the Hebrew prophets was the Chinese philosopher Lao-tzu. We know little of this elusive man, but the *Tao-te Ching* (or *The Way*), a small book, is attributed to him; this work became the foundation of Taoism. Some of its lyrics sound surprisingly contemporary:

When the courts are decked in splendor
weeds choke the fields
and the granaries are bare
When the gentry wears embroidered robes
hiding sharpened swords
gorge themselves on fancy foods
own more than they can ever use
They are the worst of brigands
They have surely lost the way.

(Lao-tzu, 1985 translation from St. Martin's Press)

Whatever else Lao-tzu was, he was a radical. Yet Asian thinking concerning what constitutes a just social order was as divided as social thought on this subject in the Middle East and the Mediterranean. Around 500 B.C., an Indian prince named Siddhartha Gautama, in spite of all his royal privilege and training in caste ideology, became miserable as he pondered the state of humanity and the misery of the poor. He fasted and meditated until he reached the enlightenment that earned him the title of the Buddha. He taught that liberation from suffering means giving up desire and that right living means moderation in all things, caring for all things, and the giving of alms. He asserted that the highest calling is the voluntary poverty of the monk. The prince had become a radical. His conservative counterpart was a Chinese bureaucrat and adviser known to Westerners as Confucius. Confucius believed in justice, duty, and order, but his just order was extremely hierarchical. Foremost was duty to the family and respect for elders, especially elder males or patriarchs. The emperor was the ultimate patriarch, a wise father figure who did what was right but also enjoyed unquestioned authority and privilege. According to Confucius, in a good society each individual knows his or her place and does not challenge the Way of Heaven. Confucius may have shared some ideas with his elder countryman Lao-tzu, but for Confucius the divine order was fundamentally conservative.

The teachings of both Confucius and the Buddha have had tremendous influence across much of Asia. The fact that social equality has not necessarily been any more common in Buddhist societies than in Confucian societies reminds us that leaders often alter the tenets of great thinkers to suit their own purposes. At the same time, many individuals have used religious tenets to challenge the existing order and repressive power. For example, Buddhist principles have inspired followers of the 14th Dalai Lama of Tibet in his struggles against Chinese occupation as well as followers of Nobel Prize winner Daw Aung San Suu Kyi in her struggles against the repressive military rulers of Myanmar (formerly known as Burma).

A century after Confucius and Lao-tzu, a similar debate in views took place between a great teacher and his star pupil. The professor was clearly a radical, but his protégé was to become a moderate conservative. They lived in ancient Athens, a democracy that gave voice to male citizens but was clearly divided into privileged males and cloistered females, free citizens and slaves, rich and poor. Plato, the radical, looked at his Athens and saw in it the picture of all the Greek city-states, and indeed all state societies:

> For any state, however small, is in fact divided into two, one the state of the poor, the other of the rich; these are at war with one another.
>
> (*The Republic,* bk. 4, translation by Benjamin Jowett)

No more succinct and vigorous statement of class struggle would come until the time of Karl Marx. Plato had a simple but compelling theory of social inequality: Whatever their commitments as citizens to the welfare of the state, all parents tend to be partial to their own children and to give them special advantages. This allows these children to prosper and in turn pass on even greater advantage to *their* children. In time, the divides separating families become both large and fixed, resulting in a class of "noble" birth and a class of "common" birth. Plato's solution to the inequality this causes was the communal raising of children, apart from their families—a children's society of equals in which the only way individuals could excel would be through their own abilities. Plato was a communist. His ideas on forbidding family privilege must have seemed as radical in his age as the similar ideas of Marx and Engels did in the nineteenth century. They are also, however, the basis of the ideal of universal public education, which is gradually being embraced by the entire modern world. In his greatest work, *The Republic,* Plato envisioned his ideal state, one in which no inequalities exist except those based on personal talent and merit. In such a state the wisest would rule as philosopher-kings, looking after the interests of all the people. They would have great power but no great wealth or privilege; presumably, they would be so wise and altruistic that they wouldn't care about such things.

Plato never wielded much real political influence; he was probably too radical even for Athens. Yet one of his students certainly had influence. Aristotle rose from Plato's tutelage to become what medieval scholars would call the sage of the ages, serving as tutor and adviser to the empire builder of the age, Alexander the Great. But Aristotle never advised Alexander to build his empire on the model of Plato's *Republic,* for Aristotle believed in the same idea of a natural order of inequality that the Hindus and the Babylonians had before him:

It is clear that some men are by nature free and others slaves, and that for these latter slavery is both expedient and right.

("On Slavery," in *The Politics,* translation by Benjamin Jowett)

The sage of the ages was clearly a conservative. To be fair, Aristotle did not believe a society should be marked by extremes of wealth and poverty; rather, he recommended a golden mean between these extremes. For Aristotle, however, inequality was rooted in human nature. The Romans, who succeeded the Greeks in dominating the Mediterranean, built their empire on this Aristotelian view of the world, as had Alexander. Like many others, the Romans also gave their ideology of inequality a "racial" basis that could justify slavery. The influential Roman orator and counselor Cicero warned his friend Atticus: "Do not obtain your slaves from Britain because they are so stupid and so utterly incapable of being taught that they are not fit to form a part of the household of Athens."

The Christian Challenge

Roman ideals of order faced at least one memorable challenge. It came from a tradesman's son and his followers in the remote province of Galilee. When they confronted the existing social order, Jesus, his brother, James, and especially his Greek biographer, Luke, sounded quite radical. Luke records Jesus as telling his followers, "Blessed are you poor, for yours is the kingdom of God," while warning, "Woe to you that are rich, you have already received it all."

Jesus warned that it is easier for a camel to pass through the eye of a needle than for a rich man to enter the kingdom of heaven, and he told at least one wealthy man who wanted to follow him to first give all his money to the poor. Jesus was fond of reminding his listeners that God has chosen the lowest outcasts to be rich in faith, and that in a time to come those who are last will be first. As leader of the early church, his brother, James, seems to have encouraged this same approach:

Has not God chosen those who are poor . . .? But you have insulted the poor. Is it not the rich who are exploiting you? Are they not the ones who are dragging you into court?

(James 2:5–6, New International Version)

It is not surprising that Jesus and most of his early followers did not win the praise and favor of the rulers, whether political or religious, of the time. Jesus and his followers practiced communal sharing and challenged the existing order; they were radicals. At least one of Jesus' followers, however, appears to have favored a more moderate approach. Lenski (1966) calls the apostle Paul a conservative. Some of Paul's ideas on the divine order, in fact, sound quite radical. He wrote to one of his churches, "for before God there is neither Jew nor Gentile, male nor female, slave nor free." Yet Paul, a Greek-speaking Jew who was born to some privilege as a Roman citizen, encouraged his followers to accommodate and support the existing order. He told them they should pray for rulers rather than denouncing them, because rulers are God's instruments for keeping the peace. It was this Paul, the conservative, rather than the man who worked alongside women and slaves, who would come to be most cited by the established Christian church. It is perhaps not surprising that once the church became an official institution in the empire, with its own access to power and privilege, the most conservative passages of Paul's view of order—such as "Slaves obey your masters"—would become the key tenets. Still, throughout the period of early Christianity there were those, such as the Desert Fathers, who clung to the more neglected passages, such as "One cannot serve both God and wealth," and abandoned all luxury to live harsh lives in remote regions.

This tension between radical and conservative Christianity continued throughout the Middle Ages, just as the tension between radical and conservative philosophies tugged back and forth across Asia. The dominant view of medieval theology was decidedly conservative. In the twelfth century, John of Salisbury revived the image of the body, now the body of Christ, to explain social inequality: The prince is the head, the senate the heart, the soldiers and officials the hands, and the common people the feet, and so they rightfully work in the soil.

Yet throughout this time there were always opposing voices, which, although they rarely swayed powerful popes, kings, or emperors, did draw their own followings. St. Francis, born to considerable wealth in Assisi, Italy, gave away his inheritance to live a life of wandering poverty, preaching a gospel for the poor. He was beloved by poor villagers in Italy and argued for persuasion over conquest during the Crusades. The Roman Catholic Church came close to excommunicating him, but instead it eventually embraced his devotion, even if not all parts of his lifestyle. Less able to stay within the bounds of official authority, the followers of Peter Waldo lived communally in the mountains of Italy, denounced the wealth of the church, and were eventually severely repressed. They were simply too radical, not just in their lifestyle, as Francis was, but in their social demands, for the church to accept them.

Eventually, other groups broke from the Roman Catholic Church. The theology of the Protestant reformers may have seemed radical to their times, but most of their social philosophy was not. Martin Luther's call for a priesthood of all believers had radical implications that would alter northern Europe. Yet Luther welcomed the protection of German princes, and when peasants rose in revolt, Luther denounced their rage. Likewise, many of the Calvinists of the Netherlands and of Scotland were emerging middle-class entrepreneurs who would alter the social structures of their societies. Yet

Calvin, like Luther, took his cues on social order from Paul, endorsing respect for rulers and sanctioned authorities and disdaining social upheaval. Sociologist Max Weber ([1905] 1997) saw in the ethics of the Protestant reformers the beginnings of the demise of old medieval divisions between nobility and peasantry. But, Weber believed, theirs was the new spirit of capitalism that also embraced inequality—so long as it was "earned" by hard work and reinvested for more profit rather than squandered in personal excess. One group differed from this pattern, the so-called Anabaptists of what became known as the radical reformation. They rejected church hierarchies in favor of a brotherhood of believers committed to humility, simplicity, and nonviolence. Even though as pacifists the members of this group posed no threat of armed rebellion, both Roman Catholic authorities and many of the other reformers bitterly repressed them. Disputes erupted over baptism, but it may also have been that the Anabaptists' vision was simply too radical. The successors to these early radicals include the Mennonites and the Brethren as well as the simple-living Amish and the communal Hutterites. Others who have reclaimed some of the same ideals have included the Society of Friends (Quakers), the first American group to denounce slavery vigorously, and other brotherhoods and sisterhoods such as the Shakers, who exulted in communal simplicity in the now famous hymn that includes these lines:

> 'Tis the gift to be simple,
> 'Tis the gift to be free,
> 'Tis the gift to come down
> Where we ought to be.

Radical thinking reached England by the seventeenth century, also in religious context. The Levelers were so called for their desire to equalize, or "level," society. Their leaders argued that control by a landed elite was neither Godly nor English. Sang the Leveler priest John Ball:

> When Adam delved and Eve span
> Who then was the gentleman?

Gerrard Winstanley argued that social inequality had been imposed on the English by their Norman conquerors, whose descendants still oppressed the British commoner. Jailed and repressed, the group's membership declined over time, but the ideas of the Levelers influenced others. John Wesley, the founder of Methodism, preached social order and respect for authority. But he also preached to the poorest segments of society and took great interest in their welfare as well as their conversion. Other evangelical reformers came in his wake, also challenging social divisions. Among them was William Wilberforce, who led the drive to abolish slavery and British participation in the slave trade in addition to seeking reforms in prisons, debtors' prisons, and orphanages. These were reformers rather than true radicals, although they must have seemed radical to others in their times.

The Social Contract

By the eighteenth century in Europe, however, the arguments for social change tended to draw less on the Bible than on a new understanding of a social contract that included the rights of all. The emphasis was on political rather than economic reform, and so legal rights were the prime concern. John Locke, who was English, and Jean-Jacques Rousseau, who was French, argued that rulers' political authority comes from the consent of the governed rather than from divine right. These thinkers' ideas for reform ultimately had radical implications. They became the basis of the 1776 American Declaration of Independence and of the 1789 U.S. Constitution, with its Bill of Rights. They were also the foundation of the subsequent French Revolution, with its more radical cry of "Liberty, fraternity, and equality!"

Two great documents of reform were written in 1776. The first was the American Declaration, which includes Thomas Jefferson's assertion that "all Men are created equal, that they are endowed by their Creator with certain unalienable Rights, that among these are Life, Liberty, and the Pursuit of Happiness—That to secure these Rights, governments are instituted among Men, deriving their just Powers from the Consent of the Governed." It is true that the Declaration never mentions women, in the rhetorical custom of the time, and that Jefferson was attended by slaves as he wrote these sentences, although he personally wrestled with the issue of slavery and wanted to include a statement against it in the Declaration. He considered including a right to property in his list of rights but settled on the pursuit of happiness as a generally understood reference to free economic activity. In the same year, a more purely economic document came from a Scottish philosopher, Adam Smith, in his *Inquiry into the Nature and Causes of the Wealth of Nations*. Against the strong economic control wielded by kings of the time, Smith argued for unfettered free trade and commerce to meet the demands of consumers. If this was done, he asserted, the "invisible hand" of the market would balance competing individual demands to produce the greatest good for all. This idea ultimately had enormous influence, setting the basis for classical economics and what became known as Liberalism. Against a world ruled by wealth-amassing royal domains, Smith envisioned a world of free trade, free markets, and free competition among firms that is still at the heart of global capitalism. Both Jefferson and Smith believed that by limiting royal power they were setting the stage for nations of free, prosperous, and more equal citizens. Radical in their day, these ideas would be incorporated into a "reformed" conservative thesis in which companies, and ultimately corporations, instead of crowns would preserve order and the common good.

The primary emphasis on legal and political rights rather than economic rights and equity distinguished eighteenth-century thinkers from those who

Radical Antithesis	Conservative Thesis
	Code of Hammurabi 1400 BC
	Hindu castes
Hebrew Prophets 800–600 BC	
Lao-tzu 600 BC	
Buddha 500 BC	Confucius 500 BC
Plato 400 BC	Aristotle 350 BC
Jesus and James 30 AD	
	Apostle Paul 60 AD
Desert Fathers 100 AD	
	Medieval Theology (John of Salisbury) 12th c.
St. Francis of Assisi 13th c.	
Waldensians 13–14th c.	
	Luther and Calvin 16th c.
Anabaptist "radical reformers"	
Levellers (Gerrard Winstanley) 17th c.	
Locke and Rousseau 18th c.	Adam Smith 18th c.
Karl Marx 19th c.	Gaetano Mosca 19th c.
Max Weber 19–20th c.	Social Darwinism early 20th c.
Conflict Theory	Functionalism

Exhibit 1.1   The Great Debate

followed in the nineteenth century. Nineteenth-century socialists took up some of the earlier rallying cries but wanted to go beyond these "false revolutions" to a new, more sweeping revolution that would utterly change the economic foundation of society. These were the true radicals (see Exhibit 1.1). The most exacting and prolific spokespersons for this movement were Karl Marx and his collaborator, Friedrich Engels.

The Sociological Debate

Karl Marx and Class Conflict

The prolific collaboration between Marx and Engels around the middle of the nineteenth century marks the entrance of a clearly social science position into the great debate on inequality. Adam Smith laid the foundations for classical economics, but he was a philosopher who was still largely working in social philosophy. Likewise, John Locke was one of the founding thinkers in political science, but he himself was also a philosopher more interested in the exchange of ideas than in the examination of data. Marx, in contrast, although well trained in philosophy, called himself a political economist and was eager to draw on both historical-comparative and quantitative data to support his positions. The data at his disposal were not always the most accurate, but bureaucratic governments were increasingly making vital statistics available, and the vast library of the British museum was collecting the findings of investigations conducted in many disciplines. Together, these developments allowed Marx to enter the debate as a social scientist and make major contributions to political science, to economics, and, ultimately, to the emerging discipline that became known as sociology.

Marx's ideas are difficult to assess in part because of Marx's enormous influence. No other social scientist has ever come close to having his or her theories become the basis of whole societies with a combined population of more than a billion people. Herein lies the difficulty. With other theorists, it is possible to note both those elements of their work that have stood the test of time and those that have not. This is difficult in Marx's case because for much of the twentieth century, he was so honored in the communist world that his ideas could not be questioned, and he was so vilified in parts of the noncommunist world that full and fair consideration of his ideas was impossible. The ideas behind the icon, both those that were amazingly accurate and those that were clearly inaccurate, are far more interesting than the stale debate between world powers that became the Cold War. The thaw in that war of words has created new interest in Marx just as the societies whose political structures bear his name are collapsing or abandoning their attachment to his ideas. Could it be, John Cassidy asks in a 1997 *New Yorker* article, that Marx, who was singularly wrong about the prospects for socialism, could have been absolutely right about the problems of capitalism?

Marx believed that he was writing not just a history of capitalism but a history of civilization itself. Like most German philosophy students of his day, he had been greatly influenced by the philosopher Hegel, who held an interesting idea about ideas. One view of how new ideas develop is that they grow as new thinkers come along and extend and refine old ideas. Hegel's view was different. He asserted that someone puts out an idea, and then someone else as likely as not comes along and says, "No, you're wrong." Ideas are not like a growing plant; rather, they come from vigorous debate. Hegel called the debate between an assertion, or thesis, and its opposite, or antithesis, a **dialectic**, and he believed that the dialectic is the driving force in the history of ideas. A thesis is offered and becomes the dominant view until it is challenged by an antithesis. A debate ensues, and out of this comes a synthesis, a blending of ideas. Once accepted, this synthesis becomes the new thesis and the process repeats.

We have, in very Hegelian fashion, just examined a dialectic on inequality between a conservative thesis and a radical antithesis. Hegel would be pleased. Marx, however, would want to change the terms of the debate. He once wrote that he was going to turn Hegel on his head. What Marx meant was that he accepted Hegel's dialectic, the battle between opposing positions, but Marx believed that the real dialectic was not the struggle between ideas but the struggle between economic classes. In Marx's view, history is driven by material circumstances and economic relations, not by abstract ideas. Ideology, a system of ideas, directs people's behavior, but this ideology is created by the ruling classes to justify their position. In Marx's phrase, "In any age the ruling ideas are the ideas of the ruling class." People can, however, come to reject those ideas when they become aware of their oppression, or when the system itself is on the verge of collapse, and this, according to Marx, is the great dialectic. All history, Marx asserted, is the history of class struggle.

Marx looked at the tumultuous state of Europe in the midst of the Industrial Revolution (and many impending or threatening social revolutions) and contended that the basis of any society is its mode of production, the way it secures its livelihood. The concept of the mode of production has two components, one physical and one human. The physical component includes the means of production, essentially the technology of the time. Marx described the human component by using his key phrase the social relations of production, which refers to the positions of groups of people, social classes, in the economic process. These groups can take many forms, but essentially, Marx believed, there are two classes: those who control the means of production, the rulers, and those who work the means of production, the ruled. Every society needs both, but the tensions between them, the class conflict, always brings the existing societal order down to be replaced by something new. This new society has its new rulers, who need and create, "call out," a new class of the ruled. And the process repeats.

Marx called the first stage in this great struggle *primitive communism*. He drew on the sketchy anthropology of his day to envision a time when fairly equal bands and tribes existed in societies where the main social institution of production was the family. This harmonious state was destroyed by the introduction of the great evil: private property. It was Marx's collaborator and frequent coauthor, Engels, who suggested how this might have begun. Engels speculated that men began to treat their wives and children as their property. Men ruled and women served, and so the first class division was begun with property, patriarchy, and gender conflict. Some of Engels's description of this process rests on shaky anthropological ground, but nonetheless he laid a foundation for a feminist view of the origins of social inequality.

The expansion of private property and eventually private landholding created the great ancient empires, such as Plato's Greece and Cicero's Rome. These were based on new and growing divisions between town and country and between emerging social classes, but most notably between property-owning citizens and slaves. The collapse of these empires gave rise to medieval feudalism and two great classes: landowning nobles and land-working peasants. Other classes helped bolster the position of the ruling nobility: Through the church, the clergy provided the justifying ideology, and knights and soldiers provided the might of coercion for the unconvinced. Amid growing struggles between nobles and peasants, a small new class gained prominence, that of capitalist merchants. The members of this new group, whom Marx called the bourgeoisie, were radical in their destruction of the old feudal order but ultimately conservative as they came to power as the new ruling elite. The basis of their wealth was not the land but urban production. As this became urban industrial production, they had at their disposal a new means of wealth and they created a new subservient class, their workers, the proletariat. The urban industrial proletariat, factory and mill workers, were the new oppressed, with "nothing to sell but their labor."

For Marx, capitalism was a new chapter in an ancient story. It was more productive and generated more wealth than any previous societal form, but it also generated more misery. Each form of society creates its own problems and contradictions, and the mode of production of industrial capitalism is marked by its own unique aspects. These include the following:

- *Wealth accumulation:* "Accumulate! This is Moses and the Prophets to the capitalists," Marx wrote. Industrial capitalism unleashes tremendous productive power and allows for great accumulation of wealth. Marx saw capitalism as a necessary evil, something that was necessary until the world had enough productive capacity and accumulated wealth to redistribute.

- *Narrowing of the class structure:* The class structure of capitalism, like that of all the societal forms that preceded it, is more complex than a simple two-class system—owners and workers, bourgeoisie and proletariat—but the forces of capitalism eventually drive almost everyone into these two classes. Rural landowners become less important and small independent producers (petite bourgeoisie) are driven out of business by large capitalists.

- *Homogenization of labor:* Under older systems, the peasants labored apart or in family units and were slow to see their common interests. In the towns, the crafts guilds all proudly guarded their own specialties. Under industrial capitalism, workers are "deskilled," turned into highly replaceable parts of the factory production. And they are all brought together on the factory floor. These two factors, common skills and common ground, make it easy for capitalists to control the workers. Marx believed that these factors would also ultimately make it easier for workers to see their common interest and join forces to overturn the system.

- *Constant crisis of profit:* Capitalists are in an intense competitive struggle that drives them to try to increase production while cutting costs. This drives wages down to a subsistence level—that is, capitalists pay their workers no more than they must to allow them to survive and keep working.
- *Alienation:* Workers take no satisfaction in being mere cogs in a machine that is making products they cannot afford and may never even see. Factory workers are alienated from the products they make, from nature, and from their own human nature, which longs to take pride in meaningful work.

The combined effect of the aspects of capitalism described above is a great contradiction: Workers under industrial capitalism make more money than ever before but have less. As the realization of this contradiction strikes them, they are ripe for revolution. Eventually, especially if they read Marx and Engels's pamphlets, they will gain **class consciousness.** They will become a "class for themselves," realizing that they are in a struggle not against each other but ultimately against the ruling class. Capitalists can forestall this realization by trying to hide the nature of system, telling workers that they need only work harder or better to improve their lives. Capitalists can resort to coercion, using the military or the state police against the workers. But ultimately, as the capitalists become richer and fewer, and the workers become ever more numerous and ever more miserable, the system must collapse. When it does, the stage is set for the next phase: socialism, a system of collective production and just distribution that overturns the class structure. Here the prior process of history stops. Given that history is the history of class struggles, and class struggle is the force that ultimately brings down each society, it stands to reason that a classless society with no class struggle will stand forever. For Marx, true socialism is the final stage of economic history.

In the meantime, Marx encouraged his followers to work with sincere reformers wherever they could. Thus these radicals promoted practical ideas that no longer seem radical: minimum wage laws, worker safety laws, the end of the 16-hour workday and the seven-day workweek, the abolition of child labor, and the creation of unions. Marx, however, did not believe the capitalist system could be fully reformed; for Marx, capitalism is corrupt at its heart. Revolutions that change only governments without overturning the nature of the economy are ultimately false revolutions, the French and American revolutions included. Yet Marx believed that the efforts of the reformers were sincere and could be supported as first steps. Eventually it would become obvious to them that they could never tame the beast of capitalism; they would have to slay it.

Marx's grand revolution never came—at least it has not come yet. The revolutions that convulsed Europe in 1848 as Marx and Engels worked on the *Communist Manifesto* were put down by the force of repressive states.

The revolutions that would succeed in later years—in Russia, China, Cuba, and Nicaragua, among others—were all closer to old-style peasant revolts. Many of these were led by educated revolutionaries, but they occurred in largely **agrarian societies** as revolts against landlords. Marx looked for true revolution in the most advanced capitalist countries, including Germany and Great Britain, and he was particularly hopeful about the United States. What happened?

In part, the changes brought about by social reformers, sometimes with the support of Marxist socialists, alleviated the worst misery that Marx had witnessed. Gradually, the most unsafe workplace conditions were improved, workdays and workweeks were shortened, and child labor was curtailed. Unions gained growing clout. Further, Marx could not have anticipated how continually and quickly industry would make technological advances. New productive capacity allowed capitalists to cut costs without cutting wages. New technologies also required the employment of a whole new group of technicians and engineers—and, later, programmers and analysts—who had new skills to sell and could command higher wages. Even as the middle class of small, independent producers, the petite bourgeoisie, was declining, a new middle class of salaried professionals was emerging.

Although Marx was clearly aware of the importance of technological change and continually critiqued industrial capitalism, his focus was always on the social relations of capitalism rather than on the social relations of industrialism. Could it be that the mass-production process of full-scale industrialization was inherently alienating, whether it was done for capitalist owners or a socialist government? Marx was accurate in describing the plight of the workers of his day, yet in hindsight he seems to have been greatly overoptimistic about a socialist system's ability to address that plight.

Marx was clearly wrong in some of his predictions, but he has not been retired from the great debate. New generations of neo-Marxists continue to rediscover and refine his ideas. This group plays a key role in what has become known as the *conflict position* in sociology, of which Marx must clearly be seen as a founding thinker. Many in the conflict school of thought believe Marx was fundamentally right in viewing conflict in general, and class conflict in particular, as the driving force in society and social change. They differ with Marx only concerning the nature of that conflict.

Conflict theorists such as Ralf Dahrendorf contend that Marx was right about the tension in the social relations of production but wrong in seeing this tension as based solely on ownership of property. Dahrendorf (1959) asserts that the real issue is authority relations: who has the power to command and who must take the orders. Property, in this view, is only one basis of authority. A top corporate executive may have great authority even without owning a majority interest in the company. A government or military leader, even a communist bureaucrat, may have authority and use it abusively without actually owning the productive forces being commanded. Erik Olin Wright and Luca Perrone (1977) have demonstrated that Marxist class categories are good predictors of income if a third category, managers (those who have authority without property), is added to the categories of owners and workers.

Others have noted that capitalism has proven more adaptable than Marx realized it could be. Marx described the perils of competitive capitalism. Some neo-Marxist conflict theorists, such as Michael Burawoy (1979), contend that in fact what we now have is monopoly capitalism. In this system the heads of major corporations and financial institutions can coordinate their actions and control their competition to ensure profits while still offering workers enough to secure their consent. In these theorists' view, the workers are indeed consenting to their own exploitation as they work to secure bonuses and benefits, but the system goes on because these perks hide the exploitative nature of the system.

One of the most interesting extensions of Marx's thinking comes from the most famous Marxist of all, Vladimir Lenin, and Lenin's intellectual contemporary Nikolay Bukharin. Lenin ([1917] 1948) and Bukharin ([1921] 1924, [1917] 1973) contended that Marx was essentially right but only beginning to understand the full nature of global capitalism. Britain could have what Lenin called a "laboring aristocracy" of well-paid labor only because the miserable subsistence-level workers who were really supporting the system were located somewhere else, such as Calcutta, India. Capitalist exploitation had moved from the national to the international level, and the only answer was global revolution and international communism. Lenin believed that in the Russian Revolution *he* was firing the shot that would be heard around the world. Russian communism under Stalin turned inward and became nationalistic, but some in this line of thinking believe that the only true revolution must be international. Only when global capitalism is replaced by international socialism, ideally of the humane and democratic form that Marx dreamed of, will the misery and exploitation end (Wallerstein 1974). This is the foundation of the international conflict perspectives that have become known as dependency theory and world systems theory. Dependency theorists argue that poor nations are poor because they are still dependent on the First World nations, many of which were their old colonial masters. The world systems approach extends this understanding to look at the way the world operates as a single economic unit with a privileged core and an impoverished periphery.

Max Weber and Life Chances

Max Weber, a founding thinker in the emerging field of sociology at the beginning of the twentieth century, was writing in Germany at a time when the ideas of the late Karl Marx were much debated. Weber accepted many of Marx's ideas: the centrality of economics to all other human affairs, the importance of property relations in making social classes, and the importance of social conflict in creating social change. Weber, however, sought to expand and refine Marx's ideas to fit more accurately the realities he observed and analyzed. In Weber's view, a person's social class is defined by that individual's life chances in the marketplace. Ownership of property matters, but so do authority and expertise, particularly what the person can command based on these assets. The real divisions are between the powerful and the powerless, with gradations in between. Further, power is exercised in different realms: the economic realm, the social realm, and the political realm. In formulating these ideas, Weber often moved among what are now the separate disciplines of economics, sociology, and political science, respectively.

Power in the economic realm is **social class.** It is vested in possession of goods and opportunities: what one can sell in the commodity markets (investments) and what one can sell in the labor markets (skills and expertise). Weber's emphasis on the marketplace as the arena for power struggles continues to fit well with what we see in the often-contentious market-driven economy that is part of U.S. society.

Power in the social realm is status honor, or **prestige.** It is vested in respect and respectability as well as just plain showing off. According to Weber, "Classes are stratified according to their relations to the production and acquisition of goods; whereas 'status groups' are stratified according to the principles of their *consumption* of goods as represented by special 'styles of life'" (in Gerth and Mills 1946:193). Fine clothes and fine cars are a part of status honor, as are one's family background and family name, residence, and reputation. Status groups are communities in which the members recognize one another and common sets of symbols or indicators of **status.** What constitutes prestige varies greatly across communities. The distinguished sociologist who commands great respect and deference from other sociologists at a professional conference may be largely unknown and undistinguished outside of the discipline. A gang lord who commands great respect within a particular community may be reviled as a thug outside of that community. Weber's emphasis on what we now call *lifestyle* is also very contemporary and fits well with our consumption-oriented and prestige-conscious society.

Power in the legal realm is what Weber called "party." A political party is clearly a community based on gaining power through legal authority. Weber's term, however, may also be used for a labor union, a student union, a social action group, a lobbying organization, or a political action committee. Any group that is involved in struggle to use the legal realm to gain advantage and position is an example of the kinds of groups that Weber called "parties." "Parties," he wrote, "live in a house of power" (in Gerth and Mills 1946:194).

Weber acknowledged that the three realms described above are not isolated spheres; rather, they are constantly interacting. Despite this, he believed that they are distinct. The pope may command great social honor within some communities while possessing little personal wealth and limited legal authority. A political boss from a poor family background may wield great political power without having any obvious personal wealth, and perhaps may have a mixed and dubious reputation. Yet Weber acknowledged that if one of these realms is dominant, Marx was likely right in looking to social class. Command of great wealth can be used to gain prestige and buy influence if not outright power. Again, his assessment sounds very contemporary.

Whereas Marx emphasized struggles between classes that were largely fixed in place, as social classes have been over most of history, Weber was writing at a time of greater class mobility: As the Industrial Revolution matured, some former members of the working class were entering the middle classes, some in the middle class were getting fabulously rich, some in the upper classes were trying to protect their position of old wealth from the "new rich," and some people seemed to be losing ground altogether. Weber thus focused more on the up-and-down nature of social mobility. In particular, he stressed the idea of **social closure**, or monopolization. Groups that have attained positions of power, prestige, and privilege try to close off access to other groups; that is, they attempt to monopolize these positions. In a sense, power, prestige, and privilege are limited goods. If all are prestigious, then no one is *really* prestigious; if all are powerful, then no one can be *really* powerful. Against this backdrop of monopolization, outside groups are continually trying to usurp power, prestige, and privilege, trying to claim these goods for themselves and win social acceptance of their new standing (Weber [1920] 1964, [1922] 1979; see also Parkin 1979).

Whereas Marx seemed to sympathize with the struggles of the exploited, Weber wrote about the struggles of the excluded. Marx described the conflicts between owners and workers, landlords and landless, that have wracked societies and continue to divide our own. Weber anticipated the rivalries that continue to rage in our times: between political parties, factions, and points of view; between racial and ethnic groups; between conservative and liberal attitudes toward lifestyles and values. Marx and Weber agreed on this common dynamic: Social interaction is filled with conflict, social organization is built on conflict, and social change is the result of conflict. Both were conflict theorists.

Émile Durkheim and the Search for Order

Not all of the social scientists working in the emerging disciplines in the early twentieth century were convinced that the underlying issue of society is conflict. Although they recognized the reality of conflict, they were more interested in the question of how a society maintains order. Why doesn't it fly apart, becoming a battle of all against all? These theorists did not necessarily favor great inequalities, but, like Aristotle and philosophical conservatives, they believed that stratification is a part of maintaining a functioning social order. At times they expressed this in terms of that favorite analogy: society as a body with differentiated parts.

The most profound and influential early thinker in this line was the French sociologist Émile Durkheim. A contemporary of Max Weber, Durkheim was especially interested in the issue of social solidarity. How did societies first come together, and, amid the changes of urban industrial society then gripping France, how could they continue to hold together and function? A central concept for Durkheim was the **division of labor,** the way tasks are ever more likely to be divided into the domains of specialists. Simple societies, according to Durkheim ([1895] 1964), have "mechanical solidarity," the solidarity that comes from shared experience in which everyone works together on common tasks. This solidarity, which can be reinforced by religion and ritual, forms the basis of social cohesion. Modern societies have seen a shift to what Durkheim called "organic solidarity." Like the organs of the body, all persons in a society have their own specialized tasks, and each individual needs all the others for survival. We hold together as a society because we realize that few of us could make it alone; we are dependent on all the other "organs" to play their part. Durkheim was concerned with social evolution and the ways in which societies and their members cope with the changes around them. His focus on social order and the functions of social differentiation, the division of labor, became the basis of a largely conservative line of theory that was dubbed *functionalism.*

Some American theorists were even more explicit in contending that inequality is fundamental to a working society. Charles Sumner, with little of Durkheim's sophistication, seized on the growing interest in Charles Darwin's theory of evolution to stress what he termed "social Darwinism." He explained the great inequalities and social struggles that marked the Industrial Revolution as merely social examples of the survival of the fittest. The strongest, brightest, and most ambitious (some might say most ruthless) moved to the top, where they could command further progress, while the weakest and least able fell to the bottom. The actual links between Sumner's theory and Darwinian theory were thin and forced, but the approach provided a veneer of scientific-sounding explanation to the harsh realities of wealth and poverty at the turn of the century.

The growth of the social sciences brought new data and new theories to an old debate. They intensified rather than resolved this debate, however, and set the stage for the sociological debate on inequality that came into focus in the middle of the twentieth century.

Conflict and Functionalist Approaches to the Debate

The intellectual legacy of Marx and Weber, already well established in Europe, became central to American conflict sociology through the work of Ralf Dahrendorf on authority relations and the work of C. Wright Mills on changing American classes and power elites. The Durkheimian legacy became American functionalism through the extensive work of Talcott Parsons and, later, Robert

Merton. The essence of the debate between these schools of thought was captured in a midcentury exchange published in the *American Sociological Review* in which Kingsley Davis and Wilbert Moore presented the essential functionalist statement on inequality, and Melvin Tumin countered with rejoinders based in the conflict tradition. In many ways, the points made in this exchange provide a systematic outline of the conservative thesis and the radical antithesis.

Davis and Moore (1945) began with the simple assertion that to maintain a working social order, a modern society must do two things: It must place people in the division of labor, and it must motivate them to work hard in that position. They argued that social stratification does both. Differential rewards are needed to compensate those people who make sacrifices to gain an education and work to make it to the top. The competitive struggle to reach the top ensures that everyone works hard, hoping for advancement, and that the most talented should eventually garner the most powerful positions, where they can accomplish the most good. Stratification is universal, occurring in all societies, because it is necessary and inevitable, resulting from the need for a working social order. It is equitable insofar as the competition is fair, and it ultimately benefits everyone by creating the most efficient, most productive society. Although these ideas are now 50 years old, they could have been drawn from yesterday's campaign speeches. In fact, they may summarize many of the ideas you offered in response to my questions at the beginning of this chapter.

But is this system fair? Is it truly efficient and productive? Is it inevitable? Tumin (1953) drew on the conflict tradition to deny all these things. Stratification systems may actually limit the discovery of talent, he argued, because those without access to resources such as fine schools may never be able to develop and display their talents. Many of the people in the "best" schools and in the most powerful positions are the children of people who have previously attended those schools and held those positions; what of the talented poor who may never get a chance to reach the top? Further, is working toward the top really a sacrifice? Given the choice of attending an elite college with a beautiful campus and then moving from one executive suite to another or going directly into the workforce to help support oneself and one's family through backbreaking unskilled labor, how many would not prefer the former, quite apart from the higher income to be gained? Further, "sacrifices" such as college tuition may be made by family members and not directly by the persons benefiting. Tumin argued that, quite apart from creating social solidarity and consensus, inequality is likely to create hostile parties who distrust one another. The losers in the great game are likely to be discouraged, disgruntled, alienated, and openly hostile to the system. It is neither easy nor "efficient" for a society to control such hostile factions, nor are the losers likely to be highly productive. Certainly, Tumin contended, there must be other, better ways to motivate people.

Davis and Moore responded that Tumin was bringing in secondary issues. The role of family and inheritance is not a fundamental part of stratification.

A system such as our own could be reformed by laws encouraging equal opportunity. Further, Davis and Moore contended that the conflict approach is ideological, arguing for what ought to be and not describing what is, and that it is contrafactual, flying in the face of the existing evidence on all actual working societies. Tumin responded to this by saying that inequality may be universal, but does that mean it is necessarily functional and indispensable? Other evils have also been universal. Tumin contended, like Plato and Marx before him, that the role of family and inheritance is not secondary; rather, it is a crucial part of a stratified system. Finally, he asked, isn't all of this ideological? The analyses on both sides of this debate were shaped by their authors' views of a social ideal, and their arguments were created to justify particular social patterns.

The debate in the *American Sociological Review* broke off at this point, but the ideas are going with us into the twenty-first century, and they have clearly become wedded to ideologies. The "conservative thesis" of the new right in American politics and much of the rest of the world echoes Davis and Moore's arguments: Inequality motivates hard work, competition, and efficiency. The antithesis from the left echoes Tumin's assertions: Inequality erodes opportunity, perpetuates privilege, and undermines motivation and hard work while it perpetuates inefficiencies. From classrooms to campaigns, the debate continues.

Moving Beyond the Debate: A New Synthesis

Gerhard Lenski (1966) sought to lay out a new theory of stratification, a synthesis of the functionalist and conflict views. Lenski, like Marx, wanted to show how patterns of stratification had shifted over different societies. He had access to better anthropology and historical-comparative sociology than Marx did, and he focused his attention on the technology of production, what Marx called the mode of production. Lenski's work addressed a variation on Weber's three dimensions of class, status, and party, which Lenski labeled *privilege, prestige,* and *power.* Clearly conversant with the ideas of conflict theorists, Lenski focused on societal evolution in a manner similar to Durkheim and later functionalists. He called his approach *ecological-evolutionary theory.*

As Lenski surveyed the sweep of human society from the simplest hunter-gatherers through simple horticultural farmers to vast agrarian empires and on to industrial states, he found common trends at work. The expansion of technology and a growing division of labor made each stage in social development more powerful but also made certain individuals and families within those societies more powerful. He concluded, with the functionalists, that some measure of social inequality is inevitable in complex societies, given the multitude of tasks and social positions that exist in such societies. Yet he argued, in line with the conflict theorists, that the level of inequality in complex societies is always higher than necessary, as powerful and well-placed individuals used their social power to increase their prestige and commandeer greater privilege.

Lenski's work is often cited, but his approach, filled with historical complexities, has not been widely expanded upon in the great debate. There is one very simple and important kernel of a theory of stratification in his work, however: Although inequality may begin in differences in human abilities, it is primarily a social rather than a natural construction.

Functionalist theories of inequality place the roots of social stratification in fundamental human differences in talents, abilities, and, possibly, motivation. Marxist theories question this underlying assumption, placing the roots of inequality in property relations, the social relations of production. Anthropological evidence suggests that inequality is not based on the fact that humans have different talents and levels of ability (Harris 1989; Diamond 1997). Hunter-gatherers have different levels of ability, but because individuals in hunter-gatherer societies all share and work together, they remain essentially equal. The acceptance of inequality begins when someone can claim a position of social power, a central position in a network of exchange that can be exploited for personal as well as clan gain. The families of key "big men" who redistribute resources may receive more than others, gaining in both prestige and privilege. Inequality is thus based not so much on differences in human ability as on differences in social position within a network of exchange. Unlike in the functionalist model, gain depends on social position rather than mere talent, and unlike in the Marxist model, the key role is not in the production but in the distribution and redistribution of goods. Privilege goes not to the exceptionally talented but to the exceptionally well connected.

The failure of Marxist states lies in the fact that they revised the social relations of production but did not substantially alter the privileged positions of distribution and redistribution. Such a model is also important for understanding contemporary social inequalities. Despite the so-called triumph of markets, these are not face-to-face markets but rather markets increasingly mediated by global redistributors who are able to garner privileged positions in global networks of exchange. New technologies have created possibilities for new and broader opportunities, but they have also created new concentrations of power. As Max Weber noted a century ago, those benefiting from this power have often used it to guard their position by means of monopolization and social closure. The current prospects for a more equitable global economy hinge on humankind's ability to limit concentrations of distributive power and open multifaceted avenues of information and opportunity through more open social networks.

KEY POINTS

- The debate about whether inequality is just and necessary has been ongoing since the establishment of the earliest civilizations.
- The *conservative thesis* represents the dominant thinking that social inequality reflects basic differences among people in creation, ability, or worth and is necessary to the orderly functioning of society.

- Challengers to unequal systems have offered the *radical antithesis,* the argument that great social inequality is fundamentally unjust and ultimately destructive to societies.
- The philosophical debate concerning inequality was taken up by social scientists in the 1800s. Karl Marx developed an approach to understanding history that is based on conflict between social classes. Capitalism takes this conflict to new intensity in the struggle between owners and workers. Marx contended that workers could fully secure their rights only through collective control of the means of production.
- Max Weber continued Marx's emphasis on social conflict but asserted that the struggles are rooted in three different dimensions of unequal social power: social class or economic power, status honor or social prestige, and "party" or political and legal power.
- Émile Durkheim emphasized the importance of a complex division of labor for modern societies, arguing that social stability depends on a society's filling a multitude of interdependent positions. Functionalist sociologists built on Durkheim's ideas to contend that social inequality serves an important function for society by helping to place people in this division of labor and motivating them to work hard.
- Conflict theorists drew on the work of Marx and Weber to point out how large social inequalities can function to create social unrest, overlook abilities, and discourage workers while promoting social misery.
- Gerhard Lenski tried to synthesize conflict and functionalist ideas in looking at how privilege, power, and prestige emerge from different types of societies. The debate on what constitutes a just social order and how such an order can best be achieved continues.

FOR REVIEW AND DISCUSSION

1. What arguments have been offered in support of the social benefits of inequality? What counterarguments have been offered to challenge these supposed benefits? How have these arguments formed the basis of conflict and functionalist views of social inequality?
2. In what ways are Marx and Weber in agreement on the causes and nature of social inequality? In what ways do their views of stratification and class formation differ?
3. Is social inequality desirable for society? Defend your view with arguments from historical and sociological viewpoints discussed in this chapter.

MAKING CONNECTIONS

In the Media

The conservative position on social inequality has gained new momentum in recent decades through many social and political groups. The radical antithesis also continues in the arguments of many progressive (many prefer this term to the extremist-sounding *radical* or the often ambiguous *liberal*) groups. To get a sense of how the debates discussed in this chapter continue into the present, try one of the following:

- Look at the coverage of issues related to the economy and society in a publication associated with conservative political opinion, such as the *National Review*. Compare the ideas presented there with those in a publication associated with progressive political opinion, such as the *New Republic*. Many campus libraries carry both the *National Review* and *New Republic* or have them available online. These magazines are targeted toward educated readers and are not particularly extremist, but they do have their own definite points of view.

- Compare the newspaper columns of a respected columnist associated with conservative opinions, such as George Will, with those of a respected columnist associated with progressive opinions, such as Molly Ivins. You can find back issues of major newspapers and search for authors and topics quickly using the Internet (for example, see www.nytimes.com, www.chicagotribune.com, www.latimes.com, or www.washingtonpost.com).

- Compare the perspectives on social and economic issues, as well as the use of religious tradition, in publications of groups on the religious right, such as the Christian Coalition and Focus on the Family, with those in publications of "radical discipleship" groups that generally take progressive stands on economic issues, such as Sojourners. What positions, concerns, values, and emphases do these groups have in common? How do their positions differ? How does each side support and defend its positions?

2 The Global Divide

Inequality across Societies

_____ **Worlds Coming Apart and Coming Together**

The explosion rocked the entire city. Stunned onlookers watched the crumbling walls and then turned to run, covering their faces as great billows of smoke rushed outward and threatened to engulf them. As the smoke cleared, the onlookers turned back to witness a scene of complete destruction. Some wept. Others cheered. This was neither an act of war nor terrorism; these were the tremors of economic change. The Uniroyal plant in the small city next to mine had come down in a great planned "implosion." This riverside site had once been home to the Ball Band factory. Only the most senior readers will remember Ball Bands, but they were one of the first prized athletic shoes in the United States—the Nike Air Jordans of their day. In time, Ball Bands gave way to new national competition, and then the market for domestically produced athletic shoes almost disappeared as competitors such as Nike dominated the market with shoes made in East Asia and aggressively marketed to American urban youth, whose fashion tastes have in turn influenced much of the world's urban youth culture. Ball Band was bought out by the national giant Uniroyal, which manufactured various products made of synthetic rubber and plastic at the site. One by one these operations were transferred overseas, where light products needing inexpensive resources but high labor demands could be made much more cheaply. "I'm sorry to see it go, it's the end of an era for this city," said one onlooker. "I think it's a good thing," offered another, "it's the beginning of new development."

The city is Mishawaka, Indiana, better known as one of the "homes" of *The Late Show with David Letterman* than as an industrial giant. It is a small city, population approximately 40,000, one of several such cities that are strung out like beads (I hesitate to claim "pearls") along the St. Joseph River as it winds along the border between Indiana and Michigan before emptying into Lake Michigan. On the lake, a few old tankers still head for

U.S. Steel in Gary, and brand-new French and German cruise ships take passengers from northern Michigan to the new attractions of waterfront Chicago. The ships don't venture upstream, but if they did, passengers would glimpse the social and economic changes taking place in the American heartland. First port of call is Benton Harbor–St. Joseph, Michigan, twin cities divided by the river. They are far from identical twins, however: St. Joseph is almost entirely white, and Benton Harbor is largely black. This is not an accident of geography, of course. When black workers began coming to work in the growing industrial region of western Michigan early in the twentieth century, they were not welcome in St. Joseph, so the black population became concentrated in the cheaper, less desirable areas across the river. Today segregation remains so severe that many white residents in St. Joseph have never been to Benton Harbor, and some black residents in Benton Harbor have never crossed the bridge into St. Joseph. Alex Kotlowitz describes this amazing divide of two small communities in his book *The Other Side of the River* (1998). Benton Harbor has a national distinction: It is the single-parent capital of the nation. Eight out of ten children in Benton Harbor are born to single mothers. Three-quarters of Benton Harbor residents are poor. When industry left the area, so did options for the people who live in Benton Harbor, especially for black men. They are not welcome in the upscale residential and tourist community and economy of St. Joseph. Even those who can find employment in the restaurants and stores that line Interstate Highway 94 as it curls around the city don't earn enough to support families. Seeking higher wages means leaving the community, and many have been left behind.

Continuing upstream, small Michigan cities surrounded by orchards boast new industrial parks built with the help of state-sponsored tax abatements to try to lure industry north from Indiana. Further upstream and crossing into Indiana, the city of South Bend does not have much industry left to lose. It experienced the "rust belt" phenomenon early, when automaker Studebaker closed its large plant. Other heavy industry soon followed, leaving a corridor of hollow shells of dirty brick buildings unsuited to modern global industry. The city has made its own comeback: new sports stadium, new College Football Hall of Fame, new industrial park on the outskirts of town, and new jobs in a growing service sector supported by regional hospitals as well as regional and national universities. The only large industrial employer to remain was bought out by a large national firm that was in turn bought out by a larger multinational German-owned firm. This corporation still employs engineers, designers, and accountants, but few entry-level industrial workers. Unemployment is very low in South Bend, but so are prevailing wages. Low-skill, minimal-credential employment is dominated by low-level service jobs, as indicated by vacancies for nurse's aides in hospitals, custodians in office buildings, and clerks in liquor stores and gas stations. The prevailing wage is just over $6 an hour, and the local United Way estimates that with just one dependent, a worker would need closer to

$10 an hour to be self-sufficient. The only new construction downtown is a million-dollar addition to double the capacity of the Center for the Homeless. Many low-income whites live in South Bend, but the neighborhoods that once housed predominantly Polish American industrial workers are increasingly Hispanic, sharing the space with a large African American population. All large retail has left the city and is now overwhelmingly concentrated along a "miracle mile" strip of big-box chain retailers that ends in a regional shopping mall outside the city limits.

Further upstream, the city leaders in Mishawaka hope that the vacant land on which the Uniroyal plant stood will attract new development downtown. Jobs are now dispersed, and no public transportation reaches across the entire metropolitan region. Further upstream still is the city of Elkhart, which has one of the lowest unemployment rates in the nation thanks to a booming recreational vehicle industry. This is "new industry," however: constantly changing and often seasonal, nonunion, and paying wages of around $8 an hour with few benefits; in real dollars, this is about one-third the rate of pay enjoyed in the past by workers employed with the old unionized automakers. Word that jobs are plentiful in Elkhart has drawn a chain migration of Latino workers to the city, both Mexican Americans who are longtime U.S. residents and newcomers of uncertain legal status.

This tale of a string of cities is neither about perpetual decline nor about a chain of prosperity, although both have been claimed. It is, as Dickens wrote, the best of times and the worst of times, and it just matters who, and where, you are. White, black, and Latino workers in this area may have little in common, but their children have found common ground—shoes. They all wear popular brands of athletic shoes, none of which are made in the United States. Searching out the source of all those shoes takes us to another river valley, quite different from the first.

Athletic shoes are now made in export zones outside of Jakarta, Indonesia, and Penang, Malaysia, and increasingly in the great export-oriented zones of China. The largest of these zones centers on Hong Kong, the world's most crowded and maybe most bustling city, which thrived as a British treaty port and colony and is now again part of China. Hong Kong shares the mouth of the Pearl River with Guangzhou, the city Westerners know best as Canton. This was China's window on foreign trade for centuries and is again the center of China's trade policies. Hong Kong and Guangzhou are the anchors of a valley of a dozen cities, at least half of which have populations of more than a million, dedicated to growth through export manufacturing. From shoes to toys, almost anything that is light and transportable—especially if it is made of molded plastic, glued nylon, and rubber and easily shipped—is manufactured in China, often in this valley.

The Pearl River delta supplies the entire world with Barbie dolls and Mickey Mouse. Some 80,000 Hong Kong–related enterprises have factories in the area, and 1,800 other Chinese companies have come here, along with thousands of Taiwanese, Japanese, U.S., and European firms (Edwards 1997).

High-rise office buildings gleam throughout the night, not just in Hong Kong but in all of these cities, as the manufacturing companies try to manage it all. More and more look-alike concrete high-rise apartments crowd shoulder to shoulder to house all the workers needed. Potential workers must seek permits to come here, and coils of razor wire and vigilant police keep out those who don't have permits. Huge new shopping malls and amusement parks cater to the new middle class, but most Chinese who come here are seeking work. Young women wearing paper masks stand closely side by side, gluing soles on Reeboks. Increasingly, men are trying to find work here as well to escape the rural poverty of neighboring regions.

Fu came to the Pearl River cities from rural China in hopes of finding better wages in industrial employment. He went to work molding bits of red and white plastic into jolly Santa Claus faces for export. Fu was glad to have steady work, but he himself was anything but jolly. He had to turn out box after box of plastic Santas, working at breakneck speed with a stamping machine that frequently malfunctioned because it was old and poorly maintained. His bosses ignored his complaints about the faulty machine. One day, as he reached to scoop another Santa out of the machine and into a box, the machine press came down on his arm and severed his hand. Fu's recovery was slow and painful, and he could no longer work in any of the factories where the work required speed and dexterity. His employer, a Taiwanese industrialist with offices in Hong Kong, offered him a small "severance package" ("severance" taking on a new meaning in this case). Fu refused the offer and turned to Jo, an attorney who specializes in worker injuries. Jo assured Fu that the loss of his hand, which was clearly caused by company negligence, was worth much more than the small settlement the employer had offered. When Fu returned to the factory and confronted the employer, the man had Fu locked in a room where he was told he would stay until he agreed to drop his case. Jo, well acquainted with the intimidation that employers often apply in such cases, went to the factory with some local police officers and secured Fu's release. Fu did sue and won several times the original offer. Although this was still a tiny amount compared with the awards in U.S. litigation cases, it was enough for Fu to return to school and study for an occupation that uses mental rather than manual dexterity. He is thinking about becoming a lawyer.

Increasingly, however, workers like Fu and even lawyers like Jo must be careful how they press their cases. Taiwanese firms continue to manufacture plastic products in the Pearl River delta, but they no longer make blue jeans there. They have moved that work to Nicaragua. The country where Sandinista revolutionaries once battled U.S.-backed Contras is now home to a large "free zone." Some 25,000 workers, mostly young women, now work in factories in Nicaragua for wages of about 70 cents an hour, half the rate paid for such jobs in Mexico. They work large amounts of forced overtime, often until midnight, sometimes seven days a week. In the late 1990s, one of these workers, Gladys Manzanares, mother of six and grandmother of two, decided to form a union, and, after months of organizing and struggle, she succeeded. The new union negotiated a contract with the Taiwanese factory managers that included protective filters for the workers to wear amid the cotton dust, paid pregnancy leave, and the right for workers to have coffee on the job. A few years later, new factory owners, also Taiwanese, examined the contract and began to move against the union. They attempted to buy Manzanares out, offering her $25,000 to abandon the union—more than she will earn in a lifetime—but she refused and lost her job along with other union leaders. Nicaraguans, emboldened by the Sandinista years, are used to fighting for their rights, but this time they got no help from the government. Some say it is because the factory workers already earn as much as nurses and police officers in poverty-stricken Nicaragua. Others wonder about the beautiful new buildings glistening in Managua, the country's capital, buildings for the Nicaraguan government financed by Taiwan. Was this a gesture of friendship or an act of manipulation? The firms themselves had threatened to leave Nicaragua, but now with the union gone they will stay. Human rights groups have accused the government of collusion and have turned to international consumers, threatening a boycott of the brands made in the Nicaragua factories, which are sold at major retailers in the United States such as Target and Kohl's. Labor leaders in the United States have also provided Nicaraguan workers with some money for renewed organizing, but they too must move carefully; there are few unions in neighboring Honduras.

The cases related above are just two of many that could be told about the effects of our globalizing economy on workers, industries, and urban regions. I have chosen these cases particularly because of the complexities and ambiguities they raise. Many observers have provided accounts of the effects of deindustrialization in the United States. In his 1989 film *Roger & Me,* for example, Michael Moore provides a captivating account of the utter devastation of the economy of Flint, Michigan, along with the lives of many workers, caused by the massive closings of General Motors plants, a process that continues. But what of Ball Band/Uniroyal? If you had been standing on that riverside when the building imploded, after you wiped the soot from your eyes, would you have wept or cheered? Is this a tragic loss for the community or a new beginning? Has the city lost its economic base or merely traded ugly buildings and hot sweaty jobs for new opportunities and better jobs?

The Chinese situation has similar ambiguities. Is the economic boom in the Pearl River valley the driving force behind a new, prosperous, modern China, or the latest page in a history that includes five centuries of exploitation of Chinese workers by foreign interests? Certainly the irony, and ultimately the tragedy, of Fu's risking his body to turn out ever more trivial trinkets says little about the ability of global capitalism to advance balanced development and human priorities. Many have written forcefully but the abusive conditions in international sweatshops (see Peña 1997; Ehrenreich and Fuentes 1981), yet Fu was eager to leave the poverty of the countryside for urban opportunity. The treatment Fu received from his employer was deplorable, but this is not just a simple story about foreign exploiters and national police states. Jo himself notes that in his labor rights work, he often finds that the plants with the best safety records are the largest operations controlled by U.S. and Japanese multinationals. Often the worst abuses occur in plants owned by Taiwanese and Hong Kong partnerships that subcontract to larger firms. This is true elsewhere as well: Often the greatest human rights abuses perpetrated in Southeast and South Asia occur in operations controlled by local employers or regional middlemen, who frequently employ young women and even children at the lowest possible wages to work under the worst conditions as the employers seek competitive advantage. The workers' ability to seek legal recourse is also interesting. China is frequently noted for its repression of civil and human rights, and some observers offer this repression as a reason the United States should cease trade relations with China. Certainly, Fu was horribly mistreated by a controlling employer, but note that he was able to hire an attorney, get police protection, sue in court, and win a settlement. China, in fact, has quite good labor protection laws on the books; these laws are often not aggressively enforced, however, especially in smaller plants, unless lawyers like Jo can prepare and advance the workers' cases.

Is this a story about progress or betrayal? Is it about worker oppression or democratic progress? Perhaps the same questions could be asked about the Chinese, Nicaraguan, and U.S. examples. As enterprises become ever more linked around the world, are we witnessing a chain of prosperity or a race to the bottom?

Those who take the former view see a chain reaction that in time will bring modernity and prosperity to ever greater numbers of people. After World War II, the United States was victorious and embarking on a period of prosperity. Japan, in contrast, was defeated and falling into hard economic times. Japan's response was first to become a manufacturer of less expensive consumer goods using the relatively cheap labor available. It then became a "middleman" for the production and export of such goods, and eventually Japanese firms gained full corporate control over the goods' production. As Japanese wages rose, the production itself shifted to new places with cheaper labor: South Korea, Hong Kong, Taiwan, and Singapore. These places became primary producers of inexpensive consumer goods, then middlemen in transactions between new sources of cheap labor and the United States, Japan, and Europe, and now firms headquartered in these countries are controlling the trade. Singapore has experienced so much growth that it now has a higher gross national income per capita than Great Britain, its former colonial master (World Bank 2002). Singapore also has a very high cost of living, a great deal of urban congestion, and problems with pollution. Labor-intensive production has again shifted, this time to Indonesia, Malaysia, and Thailand. In these places entrepreneurs now hope to grow rich and maybe shift production onward: to Bangladesh, India, Pakistan, and locations in the Indian Ocean. The chain continues and alters everything in its wake. What about all those software engineers in India? Are they poised to lead South Asia into the postindustrial information age without ever fully entering the industrial age? When they look offshore for their production, where will they look—Africa? Can the chain continue? Will the next decades find Ugandan businessmen in air-conditioned offices managing production by low-wage laborers in Congo? Can this go on until all nations are prosperous? Or is it all a great big pyramid scheme that must eventually end? A pyramid scheme works only as long as new people can be brought into the system, and ultimately the gains of the winners are paid for by the losses of the latecomers. Such a scheme produces a few big winners and ends with many more losers. Is this the real nature of the global chain of prosperity? Or will technological advances make this process obsolete? Will those plastic Santas ever be able to mix and mold themselves?

The other, less optimistic view of global work has been dubbed the race to the bottom. This can occur, and has occurred, even within large countries. For example, individual U.S. states often compete with one another to attract new, mobile industries, and states often "steal" companies from one another. When a state rewards a firm with big tax breaks for locating within its borders, the state gains new jobs but not additional tax revenue to support public institutions, such as schools. States also compete for firms by offering a "business-friendly" environment: low taxes, low wages, few unions, few labor or environmental regulations, and so forth. The danger is that in their efforts to keep jobs and attract new ones, states may be tempted to give away all the things that might help build a better quality of life for their residents. Workers may get jobs, but with minimal wages, benefits, and protections. States that refuse to play this game may simply lose businesses altogether. One way the United States can protect against this process is to enact national laws that set minimum standards for wages, environmental protection, and so forth.

This problem grows when the race goes international. Mobile firms can move from low-wage countries to countries where wages are even lower, from countries with few regulations to countries with no regulations at all. Firms may bypass nations with progressive policies altogether, and countries that are eager to grow will be tempted to offer more and more to gain the attention of multinational firms. Desperate countries, such as Nicaragua, may give away a great deal. Some may try to attract firms by emphasizing the particular resources they have to offer. For example, Singapore (and, increasingly, Mexico) has promoted the availability of an educated and English-speaking workforce. (Ironically, as more countries improve their educational levels, this is becoming less of a selling point.) Further, no international government exists to set minimum standards and prevent the race from truly reaching the bottom: workers laboring for bare subsistence pay with little or no protection, as Marx envisioned. Is this the world of work we will see in the coming decades?

The Double Divide

The world is getting smaller, but its inhabitants are not necessarily growing closer to one another. Compared with those in the past, the gulfs that divide people now are less likely to be geographic and more likely to be social and economic. In an excerpt from his book *The Global Soul,* frequent flyer Pico Iyer (2000) writes of his recent experience:

> I woke up one morning recently in sleepy never-never Laos (where the center of the capital is unpaved red dirt and a fountain without water) and went to a movie that same evening in the Westside Pavilion in Los Angeles, where a Korean at a croissanterie made an iced cappuccino for a young Japanese boy surrounded by the East Wind snack bar and Panda Express, Fajita Flats and the Hana Grill; two weeks later I woke up in placid, acupuncture-loving Santa Barbara and went to sleep that night in the broken heart of Manila, where children were piled up like rags on the pedestrian overpasses and girls scarcely in their teens combed, combed their long hair before lying down to sleep under sheets of cellophane on the central dividers of city streets. (P. 75)

The Gap between Nations

Each year, the United Nations Development Program issues its *Human Development Report.* These yearly reports reveal a mind-boggling gap between nations in income, wealth, and well-being. The richest one-fifth of the world's people consume 86% of all the world's goods and services, while the poorest one-fifth consume just 1.3%. U.S. citizens alone spend $8 billion a year on cosmetics, $2 billion more than the estimated amount needed annually to provide basic education for everyone in the world. Not to be outdone in consumption, Europeans spend $11 billion a year on ice cream, $2 billion more than would be needed to provide clean water and safe sewers for the rest of the world's population (see, e.g., United Nations Development Program 1996, 1998, 2000, 2003, 2004).

The gap between nations is greater than the gap within any single country (United Nations Development Program 2003). The richest 5% of the world's people receive more than 100 times the income of the poorest 5%. In fact, the richest 1% alone receive more income than the poorest half of the world's people. Just 25 people, the 25 richest Americans, have a combined income almost as great as the combined income of 2 billion of the world's poor.

Some observers have argued that scholars should use other measures in addition to income when measuring poverty and inequality (Sen 1999). For example, some Arab states have been able to use oil wealth to increase overall national incomes, but these nations still tend to lag far behind in measures of well-being. To compare incomes, the United Nations uses a purchasing power parity index that adjusts the per capita national income for cost of living (it does not adjust for internal inequalities). By this measure, the highest-income country in the world is the United States, because many other high-income countries—Switzerland and Japan, for example—also have

HDI rank	Life expectancy at birth (years) 2002	Combined gross enrollment ratio for primary, secondary, and tertiary schools (%) 2001/02	GDP per capita (PPP US$) 2002	Human development index (HDI) value 2002
High human development				
1 Norway	78.9	98	36,600	0.956
2 Sweden	80.0	114	26,050	0.946
3 Australia	79.1	113	28,260	0.946
4 Canada	79.3	95	29,480	0.943
5 Netherlands	78.3	99	29,100	0.942
6 Iceland	79.7	90	29,750	0.941
7 United States	77.0	92	35,750	0.939
8 Japan	81.5	84	26,940	0.938
9 Ireland	76.9	90	36,360	0.936
10 Switzerland	79.1	88	30,010	0.936
11 United Kingdom	78.1	113	26,150	0.936
12 France	78.9	91	26,920	0.932
13 Germany	78.2	88	27,100	0.925
14 Poland	73.8	90	10,560	0.850
15 Costa Rica	78.0	69	8,840	0.834
16 Mexico	73.3	74	8,970	0.802
Medium human development				
17 Russian Federation	66.7	88	8,230	0.795
18 Brazil	68.0	92	7,770	0.775
19 Saudi Arabia	72.1	57	12,650	0.768
20 China	70.9	68	4,580	0.745
21 Iran, Islamic Rep. of	70.1	69	6,690	0.732
22 Indonesia	66.6	65	3,230	0.692
23 South Africa	48.8	77	10,070	0.666
24 India	63.7	55	2,670	0.595
25 Cambodia	57.4	59	2,060	0.568
26 Bangladesh	61.1	54	1,700	0.509
Low human development				
27 Pakistan	60.8	37	1,940	0.497
28 Kenya	45.2	53	1,020	0.488
29 Nigeria	51.6	45	860	0,466
30 Haiti	49.4	52	1,610	0.463
31 Ethiopia	45.5	34	780	0.359
32 Mozambique	38.5	41	1,050	0.354
Country Averages				
Developing countries	64.6	60	4,054	0.663
Least developed countries	50.6	43	1,307	0.446
High human development	77.4	89	24,806	0.915
Medium human development	67.2	64	4,269	0.695
Low human development	49.1	40	1,184	0.438
World	66.9	64	7,804	0.729

Exhibit 2.1 International Comparisons in Income and Well-Being

Source: Data from United Nations, Human Development Report (2004).

very high costs of living. The United States does not lead the world in measures of well-being, however. A composite "Human Development Index," which includes income, health, and education, gives that place to Norway. (See Exhibit 2.1.) Incredible gaps in income are also often paralleled by huge gaps in life expectancy and education.

There is some evidence that income inequality between countries is declining (Firebaugh 2003). This is not true uniformly around the world, however. Incomes in sub-Saharan Africa have fallen absolutely, leaving much of Africa even further behind the rest of the world than in 1990. In 1820, Western Europe's per capita income was only about 3 times that of Africa; by 1992, it was more than 13 times that of Africa, and the gap continues to grow. The closing of the worldwide gap is entirely the result of gains in Asia and in two Asian countries in particular. Both China and India have seen dramatic growth in overall incomes, with the rate of growth greatest in China. Because these two countries alone account for one-third of the planet's population, their growth has a big impact on overall figures. Big gains have also been seen in Singapore and in urban Thailand. The rural-urban gap also remains large, however. Most of the gains have been in large commercial cities; rural areas, especially in India, have lagged far behind. This highlights a common problem: Comparisons between countries may mean less than comparisons between groups and regions within countries. Cosmopolitan and high-tech Mumbai (formerly known as Bombay) may flourish along with Seattle while the hill country of northern Uttar-Pradesh languishes along with the hill country of eastern Kentucky.

The Gap within Nations

The gap within nations is also staggering. Some of the countries with the greatest internal inequalities are in Latin America: in Central America, where a handful of families control most of the wealth as well the government, and in Brazil, where the fabulous star- and diamond-studded beaches of Rio de Janeiro bask in the sun just out of the shadow of towering slums that climb the steep and dramatic hillsides. These nations have the largest gaps between rich and poor in the world, but they are not alone in having wide and widening gaps. Under socialist governments, Eastern Europe had some of the modern world's most equal, although not always thriving, societies. Since the fall of communism, inequality has grown dramatically within Eastern European nations. Russia, with great resources and newly wealthy entrepreneurs alongside massive unemployment and wages below those paid in much of India, now has a greater gap between rich and poor than does the United States (see Exhibit 2.2). During the past couple of decades, the United States itself gained the unexpected distinction of being the most unequal society in the advanced industrial world. Whether inequality is determined by the share of national wealth held by the top few (the richest 1% control more than 40% of U.S. wealth) or by the size of the gap between the richest one-fifth and the

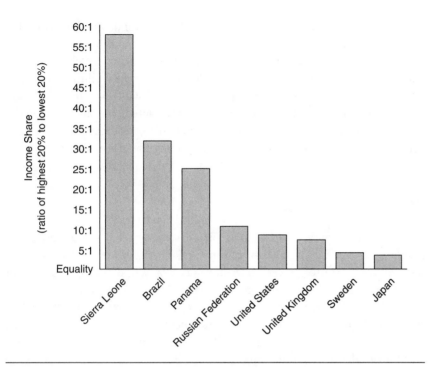

Exhibit 2.2   Inequality in Selected Countries

Source: Data from United Nations, Human Development Report (2004).

poorest one-fifth of the population, the United States is number one in inequality among wealthy advanced industrial nations.

Increasingly, it is perhaps more revealing to look at clusters of wealth—whether they be in nations, cities, multinational corporations, or wealthy individuals—than to compare nation to nation. Two cities, Singapore, independent for only a few decades, and Hong Kong, now returned to China, found themselves at the center of an exploding Asian export market and now have average incomes higher than those found in most of Europe (World Bank 2002). The assets of General Motors are larger than the total gross domestic product (GDP) of the entire nation of Indonesia, the fourth-largest country in the world. Likewise, by this measure, Exxon (United States), Royal Dutch Shell (Netherlands), and Toyota (Japan) each has more economic clout than the entire economies of Poland or the Philippines or the 130 million people of Pakistan. Despite a great expansion in the numbers of small holders in the world's stock markets, most corporate assets are held by a handful of individuals. The assets of the 3 richest people in the world exceed the combined GDPs of the 48 least developed countries. The assets of the world's 225 richest people (a group that includes 60 Americans) total more than $1 trillion, equal to the combined annual income of almost half the world's population.

What does it mean to live in a world in which the cost ($40 billion, according to always-controversial U.N. estimates) of providing basic education and health care as well as adequate food and clean water for every single person is less than 4% of the combined wealth of these 225 individuals, less in fact than the estimated wealth of one person—Bill Gates and his $60 billion of Microsoft assets? How can we make sense of a world in which the United States long claimed it could not or should not pay the $1 billion it owed to the United Nations but one person, Ted Turner, with his cable television assets, could pledge a $1 billion donation to the United Nations based on the unexpected growth of his portfolio in a single season? To begin to do so, it helps to understand some of the forces reshaping the global economy.

The Global Debate

How do we account for the gross (in all senses of the word) disparities described above? Who's to blame for poverty? Theorists' key ideas concerning these questions started to crystallize at the end of World War II, although in many ways they merely continued the ancient debate reviewed in Chapter 1.

If we live in a "new world order," it began not in 1989 with the fall of the Berlin Wall but in 1944 with the imminent fall of Berlin. The "modern world" of geopolitics began at Potsdam, where three men tried to fashion an idea of what the world would look like in the wake of world war. One presided with an iron fist over what was then the world's only Marxist state: Joseph Stalin of the Soviet Union. One presided with an iron will over the remnants of the British Empire, which he was determined to preserve: Winston Churchill. The man in the middle in the photographs, and in the debates, was *Time* magazine's statesman of the century, Franklin Roosevelt. Roosevelt argued against empires and spheres of influence and in favor of a world marked by "four freedoms": nations with freely chosen governments living as "good neighbors" in a world of free trade and mutual prosperity. Some thought that Roosevelt was too generous in meeting Stalin's demands, but everyone at Potsdam knew who would be the inevitable leader of this world: the one nation whose economy was energized rather than devastated by war, FDR's United States. By 1946, the Cold War was emerging and few agreements involved the Soviet Union. But Roosevelt's vision was expanded at Bretton Woods, New Hampshire, as the "free world" created new institutions for global prosperity: the International Monetary Fund (IMF) and the World Bank, among others. At the same time, the old empires started to crumble. Gandhi brought India successfully to independence while France struggled to hold on to Algeria and Vietnam. One by one, countries across South Asia and all across Africa became independent. Perhaps the last great empire ended in 1991 with the breakup of the Soviet Union. Two great "superstates"—the United States and China—remained to face each other as potential rivals, yet they found themselves linked by trade in spite of political differences. The world of FDR had almost fully emerged, with one exception: Two-thirds of that world was still not experiencing prosperity. What had happened?

Modernization

One answer to that question came from theorists whose work was rooted in functionalist ideas about inequality. Because one of their favorite themes was the transition from traditional society to modern society, these theorists became known as the **modernization school**. They never saw themselves as a single school of thought, but they did have some perspectives in common (see Weiner 1966). Their answer: Blame traditionalism. Poverty is the basic primordial condition of humanity: Once all societies were poor. Poor societies stay poor because they cling to traditional and inefficient attitudes, technologies, and institutions. In contrast, in the "modern world," the rise of industrial capitalism brought *modern attitudes,* such as the drive to experiment and achieve; *modern technologies,* such as machinery and electronics; and *modern institutions* to manage all this, such as financial institutions, insurers, and stock markets. Given enough time, such modernization will occur everywhere. In a widely read "non-communist manifesto," Rostow (1960) argued that poor countries slowly proceed to build the basis for mature modern economies. Once the key foundations of modernity are in place, these countries "take off" toward prosperity and a modern, high-consumption consumer economy. Although this process will take time to work in the most traditional places, eventually global capitalism and its modern corporations will carry these modern ideas, technological innovations, and efficient institutions everywhere.

Dependency

Even as these ideas were first forming, a counterargument was emerging that was rooted in conflict ideas about inequality. Because this counterargument stressed the problem of the dependency of poor nations on the whims and power of the rich, it came to be known as **dependency theory**. The first dependency theorists may have been Lenin and Bukharin, who adapted Marx's ideas to the twentieth century. Their answer to the question of poverty: Blame colonial imperialism. Worldwide industrial capitalism brings *exploitation* through unequal exchange and removal of surplus through profits, *domination* through subtle but powerful neocolonialism, and *distortion* through "disarticulated" economies that serve export needs but not the needs of local populations. Latin American thinkers were especially key in this emerging line of thought (Prebisch 1950). Latin America remained poor, they contended, because it exported raw materials at low prices to serve the needs of global industry and then had to import finished goods at high prices. Andre Gunder Frank (1967), a European transplant to Latin America, argued that poverty and deprivation is not humankind's original state. Poor societies are made, not born. They are not undeveloped but rather underdeveloped as a result of capitalist penetration. Why have poor nations been unable to resist this exploitation? First, they have been dominated by the rich nations through the neocolonial practices that replaced old-style empires, including the debt dependency that came with the World Bank and IMF, the manipulations of so-called foreign aid, and the tremendous power of foreign multinationals to coerce and co-opt national governments into compliance. Further, their economies have become so distorted that they cannot function apart from the global capitalism that is controlled by the rich centers of power, the "metropolis."

The two perspectives differed not only in their view of history but also in their view of the appropriate level at which to attack the problem (Jaffee 1998). Modernizationists saw the problems as largely internal to poor nations (Bradshaw and Wallace 1996), which would have to change their ways. Modernization theory, like functionalism, is still often termed the "mainstream school of scholarship" (Isbister 1998), especially by those who attack its ideas. Yet by the 1970s, modernization theory, like functionalism, was falling out of favor. Today, many more articles attack the theory than extend it. Modernization theory has been criticized as ethnocentric, as biased toward the West, and as measuring the world by Western standards. Modernization theory seems to have been mortally wounded by these attacks, yet its ideas are far from buried (for an adaptation and test of modernization ideas, see Inglehart and Baker 2000).

Neoliberalism

Just as modernization theory was in retreat in academic circles, many nations around the world were rediscovering the ideals of free trade and free markets. The intellectual basis for this approach comes from neoclassical economics. This approach is sometimes termed **neoliberalism**, from a time when conservatives supported royal monopolies and land-based economies and liberals favored free enterprise. In the United States, the former were largely driven out during the American Revolution, and so the latter have held sway for 200 years. Thus neoliberalism is the economic philosophy of American political conservatives (as well as many so-called moderates)—a use of terminology that adds to the confusion of many students trying to understand these concepts. This ideology was popularly dubbed "Reaganomics" in the 1980s and was also the driving policy of Margaret Thatcher and the Conservative Party in Britain at the same time. Increasingly, neoliberalism, with its roots in the economics of Adam Smith, is dominating the world. Ironically, it is the prevailing approach in the same Latin American countries where most thinkers once championed dependency theory. In fact, in a great irony, the world's only sociologist president, Fernando Henrique Cardoso of Brazil, who once coauthored a classic book on dependency in Latin America (Cardoso and Falletto 1979), has been called a neoliberal (a term he dislikes). On the international level, the IMF and the World Bank champion their own form of neoclassical economics through the ideals of **structural**

adjustment. Structural adjustment is, in effect, housecleaning. It calls on nations to reduce government spending and bureaucracy, to encourage markets, to export, and to encourage entrepreneurship as well to entice foreign investment and foreign technology.

For economists within this "mainstream" school, the evils are paternalistic politics that favor cronyism, corruption, and bloated bureaucracies; command economies that don't allow efficient supply-and-demand-driven markets; and fatalistic attitudes that result in unwillingness to risk entrepreneurial activity. In place of these evils, such economists still propose "modern" attitudes, institutions, and technology. Although this line of thinking originated in economics, neoliberals are not entirely oblivious to social problems (see, for example, World Bank 2000). They sometimes also blame racial and ethnic divisions and "tribalism," and some criticize "traditional" repression of women. Neoliberalism stresses the importance of individual rights to smooth-functioning economies, although it is often suspicious of group rights. For example, in the United States, political conservatives who champion this view favor "equal opportunity" but have intensified their attacks on affirmative action; in addition, they often oppose increases in the minimum wage, government programs for health and education, and other interference with the "free market."

World Systems

In contrast to emphasizing the need for internal reform, dependency thinkers tended to stress external causes (Bradshaw and Wallace 1996), particularly resisting external intervention: Get out of my house. This line of thought continues on in **world systems theory**, developed by Immanuel Wallerstein (1974). Wallerstein sees the "new world order" as 500 years old, beginning with the global capitalism of the emerging European powers. Since its beginnings, this modern world system has had three different core centers of power: the United Provinces (the Dutch capitalists who began this process), the United Kingdom and its various forms of empire, and now the United States and its global economic dominance. Around the core countries, which might now include other parts of Europe and maybe Japan, are scattered semiperipheral countries that serve as middlemen. This is the so-called Second World of trading states, including some newly successful entities such as Taiwan. The periphery consists of the poorest states, which bear the brunt of the oppressive system. For Wallerstein and his followers, states in the periphery are likely to remain that way until the entire system of global capitalism is overturned or radically altered.

Scholars who have deep concerns about global capitalism's ability to meet the needs of the world's poor sometimes label their perspective **political economy**. Marx preferred to refer to himself as a political economist. The term stresses the interplay of power and politics with the world of the market and exchange. Those who hold geopolitical power can use the world economy to their advantage. Some political economists are neo-Marxist, stressing the ongoing struggle between capital and labor around the world and updating Marxist concepts to fit the twenty-first century. Those in the dependency–world systems tradition begin with Marx's critique of capitalism, but they believe that the complexities of the modern world system require a new formulation that pays greater attention to international forces.

Most sides in this continuing debate agree that the key actors in the modern world are the multinational corporations, transnational lending institutions, international media, and expanding global technology and trade. They disagree on the effects of these forces. In the neoliberal view, the success of Taiwan or Singapore shows that export economies and free trade bring prosperity. In the world systems view, the core nations allowed these countries into the semiperiphery to facilitate the expansion of capitalism into new peripheral markets (such as Indonesia and Bangladesh). In one view, the structural adjustments mandated by the IMF are good examples of long-overdue housecleaning. In the other, they are further examples of policies of external domination that hurt the poor.

Is there a way to resolve this new impasse in the ancient debate? Many scholars have begun to argue that both theories have ideological blinders. Neoliberals continue to argue that poor nations must pull themselves up by their own bootstraps (even if they have to borrow the boots), and world systems theorists often see very little that poor nations themselves can do. What are the elements of a blended theory? A logical first step toward resolution is to look at both external and internal factors. Modernization and dependency may be viewed as processes that go hand in hand.

Globalization: The Ties That Bind

How is it possible to have a postindustrial economy in a society filled with industrial products? The answer lies in the phenomenon that has been termed **globalization**. Industrial products are manufactured in export zones of poor countries with low wages, then quickly shipped to wealthier consumers around the world. The process is largely controlled from the old industrial core, where smokestacks have given way to cell phone towers and factory bricks have been replaced by electronic clicks as a new service economy based on the financing, advertisement, distribution, and retailing of these products has come to dominate.

Not long ago, the word for our times was *modern:* buildings, ideas, technology, fashion, and people were all—or should be—modern. Theorists and analysts of worldwide developments spoke and wrote about "modernization." Today, everything from architecture to theory is dubbed *postmodern,* and *modernization* has disappeared from books and articles to be replaced by the new term *globalization.* Philip McMichael (2000) has labeled the period from 1945 to the mid-1970s as the "development project," when modernization views of progress dominated the international scene. Since that time, he argues, we have embarked on the "globalization project," when neoliberal free-market economics is the dominant philosophy and the favored alchemy for getting ahead. A word rarely heard before 1980, *globalization* is now in such common use that business and professional magazines feature lists of the top 20 books on the subject.

Certain people—the powerful, the greedy, the adventurous, and the curious—have always been interested in the world beyond their borders. Those with the means have sought to satisfy their curiosity and their longing for exotic luxuries by making contact with other cultures, establishing commerce with them, and controlling trade routes. Human history is the history of contact and exchange. So is globalization anything new? Yes.

Today's global economy is unlike anything humankind has known before, if only in its pace and intensity. Container ships can move the products of an entire city across the world's oceans without any individual handling of anything on board. Jumbo jets can move the equivalent of an entire village anywhere on the planet in a matter of hours. But the most important element in the rise of the global economy has been the electronics revolution. Satellite and cable carry news and ideas around the globe instantly and, through multiple media outlets, create demand for those shiploads of products, even among those who never leave home. Further, billions of dollars can be moved from one continent to another with the click of a mouse, making global control of commerce and finance easy and instantaneous.

What makes this electronically created global society such a powerful force is not just its speed but also its breadth. Global reach is no longer limited to the rich and powerful. Before I began to write this chapter, I went online and e-mailed (a very new verb) a Mexican colleague in Europe and a U.S. colleague in Sri Lanka, sent a joke to a friend in China before she leaves for Germany and to an Australian friend who is teaching in Korea, read a French newspaper, moved some savings from a money market account in New York to my checking account in Indiana, checked the prospectus of a "socially and environmentally responsible" mutual fund that wants to invest some of my retirement income in Latin America and East Asia, and checked the weather over the Pacific for a student who was leaving for Malaysia. We may not be experiencing the "global village" that Marshall McLuhan (1964) spoke of, but we all seem to be sharing one big global cybercafé and trying to figure out how it all works and what it all means.

Anthony Giddens (2000), a British social scientist who, like others, has shifted from writing about the structures of modernization to writing about the structures of globalization, calls the pace and scope of these changes a "runaway world." One of the fundamental questions of this runaway world is what it means for global inequalities, both those between nations and those within nations. The effects of globalization on inequality between nations have been complex and varied. In some ways it has led to a growing sameness, especially in the world's global cities. Whether one is in San Antonio, Texas, Nairobi, Kenya, or Kuala Lumpur, Malaysia, one sees new high-rise office buildings in the same styles and congested highways filled with a mix of the Mercedes-Benzes of the old rich, the BMWs of the new rich, the Hondas of the middle class, and, alongside, the carts and packs of the poor and the homeless. The proportions one finds in these categories, however, vary greatly. In general, the wealthy powers of the old economy have become the new nodes of commerce and control for the new economy. In many ways, the picture between nations in the global economy is much like the picture between people within nations: The rich have gotten richer, a few young, urban nations have made impressive gains, most of the poor have made few gains, and the poorest have lost absolutely.

The effects of globalization on inequality within nations have also been complex, but a pattern of growth with growing inequality seems to be emerging. This is partly a direct effect of changes in technology: Those who control and manage productive technology reap profits as a result of the greater productivity the technology affords. Workers, on the other hand, may lose their jobs to new technology or find that their skills are replaced or rendered redundant by automation, both mechanical and electronic. They are displaced or "deskilled" just as workers were in an earlier industrial era.

The **new international division of labor** that comes with globalization can also lead to expanded inequality, particularly within nations. Those who can control global markets and global production can garner a world of profits, benefiting from both the cheapest inputs of labor and resources and the largest markets afforded by the entire planet. In contrast, those whose skills are now replicated by workers around the world find themselves competing for jobs and wages with the entire planet. In this sense, the global capitalism of the twenty-first century has had effects similar to the national and international capitalism of the nineteenth century, but these effects have been greatly magnified. Spanning many different lands with many different laws and practices, they are also much more difficult to control.

These effects can be seen across the rich nations of the so-called First World and the poor nations of the Third World, and especially in the new players in this new world order (or disorder). India holds the Western imagination as the land where Mother Teresa ministered to the homeless, dying poor. Indeed, India has more beggars than any other nation in the world. But it also has more software engineers. That's right—more than Seattle, California's Silicon Valley, Texas, Boston, and beyond in the United States combined. Many have left India in a great brain drain, making Indians the best-educated immigrant group in the United States. But many more can now "telecommute," selling their skills to firms around the world. This means that even skilled, white-collar jobs can now be easily outsourced. When my home computer breaks, a toll-free telephone number puts me in touch with a polite and helpful person who will reveal only that he is in the SBC "global response center," which presumably exists somewhere in transnational cyberspace. Yet despite careful training, every helpful staff member with whom I speak has an unmistakable South Asian accent.

A student mentions in passing that her husband's small local firm is **outsourcing** the engineering for a new contract—to India, for rates around $12 an hour rather than the $50 to $70 an hour Americans would be paid for the same work. In this, no one is secure. An e-mail advertisement tells me that for much less than I would pay even work-study U.S. students, I can outsource interview transcription to India—send away audiotapes and have the transcripts delivered to me electronically. Publishers regularly outsource work on the different stages of book production. One of my first manuscripts was copyedited by an unemployed Hollywood scriptwriter. Often in the 1990s, American book publishers used printing companies located in Hong Kong. This book was copyedited by a freelance editor in the United States, then computer typeset in the growing electronic hub of Chennai, India (the British formerly knew this as the textile trade center of Madras). Americans have gotten used to the tremendous numbers of products that are now produced in China and the impact of this phenomenon on the U.S. working class. A growing issue is the impact that a well-educated and low-paid labor force in India could have on job security for the U.S. middle class.

The first country to move into white-collar outsourcing in a major way was Ireland. On the periphery of Europe, Ireland offered an English-speaking labor force that would work for low wages and had good technical infrastructure. Soon many companies realized that it was cheaper to send electronic records across the Atlantic to Shannon or Dublin to be processed. Insurance companies led the way, but eventually even municipalities also tried cutting costs in this manner. Ireland has been quite successful with this, so successful that wages are rising there, and this option is no longer so appealing to potential outsourcing companies. The first move toward outsourcing white-collar work occurred at a time when data-processing jobs were so plentiful in the United States that most U.S. workers didn't worry about the competition. Now, with the U.S. labor market again tight and a huge wage gap between the United States and India, white-collar outsourcing is back as a concern. And with technological advances, the possibilities for outsourcing continue to grow.

Coming home after a late-night party, two college students have an auto accident and are rushed to a nearby emergency room for X-rays. Already waiting in line is someone who needs a chest X-ray to check for possible pneumonia. The X-ray technician works through the night, but the radiologists, overworked from the growing numbers of complex diagnostic procedures they must perform, have all long since gone home to bed. They can rest easy. X-ray results are increasingly digitized rather than put on film, and digits can circle the globe at the speed of light. The results of the students' X-rays and those of the other patient are sent electronically halfway around the world to Hyderabad, India, where it is the middle of the day, and an Indian physician reads the results and e-mails back a diagnosis. The next day, one of the hospital's radiologists, an Indian immigrant himself, reads the e-mail and confirms the results. Such global exchanges are becoming increasingly common. They meet real needs, such as those caused by shortages of specialists, and take good advantage of time differences. Is it possible that U.S. radiologists, who are among the highest-paid medical specialists, will ever find their incomes challenged by cheap foreign competition? Might Indian radiologists find their incomes growing, moving them closer to their U.S. counterparts but further from most of their fellow Indians? The possibilities are complex and intriguing.

By some estimates, India already has a larger middle class than does the United States. How can that possibly be? The middle class in India has been growing for decades, and in a country of a billion people, even if only one-fifth can claim this status (many more are poor), that is still 200 million people and growing. Sometime during the twenty-first century, India, with a population growth rate that is slowing but still higher than China's, will be the world's most populous country; it already holds more people than all of Africa. India was much slower than most of East Asia to enter fully into the global marketplace, but it has been increasing its role in the past few years, along with impressive growth rates. International comparisons show that India is no longer one of the world's poorest nations; rather, it is a "middle-income" nation, with annual per capita income of $2,670, up from an average of only $1,422 in 1995. But that is an average of extremes, and not just extremes of very rich and very poor, as Americans often assume, but of rich, middle, and poor. India has a handful of industrial and media moguls (an Indian word, in fact, for wealthy Muslim rulers), a highly educated professional and managerial middle class that is vulnerable to economic downturns but continues to grow into the hundreds of millions, and yet a mix of rural and urban poor that may top 600 million people. Some look at these numbers and see a great sea of miserable humanity, whereas others see a great pool of potential low-wage workers alongside an untapped pool of middle-class consumers. India in the twenty-first century will likely be all of these.

Mexico, which once had a highly protected economy like India's, now has more trade agreements with other countries than any other nation, including the United States. For Mexico, the North American Free Trade Agreement (NAFTA) was only the beginning of a great embrace of free trade. Mexico's economy has experienced a growth rate of between 4% and 5% since the passage of NAFTA—not phenomenal growth, but strong. It has moved from being the world's 26th largest economy to the 8th largest. Yet almost all of Mexico's post-NAFTA industrial growth has been in maquiladoras, export processing plants that assemble goods from imported parts and then ship the goods back for sale by foreign-owned multinational corporations. Although employment has increased, real wages in industry during this time of growth have declined by as much as 20%. What happened? The pattern is by now familiar. A few spectacular successes, such as Mexico's Cenex cement industry, have become global competitors, and several of the world's richest people are Mexican. Much more extensive in Mexico are foreign-owned firms such as Nissan, Ford, Volkswagen, and General Electric, which hire a handful of Mexican engineers and managers to help bolster an always-worried but nonetheless growing middle class that also includes a mix of professionals in medicine, education, law, and government. Workers in manufacturing, however, are less likely to have effective union protection and more likely to be poorly paid young women, most of whom are secondary school graduates but have few options other than equally low-paying clerical work. Beneath these are the truly poor: rural workers with declining incomes who must now compete with cheap U.S. and South American agricultural products, and informal economy workers who try to get by hawking, fixing, and reselling the many foreign-made products that fill the country (or trying to sell "pirated" versions of these products). These poor and informal economy workers may make up as much as 40% of the total Mexican workforce.

Has globalization finally brought Mexico both democracy and prosperity? Or does it merely offer new forms of economic exploitation, political domination, and social distortion? The answers to these questions matter enormously, not only to Mexico's 100 million people, but also to the billions of the world's poor, both the desperate and the hopeful, who watch and wonder.

_____ Immigration: Seeking to Cross the Divide

To an economist, the economics of immigration are quite simple: Given the big gaps in wages and demand for workers around the world, it makes sense for people to move from low-wage, low-demand areas to high-wage, high-demand areas. Yet many people in low-wage, low-demand areas are reluctant to leave their families, their homes, and their familiar cultures, and many high-wage, high-demand locations are reluctant to welcome newcomers. **Immigration** involves leaving one's country of origin to live in another country. **Labor migration** is a temporary move in search of work, with the intention to return home. Kuwait has many **guest workers** in its oil fields. Such workers do not get Kuwaiti citizenship; rather, they work and then return to their original homes in Egypt, Jordan, the Philippines, or wherever. Germany has long had many Turkish guest workers; it was never assumed that they would become German citizens.

The United States is a nation of immigrants. A few arrived some 10,000 or more years ago, but most others within the past 400 years, and many much more recently than that (see Exhibit 2.3). Yet, at times, the United States has also sought labor migrants. The Chinese workers who were imported to help build railroads in the 1800s did not bring their families and were not expected to stay. They were to do the work they were needed to do and then go home. In the 1900s, Mexican workers were offered the same arrangement at various times. More recently, U.S. President George W. Bush has proposed establishing a new guest worker arrangement with Mexico to meet the demands of certain U.S. industries and agribusiness, but he has not offered these workers a guaranteed route to citizenship.

Immigration, labor migration, and outsourcing are intertwined in a peculiar cost accounting of micro- and macroeconomics. Places with long-term labor needs may encourage immigration, especially from certain locations deemed culturally or ethnically desirable. Cyclical demands may lead to calls for temporary labor migration, so people can be sent home when the work is done. Sometimes it is cheaper to move people, and sometimes it is cheaper to move products. For a long time in the United States, fine stitching on garments was done by immigrant labor in major cities, first by European immigrants, then by Asian and Latin American immigrants. This practice still persists in certain locations in New York and Los Angeles, where quick turnaround on orders is essential. For more routine sewing, it is now cheaper to have the work done in Asia or Latin America and to move the garments rather than the garment workers across borders. The largest of the U.S. guest worker programs for Mexicans ended as U.S. factories found it cheaper to move their operations over the border into Mexico than to bring Mexican workers into the United States. In agriculture, Mexican American migrant laborers in California must now compete with low-cost farm produce coming from northern Mexico. The irony is not lost on the struggling workers:

> *El Paso, Texas:* Ernestina Miranda left Mexico for the United States in 1979 in the trunk of a car. She found a job sewing blue jeans in one of the dozens of clothing factories here. Work was steady, six days a week, 12 hours a day. She married and bought a trailer—without running water or electricity—on a plot of land. She was awarded citizenship in the late 1980's. Now, those blue jeans jobs that brought Mrs. Miranda and thousands of others like her north have gone south, to Mexico.
>
> "My American dream has turned into a nightmare," she said, over a glass of strawberry Kool-Aid in her listing trailer. Until recently, she had made a life on $7.50 an hour. She has become a temporary worker in a plastics plant that used to be based in Michigan, earning minimum wage, no benefits, no security. Her husband, Miguel, is unemployed. The mortgage on the slapdash home is in peril. "I worry about the future," she said, echoing the sentiment of blue-collar and increasingly of white-collar workers from Los Angeles to Detroit, people who find their jobs being shipped to countries where wages are a small fraction of theirs. (LeDuff 2004)

The greatest demand now for guest labor, whether official or illegal, is in those service areas that are hard to outsource: gardening, housekeeping, and child care. Immigrants find work near the bottom of the wage scale as taxi drivers, kitchen workers, and health care aides, and high on the wage scale

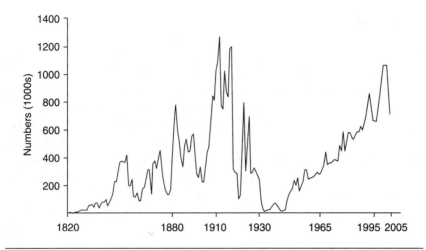

Exhibit 2.3a Immigration to the United States

Rank	Country	Number Entering United States Legally
1	Mexico	131,575
2	China	36,884
3	India	36,482
4	Philippines	34,466
5	Former Soviet Union	30,163
6	Dominican Republic	20,387
7	Vietnam	17,649
8	Cuba	17,375
9	Jamaica	15,146
10	El Salvador	14,590
11	South Korea	14,268
12	Haiti	13,449
13	Pakistan	13,094
14	Colombia	11,836
15	Canada	10,190

Exhibit 2.3b Immigration to the United States, 1998

Source: U.S. Immigration and Naturalization Service, 1997, 2000, 2004.

as technical and medical professionals in high-demand areas. Much of the traditional industrial draw is no longer available. At this point immigration into the United States and Western Europe continues to be strong, accounting for almost all of the growth in these areas. Yet as products and information move about the globe ever more quickly, the main need for immigrants and labor migrants will likely be to fill the often low-wage sector of hands-on service. In Europe, and maybe now in Japan, an aging population may be dependent on this guest labor for personal and medical needs, and in the United States professionals working long hours may also increasingly need the domestic services these workers provide.

The Market Paradox

Neoclassical economics appears quite correct about the power and appeal of the market. Yet political economists provide an important reminder that the market operates in conditions of unequal power. Just as well-connected individuals can command great returns on their efforts and excluded and marginal people must accept whatever they are offered, so it is also in the world of nations. World systems theorists note that great wealth does not automatically pour into places with great natural resources, such as Angola and the Democratic Republic of the Congo, two of the most resource-rich and income-poor countries in the world, but rather to the well-connected core cities of core countries: Tokyo, Zurich, London, Frankfurt, and New York.

The problem with markets is rooted in a great paradox: Markets work best under conditions of relative equality, but they inevitably tend to produce conditions of extreme inequality. Market solutions work best when everyone has relatively equal incomes. Food and housing, perhaps even health and education, can be "marketized" if everyone has sufficient income and information to make good choices, but these markets quickly break down when some can dominate the market while others have no say at all. The same is true for power. Markets work best in a highly competitive environment with many players of equal power all competing for that equally distributed income. From Adam Smith onward, market proponents have argued for competitive marketplaces over monopolies, whether private or public. Yet markets, at least unregulated markets, continually produce monopolies. From the British East India Company to Standard Oil to Enron and Microsoft, a few key, centrally placed players can use a combination of market power and political power to control the marketplace. The process is familiar to any player of Monopoly: the game begins as a competitive free-for-all, but early successes translate into growing dominance until one player controls it all and the others face inevitable bankruptcy.

This realization led Karl Marx to believe that a replacement for capitalist markets must be found. Some world systems theorists contend that in a globalized world, this replacement can no longer be socialism at the national level; rather, it must be some form of democratic global socialism. Many others are still trying to find ways to tame and harness the market: breaking up international monopolies, encouraging local entrepreneurial alternatives, and strengthening the role of labor in the global labor market.

The debate is often politically polarized between conservative free-trade advocates and progressive antiglobalization protesters. Yet the strategy of supporting entrepreneurship and small business while limiting the power of monopolies has appealed to free-market advocates from Adam Smith, who wrote during a great wave of globalization in the late 1700s, to Theodore Roosevelt, who campaigned for U.S. president during a great wave of globalization in the early 1900s. And even those who advocate the eventual rejection of global capitalism concede that the need in the medium and short term is to find ways to turn the market toward the meeting of human need while controlling excess concentrations of wealth and power. This is the great challenge of the twenty-first-century world economy.

KEY POINTS

- The world faces a double divide: Income inequalities are increasing both between and within nations.
- Modernization theory contends that low-come countries need modern ideas, technology, and institutions. Dependency theory and world systems theory contend that poor countries face exploitation, domination, and economic distortions because of their dependence on rich countries.
- The perspective variously known as neoliberalism, neoclassical economics, and free-market economics stresses the importance of free trade among nations and the need to reduce government interference in domestic markets.
- Political economists question the wisdom of trade liberalization in a world of vastly unequal power and access to resources.
- Globalization is radically altering the world economic system. Global interconnections are affording new opportunities to some. Global markets can increase global inequalities by increasing the revenues of large corporations while at the same time forcing wages down through increased labor competition.
- Continued wage gaps between nations have led to outsourcing, labor migration, and immigration. The most rapidly growing income inequalities are now those within nations between new elites and the excluded and marginalized classes.

FOR REVIEW AND DISCUSSION

1. What are major causes of the "double divide" between and among nations? What factors are emphasized by dependency and world systems theorists, by modernization theorists, and by neoclassical economists?
2. How has globalization changed economies and the lives of workers? Is it likely to lead to greater poverty or greater prosperity? What factors support your point of view?

MAKING CONNECTIONS

Fair Trade

Explore any "alternative traders" that operate in your community or a neighboring city. Alternative traders (sometimes called fair traders) attempt to provide alternatives to exploitative international trade relations. For example, they may purchase products directly from local producers in low-income countries and communities, often giving special attention to local cooperatives, to women's and poor people's groups, and to products that are produced in socially and environmentally responsible ways.

The products such traders purchase include clothing, coffee, and crafts, which they often resell for little or no middleman profit, so that a large portion of the retail price returns to the producers. Alternative traders with local outlets include Ten Thousand Villages (www.tenthousandvillages.com), SERRV International (www.serrv.org), Equal Exchange (www.equalexchange.com), and MarketPlace: Handiwork of India (www.marketplaceindia.org). Visit their Web sites for information on fair trade and then check for locations or outlets near you. If possible, visit some alternative trade stores, look at the items (bring your Christmas list if you like!), read the brochures, and talk with store personnel. How do alternative traders attempt to cope with the problems and inequality of the global economy? Is this a viable way to combat the destructive and exploitative aspects of world trade?

United Nations

Visit the Web sites of the United Nations (www.un.org) and the United Nations Development Program (www.undp.org). The main U.N. site is "information central," providing facts and statistics on a host of international issues. You can focus on a topic of interest, such as human rights or gender issues, if you like. The U.N. Development Program collects a wide range of data on global well-being and economic development and publishes the annual *Human Development Report*. What are the key indicators of well-being? Where is progress being made? Where are we "losing ground"?

Photo Essay: Honduras

Catherine S. Alley

The strength of those who sustain themselves with subsistence farming and money sent from those family members working in Tegucigalpa several hours away is evident in the faces of the residents of the Agalta valley of Honduras. Villages are tight-knit microcosms of families, who live near one another and help to support each other, assisting in projects, farming, and caring for children.

Living in the small town of Juticalpa affords these Honduran boys the luxury of attending school. Classrooms may be crowded and the supplies limited, but they are in fact among the privileged when compared with their more rural counterparts, who often live prohibitive distances away from the few schools.

Children help at home and with farming from an early age, and the youngest children can be found playing with whatever objects they can find. This little girl who is playing with an ab-roller—perhaps found in a box sent by a church in the United States.

Transportation for the majority of rural Hondurans is limited to their undersized horses and oxen, which are also used in farming. Running water essentially does not exist.

While visiting, "gringos" are warned not to consume any water; families wash clothes in the same rivers where they bathe and collect water to drink.

A village might have one communal outhouse. During the rainy season, it is not uncommon to see the waste of chickens and dogs in muddy yards outside of villages.

Homes are adobe. Most floors consist of packed dirt; however, it is becoming a more common mission activity to provide cement for families to use for paving the floors of their homes.

PART II

Dimensions of Inequality

3

The Gordian Knot of Race, Class, and Gender

When Alexander the Great brought his armies across Asia Minor, he was reportedly shown the Gordian knot, an intricate, tightly bound tangle of cords tied by Gordius, king of Phrygia. It was said that only the future ruler of all Asia would be capable of untying the knot. The story recounts that a frustrated Alexander finally sliced the knot open with his sword.

There are many dimensions to inequality, and all of these dimensions are interrelated. **Class, race,** and **gender** are three of inequality's core dimensions. Asking which of them is most important may be like asking which matters more in the making of a box, length or width? These dimensions are like the 9 to 11 dimensions that quantum physics imagines for our universe: tangled, intertwined, some hard to see, others hard to measure, but all affecting the makeup of the whole. We could note other dimensions as well. Age, for example, can provide both advantage and disadvantage, privileges and problems. We stereotype both ends of the age spectrum: "silly teenagers" who talk, dress, and act funny, and "silly old codgers" who talk, dress, and act funny. Age is unlike class, race, or gender, however, in that unless our lives are cut short, we all move through all age categories. Sexuality and sexual orientation also constitute a complex dimension. Debates over "gay marriage" and who qualifies as a partner for the purposes of health care, tax, and housing benefits highlight how sexuality can be a dimension of privilege or disadvantage. Stereotypes, discrimination, and vulnerability to violence are all also bound up in the sexuality dimension. Some dimensions, such as race, **ethnicity**, and religion, are frequently so bound up together that they are hard to disentangle. In this chapter we explore some of the dimensions of inequality. We can't completely untangle this knot in our social fabric, but we can at least slice into it.

Dimensions of an Unequal World

Inequality is at the core of sociology and its analysis of society. It is also at the core of your daily life experience, although you may not realize it. You may know you are broke. You may wish you were rich. You may be angry about the time you felt rebuffed as a black female—or as a white male. You may have a sense that some people's lives have been a lot easier than yours—or that some have had a much harder time. In the United States in particular, and in most of the world generally, we are continually affected by social inequalities yet we are rarely encouraged to think in those terms.

We know that many people are poor, but why are they poor? Perhaps they are just lazy. That's certainly possible—I have met some very lazy people. But, come to think of it, not all of them are poor! If you have ever worked for a "lazy" supervisor or dealt with a "lazy" professional (not among your professors, I hope!), you know that it's possible for some people to be less than diligent and still command positions of authority and high salaries. Perhaps the poor are just unlucky. Certainly luck matters a great deal in our society. You may know of people who have had "bad luck": They've lost their jobs, or are in fear of losing long-held positions, just because their companies are closing or moving. Yet when we step back to look at the numbers, we find there are a great many of these "unlucky" individuals out there, all with similar stories. Patterns that go beyond individual misfortune are clearly at work.

You may also know people who "have it made" and wonder how they got to where they are. If you ask them, most will decline to claim special talents or brilliance; instead, they're likely to say something about diligence and hard work. Hard work certainly can't hurt anyone seeking success. But then again, I know of a woman who works 12-hour days doing the backbreaking work of picking vegetables and then goes home to care for three tired and hungry children. She works hard, but she does not seem to be climbing the ladder of success. Having access to the right schools, financial resources, business and professional contacts, and particular opportunities seems to play a large role in turning hard work into hard cash. The sociological study of social inequality does not negate individual differences and efforts, but it seeks to examine patterns that go beyond individual cases, to explore differences in access and opportunity and the constraints that shape people's choices.

Sociologists are interested not only in the fact of inequality but also in how this inequality is structured. When geologists are trying to understand the structure of rock formations, they look for strata: layers with discernible borders between the levels. Sociologists look for social stratification—that is, how the inequalities in a society are sorted into identifiable layers of persons with common characteristics. Those layers are social classes. Although scholars have examined the structure of social classes since the middle of the nineteenth century, most of us rarely think in class terms. Particularly in the United States (as well as in some other countries, such as Canada and Australia), the cultural emphasis has been on the equal standing of all members of society; Americans are generally reluctant to use the language of class beyond vague and all-encompassing allusions to being "middle-class." The term *middle class* once referred quite specifically to that group that stood in the middle ground between the common working classes and the wealthy propertied classes. Today, a wide range of people willingly claim middle-class status, for it seems uppity to label oneself upper-class, and almost no one wants to admit to being lower-class, which sounds like an admission of personal failings.

Certainly a simple division of American society into distinct social classes is not easy, and the difficulty is compounded by inequalities that come with gender, race, ethnicity, nationality, and age. Yet if we look even casually at various neighborhoods, we can easily see that we are looking at clusters of very different lifestyles. We can see class distinctions in the houses and the cars, and also in the residents' attitudes and routines as well as in their preferences or tastes in everything from yard decor to Christmas lights. In many ways, members of different classes live in different, and divided, worlds.

We may be most resistant to the language of social inequality and social class when it comes to the area of our past accomplishments and future aspirations. Certainly we are where we are, in this place in life, in this university, with a particular set of prospects, because of our own abilities and hard work. The message of sociologists that our life chances (what we can hope for from our lives) and our mobility (whether we move up or down in a stratified system) are both socially conditioned and socially constrained is not likely to be a popular one—at least not until we have to explain why we failed that entrance exam or didn't get that job! We may be sensitive to personal bias ("That person was against me because of my age [or race, or gender, or clothes, or whatever]"), but most of us overlook the way the entire structure of our social system shapes our opportunities.

Let me illustrate with an example from my own experience and background. I am a sociology professor at a midsize public university in the Midwest. I have an occupational title and educational credentials that place me near the upper end of job prestige rankings—at least I get called "Dr." by my students and by telemarketers. I earn a salary that places me somewhat above the overall national average (although somewhat below the average for persons of my age, gender, and race). My social class background doesn't differ greatly from that of many of my students. My grandfather was the son of German immigrants who farmed a bit and ran a small "saloon." He did various odd jobs before marriage, and then helped run his wife's parents' struggling farm. During the Great Depression of the 1930s he worked as a night watchman in Chicago until he was injured in a fall, and then he drove a cab before returning to central Wisconsin to work in a paper mill and do some minor truck farming. Asked about his life, he would talk about hard work, about good times and hard times, about luck and perseverance and getting by. He was right about all this, of course. Yet his life was shaped in innumerable ways by his social class, as well as by his ethnicity, and these were constantly interacting with broader social and economic changes. Had his own parents been wealthier, they could have bought richer farmland where they began in Indiana rather than moving on to marginal land in central Wisconsin, and their "luck" at farming might have proven much better. Had they been able to afford a better-capitalized business than their small saloon, they might have joined a growing circle of prosperous German American small business owners in Chicago or Milwaukee. And if my grandfather's parents had brought no money at all with them from Germany, they may never have been able to buy land or a business and so would have faced even more difficult times. Had they been African American or Hispanic American, they would not have been accepted in rural central Wisconsin and would more likely have found a home in Gary, Indiana, or South Chicago. As it was, they were part of the great immigrant movement that reshaped this country early in the twentieth century, and also part of the great movement of people from agriculture into industrial and service economies.

My father grew up amid the difficult times and ethnic antagonisms of Depression-era central Chicago. When his family moved back to central Wisconsin, he helped on the farm and went to work in the paper mill. When he found he hated mill work, he tried bartending. Finally, he studied for a real estate license and subsequently sold new homes for many years in north suburban Milwaukee, prospering slightly during the growth years of the 1950s and 1960s and facing lean times during the recession years and slow growth of much of the Midwest that followed in the 1970s. When asked about his life, he too spoke of work that he liked and hated, of good times and hard times, and of hard work and perseverance. He did not speak of the real estate license as his precarious step into a growing middle class, or of his being part of a generational movement from blue-collar industry into the white-collar service sector, or of social forces such as suburbanization and the flight of industry and jobs from the Midwest and Northeast, yet his opportunities and life chances were very much influenced by these events.

As for myself, truly dismal performances in door-to-door sales and in junior high school "shop" classes convinced me early on that I was suited for neither sales nor industry. I excelled at school, especially in courses involving writing or science. A move during my fourth-grade year from an increasingly troubled urban elementary school to a substantially more rigorous and well-equipped middle-class suburban school system helped me to develop these interests. I worked odd jobs during high school to earn enough money to pay for three years at a small public university at a time when in-state tuition was quite reasonable. With very high SAT scores (I excel at the abstract and impractical), I had many offers from other colleges and universities, but I didn't believe I could afford any of them—I had neither the savings nor the savvy needed to pursue scholarships and financial aid. My father and my grandmother contributed a bit, and my flair for testing out of courses got me an undergraduate degree in the three years of college I could afford. The downside of this rush was that I graduated into an economic recession, and there was little new hiring going on. My penchant for academics suggested graduate school as a likely course. An early interest in law and politics had shifted to social science and social policy, and so, eventually, I landed at a large Ivy League university and emerged with a Ph.D. in sociology.

Certainly my experience, neither especially privileged nor especially deprived, would seem to be the result of my individual motivations and abilities. The fact that I'm not rich is explained by my choice of profession. The fact that I was the first in my family to obtain a college degree and the fact that I have a secure white-collar position (although I actually wear far fewer white collars than my father did) are both explained by my own individual set of abilities and hard work. My students who come from working-class origins strongly argue the same, and few would ever admit privilege. But note how my experience might have been different.

What if my family had been wealthy? What if my father's failed attempt to start his own business had succeeded and he had ridden a suburban building boom to great prosperity as a real estate developer? Others with a bit more investment capital and better timing had done just that. We would have lived in one of the wealthier suburbs across the river. I would still have attended a public high school, but it would have been one with exceptional facilities and programs. In place of the odd jobs I worked, I would have indulged my interest in tennis—I'm not very good, but with early private instruction I may well have been able to make my high school's tennis team. In a school where virtually every student was college bound, I would have received very good guidance counseling and would have carefully gone through the many brochures that I received for selective liberal arts colleges with beautiful buildings and beautiful female students on the cover. With the assurance that I could afford to attend these schools (I had the test scores to get in), I might have welcomed my guidance counselor's advice in choosing the best. (I would have needed this advice, given that this scenario still assumes I did not have college-educated parents to help steer my decisions.) My parents, my peers, and my counselor would likely have strongly encouraged my interest in law. Only with a high-income profession could I hope to return to live in the same exclusive suburban area with my friends and classmates. A private law practice would still allow me to work with and consult for my father's business. Alternatively, I could indulge my interests in geography and urban studies as an undergraduate, then pursue a graduate business degree, and then combine these interests as I eventually took over that business. Certainly, ambling excursions into the social sciences would have been discouraged. Some of my individual tastes and abilities would still be there, although now honed and shaped in new directions by my social situation. I might still enjoy history and social science and writing, but I might be spending evenings in the study of my large suburban home writing something like *The Seven Business Secrets of Benjamin Franklin*.

On the other hand, what if my family had been poor? Our important move from the city to the suburbs when I was 9 years old would never have taken place. I would instead have attended an urban high school in north-central Milwaukee that suffered through a decade of declining facilities and neighborhoods, mounting racial tension, and a growing drug problem. I probably still would have graduated and might still have aspired to college, but I would have had to work out attending occasional classes at the nearby university while attempting to work and contribute to the family income. Some early difficulties I had with math would likely have gone uncorrected, limiting my academic options. If I had managed to graduate from college, it would not have been in three years but in six or seven, and I don't know if I would have had the energy or enthusiasm to consider graduate school.

Consider an even thornier question: What if my family had been black? My father might well have still made his way to Milwaukee from Chicago, but he probably would never have left industrial work. He could not have left to tend bar in the late 1940s unless he was willing to work in an all-black club. He could not have gone into selling suburban new construction, as that was an all-white domain right through the 1970s. Black would-be home buyers were systematically excluded by their limited financial resources, banks, and fearful white suburbanites, and there was no place for a black real estate salesman in suburban Milwaukee. My father would likely have continued to work at one or more of the factories on the west side of the city, many of which closed or laid off workers in the industrial downturn of the 1970s and 1980s. This would have occurred just as I was reaching crucial high school and college years. I would have faced all the hurdles of the "poor" scenario, plus the school counselors of the time might not have strongly encouraged me to pursue college.

Finally, what if I had been female? This may be the hardest scenario of all to unpack, for gender assumptions and inequalities are so thoroughly built into our families, our peer interactions, the media, and institutions such as schools that they are nearly impossible to disentangle from personal characteristics. My difficulties in math, which stemmed from the move from an urban to a suburban school that was a full year ahead, so that I virtually skipped long division and fractions, would likely have been attributed to my gender and not to the move. My high school counselors would not have pushed me to overcome the deficiencies, and although I still would have easily graduated from high school, my SAT and GRE scores would have been much lower, dragged down by poor math performance. With lower scores, I would have had fewer choices regarding colleges and may not have gotten into a top graduate school. I may have benefited from being female in one area: the odd jobs I worked on and off during college. The temporary agencies I worked for automatically assigned female applicants to clerical positions, which were cleaner and paid just a bit more than the "light industrial" work that males were assigned. Even though I type far faster than I pack boxes, I never got a clerical position until finally the federal student loan office needed an office worker who could lift heavy boxes full of loan applications and files, and so decided to hire a young man. Of course, in today's economy, I would have to market my computer skills—searching those student loan databases—rather than just a strong back.

But if I had been female, other challenges would have been waiting down the road. Our first child was born while I was in graduate school, and my wife was able to take a short amount of maternity leave from work. Had that been my husband instead, he could not have taken off for "paternity" leave, and I would have had to delay completion of my graduate program, as I would have had to take on a greater portion of the child care than I did. Career success following graduation (assuming I got there) is also a complex question. Sociology is a field rapidly opening to women, and, depending on where I went, I might have done quite well as a female sociologist. Had I chosen the legal profession, I may also have done well, but as a woman I would have been able to expect to earn less than 70% of what male law school graduates earn over the course of my career.

A key theme of sociology is that who we become is the result of a complex interplay between individual characteristics and our place in society, which determines which of our characteristics are encouraged, rewarded, and constrained. The fact that these rewards and constraints are so unequally distributed makes the topic of social inequality at times very disturbing but also intensely interesting. In a now famous turn of phrase, C. Wright Mills (1959) referred to our ability to connect our personal biographies to the broader sweep of history and society and to see the connections between personal troubles and social conditions as the **sociological imagination**. I hope you won't read the chapters that follow passively, but instead try to engage and develop your sociological imagination, actively making connections between personal experiences—your own as well as those of people you know—and the stratified social structures that advance or hinder our hopes, plans, and possibilities.

Intersections of Race, Class, and Gender in the United States

Where you stand in Robeson County depends on who you are, who your family is, and to what group you belong. The county courthouse even lists veterans in a hierarchy: whites first, then Lumbee Indians, then African Americans. Where you stand at Smithfield Packing, the largest hog-butchering and pork production plant in the world, located nearby, also depends on who you are. White men hold supervisory roles:

> It must have been 1 o'clock. That's when the white man usually comes out of his glass office and stands on the scaffolding above the factory floor. He stood with his palms on the rails, his elbows out. He looked like a tower guard up there or a border patrol agent. He stood with his head cocked. One o'clock means it is getting near the end of the workday. Quota has to be met and the workload doubles. The conveyor belt always overflows with meat around 1 o'clock. So the workers double their pace, hacking pork from shoulder bones with a driven single-mindedness. They stare blankly, like mules in wooden blinders, as the butchered slabs pass by. (LeDuff 2001:97)

Most of the other white men at the plant are mechanics. A couple of white men and the Lumbee Indians make boxes. A few Indians are supervisors or have other "clean" jobs. Given that most of the local businesses are owned by whites or Indians, members of these groups may have other job options. Almost all of the newly hired black women go to the "chitterlings" room to scrape feces and worms from the hogs' intestines. Hardly elegant, but it is "sit-down work." The black men and most of the Mexicans and Mexican Americans, both men and women, go to the butchering floor. There they stand for eight and a half hours at a stretch, slashing at hog carcasses with sharp knives, trying to get all the meat from the bone and to turn out the required 32,000 pork shoulders per shift at a rate of 17 seconds per hog per worker. At the end of a shift their backs ache, their wrists ache, their hands are numb. Some say this giant plant doesn't kill just pigs—it kills the hearts, minds, and bodies of the workers. But this is the only job for miles in any direction that pays "unskilled workers" as much as $8 an hour.

Profits at this plant have doubled in the past year while wages have remained stagnant. The management is vigorously anti-union and has been accused of assigning workers to their stations based on race and gender. Yet the plant remains able to get workers, although with difficulty: Some come on release from the local prison, some are recruited from New York's immigrant communities, and some come because they've heard by word of mouth, in Mexico and beyond, that jobs are available. By far most of the newcomers are Mexicans and Mexican Americans—some in the United States legally, some with forged papers and huge debts owed to "coyotes" who slipped them across the border. The work on the plant floor is so hard that employee turnover is virtually 100%—5,000 leave in a year and 5,000 come. Increasingly, those who come are Hispanic.

LeDuff (2000) notes that African Americans in Robeson County had long hoped that their position in the plant and in the community would improve, but they now find it stagnating; they are as poor as ever. They tend not to blame the plant's management or the town leadership for this situation; rather, they blame "the Mexicans" for taking their jobs and lowering their wages. Even the anger that black workers express toward abusive management is tinged with racial expectations. As one said about a manager, "Who that cracker think he is . . . keep treating me like a Mexican and I'll beat him!" (p. 98). The strain and pain of this work revive bitter racial epithets that have otherwise disappeared from polite American speech, as weathered old white employees speak of "the tacos" and "the niggers."

At the end of a day, the workers pop pain pills, swab cuts with antiseptics, and make the long drive to trailers, cinder block houses, and wooden shacks in segregated communities: whites to Lumberton, blacks to Fayetteville, Indians to Pembroke, Mexicans to Red Springs. There are four taverns along the way, one for each racial group. Sometimes the men stop at these; most of the women, black and Mexican, must hurry home to waiting children. The surrounding counties are all poor, offering few job options. The textile mills that used to hire many locals, especially white and black women, are now gone—many to Mexico. Meanwhile, the hog-butchering business arrived, leaving union towns such as Chicago and Omaha, with their $18-an-hour wages, and finding a home here, the new "hog butcher to world" (to borrow Carl Sandburg's famous but now out-of-date line about Chicago). What remains in Chicago is the board of trade, where pork belly futures are traded at electronic speeds.

Other food production has also come south. You don't need "big shoulders" (another Sandburg description of Chicago) to butcher chicken and filet catfish, and these jobs are heavily feminized as well as racially divided:

> The catfish are trucked from the fish farms to the factory where they await the assembly line. The workers—also women, also black, also poor—are ready for them in their waders, looking like a female angler's society. But these women mean business. The fish come down the line, slippery and flopping. The sawyer grabs the fish and lops off their heads with a band saw, tossing the bodies back onto the line while the heads drop into a bucket. Down the line, women with razor-sharp filet knives make several deft cuts to eviscerate the fish and turn them into filets to be frozen. Many of the longer-term workers have lost fingers, especially to the saws. The company says they fail to follow directions and that they get careless. The women say they are overworked. They say they get tired. They say they slip in the fish guts that fill the floor. But through it all, the assembly line, like Paul Robeson's Ol' Man River, "just keeps rolling along." The line that threads between these rows of black women is operated by a tall white man who supervises from a raised control booth, adjusting the speed of the line and noting the workers' efforts. One watches and wonders: is this the face of the new South or the old South? And what of what Marx called the "social relations of production"; is this the assembly line of the future or the plantation of the past under a metal roof? (Sernau 2000:88)

These are stories about work and about food. They're also about the **industrialization** of food production, and about immigration and globalization, as well as the decline of union power. But in the day-to-day exchanges and experiences of these workers they are also stories about race, class, ethnicity, and gender, and about how these dimensions intersect in the world of work. They are stories of poverty and inequality, but also stories about the complex social relations that define the stratification system.

Sociologists trying to make sense of the complexities of social inequality have turned to the increasingly popular analytic triad of race, class, and gender. Contributors to popular anthologies and new organizations try to untangle the complex ways in which these three dimensions define inequality in U.S. society and the world. Other dimensions could be added: religion, ethnicity, sexuality, and age. This is the "new" approach to social inequality, not in that any of these are new forms of inequality but in that early "classical" theorists gave much less attention to race and gender than to class. They occasionally considered religion and nationality, but they paid scant attention to age or sexuality. Karl Marx and Friedrich Engels contended that the first class division may have come along lines of gender and age as men began to treat women and children as their personal property (see McLellan 1977). Men became the first dominant group and women the first subordinate group. Following from this, Marx and Engels developed biting critiques of middle-class ("bourgeois") family structure that in many ways anticipated feminist critiques. They were writing *The Communist Manifesto* just as the first American feminists were denouncing "domestic slavery" at a conference in Seneca Falls, New York. Marx also offered interesting observations on how capitalists could use racial divisions to keep members of the working class divided, especially in the United States. Yet both race and gender were clearly in the background in his analysis.

In describing the dimensions of stratification, Max Weber proposed the three-part division discussed in Chapter 1: class, status (prestige), and party (political power). He was interested in the organization of privilege and duty between men and women within the household and was particularly interested in cultural and religious differences, but, as they were for Marx, race and gender were secondary to his analysis.

One way to look at both race and gender, as well as the other dimensions of inequality, is as special types of status. Although many people have tried to define and describe clear racial categories, such attempts have continually foundered on the complexity and diversity of human backgrounds. Race is better understood as a social status. A racial identity or category can confer special prestige or respect within a community and may confer particular stigma and disadvantage—apart from or at least in addition to class position—in hostile communities. Gender is likewise a particular form of social and legal status that may confer privileges or barriers in addition to those of class. A wealthy woman may experience expectations and opportunities that are different from those experienced by a wealthy man, and a poor woman's experience of poverty and the prospects for upward mobility may be quite different from that of a poor man. Age may command respect or contempt, depending on the context, and ethnic heritage may be a source of pride or something to be hidden. Race and gender are also closely bound up in struggles for power. Those in power may use issues of race to attempt to divide groups that pose a threat to their power, or race may become a rallying cry for groups attempting to mobilize and challenge established power.

The difficulty in trying to analyze race, class, and gender lies in the fact that the three are in continual and complex interaction with one another and with other dimensions of inequality. There is more than just an additive effect; a poor black woman, for example, faces more than simply double or even triple disadvantage. Poor black women may be less likely than poor black men to be unemployed, with more starting-level positions available to them, but these women often also face the added burden of heavy family responsibilities. Compared with poor blacks or Latinos, poor whites, male or female, may face less discrimination in some areas (such as housing) as they seek to move out of poverty, but they may face added shame and stigma that they have somehow personally failed. It is interesting that just as some slurs for poor people of color have faded from polite usage, other slurs—such as the biting and hostile *trailer trash*—have emerged to stigmatize poor rural and suburban whites.

Race, ethnicity, and gender are social markers that confer identity and social boundaries that define a community and who may be included in or excluded from that community. They are often categories of oppression and privilege, and as such, they must concern us all. Gender is not a "woman's issue," because men must also come to terms with the ways in which their gender may confer power or privilege and may also confer isolation and unrealistic expectations, especially as it interacts with social class. Race is not a "black issue" or a "minority issue," because the members of those disparate groups—now placed together as the "majority"—must also come to terms with identity, boundaries, and perception of "whiteness" and what it can mean in either privilege or stigma, again as it interacts with social class. The stigmatization of one religion can soon begin to erode freedom of religious expression for others, and violence against gays and lesbians can quickly escalate into violence against anyone whose "differentness" violates someone else's established norms.

We can't hope to disentangle the Gordian knot of race, class, and gender completely, any more than Weber could keep his three dimensions perfectly analytically separate. Instead, the challenge is to understand more fully the interactions among oppression and privilege, dominant and subordinate positions, and inclusion and exclusion that shape our social structure.

The Development of Inequality: Race, Class, and Gender across Societies

How did the world get to this state? Was there ever a time when people related as equals? Was there ever a society in which men and women worked as equals and race and class divisions were unknown? Possibly. We must be careful about projecting too many of our hopes onto a utopian past. Yet the glimpses we get of our past, both distant and recent, and of other social arrangements, shed interesting light on the development of human societies and the accompanying development of inequality. Evidence suggests that very equal human societies have existed; in fact, such arrangements may once have been the norm. Relative equality, not massive inequality, may be part of our human origins. The story of how we got from such a place to the world we now know is not as long as you might think, and it provides a fascinating glimpse into a shared past.

Hunting and Gathering Societies

Until about 10,000 years ago, **hunting and gathering societies** contained everyone in the world. These societies were largely made up of seminomadic bands of about 50, although in lush environments some could have been larger and the bands must have interacted with one another. As a rule, hunter-gatherers have gender-divided societies: The men hunt and the women gather. With their longer legs, men have an advantage in sprinting after game, and their longer arms allow them to throw spears farther. Young women are also likely to have young children, and it's very difficult to chase down your dinner with a 2-year-old in one arm! The gender division is probably based in social needs as well as in biology, however. To provide food, these societies need both tasks—hunting and gathering—to be performed, and they don't collect résumés to fill their two-part division of labor. Because men have certain potential advantages in hunting, it makes sense to train boys to be hunters. Because gathering is also essential, it makes sense to train girls to gather.

The gender division is probably not as clear-cut as it might seem, however. Men who are out hunting often gather wild honey, birds' eggs, and probably a wide range of edible fruits and nuts. Women who are out gathering might dispatch any small animals they encounter and add them to the larder, and they may also help men to stalk and flush game. When fishing is important, men and women probably work together as teams, because successfully gathering seafood often combines the skills of both hunting and gathering. More important, however, is that both tasks are essential to the economy of the band. Men's hunting provides important protein and nutrients for the band's diet, but the women's gathering provides the most reliable and abundant food source, often between 60% and 80% of the total. As all hunters know, hunting can be unreliable business, so it's good to have more reliable staples on which to depend. Because the tasks that both men and women perform are essential to the vitality of the hunter-gatherer economy, both men and women have power bases within such societies.

Class divisions are also unheard of among hunter-gatherers. The skills and implements of hunting and gathering are equally available to everyone. Except for a healer or a storyteller or a religious specialist, who might be either male or female, there are few specialized roles. Possessions are few and cannot be hoarded because bands must carry all they possess as they travel. The hunter-gatherer economy is based on **reciprocity**, the sharing of goods. One small group of hunters might be successful one day whereas others come back empty-handed. As the successful hunters cannot eat the entire rhinoceros themselves, it makes sense for them to share their take, with the knowledge that the next time they might be on the receiving end.

In hunting and gathering societies power resides primarily in the consensus of the group, although a few leaders might emerge based on their personal charisma or ability to command respect. Age can cut both ways: Elders may be accorded added respect for their acquired knowledge of plants, seasons, stories, or incantations, but an older member's declining eyesight might require giving leadership of the hunt to someone younger. Prestige may go to the best hunter or the best storyteller, but this is not passed down from generation to generation, so there are no prestigious subgroups. Further, in a band in which all must work together for survival, prestige brings few privileges. Hunter-gatherers have little choice but to share their common lot.

Whether hunter-gatherers share a life in which all are poor or all have plenty has been a matter of debate. Hobbes wrote that life before civilization was "nasty, brutish, and short." This has often described our cartoon image of the "caveman," who himself is nasty, brutish, and short, and lives just such a life. In fact, we have evidence that the men and women of hunting and gathering bands may live well. They work less than you or I do, maybe only about 20 hours per week (Sahlins 1972). In part this is because there is only so much they can do. Their lives are tied to the rhythms of nature, and so they must wait for the returning herds or the ripening fruits. While they wait, they joke and tell stories, they mend their simple tools and temporary dwellings, they play with their children, and, it seems, they often give some energy to flirting and lovemaking. Their diets are often healthier than those of most of the world's peasants; in fact, they are quite similar to the diverse, high-fiber, organic diets based on fresh fruits and vegetables supplemented with a little lean meat that nutritionists encourage for the rest of us.

If hunter-gatherers live so well, why are there now so few of them? The first hunter-gatherers to become cultivators may have done so out of necessity when they were faced with growing populations and declining environmental abundance, whether caused by natural climate change or their own predations. Many others have succumbed to larger, more aggressive societies that needed their land and resources. This is a process that continues through the present day, so that the last true hunter-gatherers are close to their final hunt.

Horticultural and Herding Societies

Women in hunting and gathering societies know a great deal about the plants they harvest. Under pressure to provide for their bands, some may have begun to remove unwanted plants, to move and tend root crops, and eventually to place seeds in fertile ground intentionally. They became plant cultivators, or horticulturalists. **Horticultural societies** differ from hunter-gatherer societies in several key ways. They often must shift their cultivation, but they may stay in one place when they do so, such as in a village in the middle of shifting gardens. They also can produce economic surpluses and store them, thus creating commodities. In addition, a population that lives in one location grows. When a village reaches the size of several hundred inhabitants, simple reciprocity often gives way to **redistribution**, with redistributors who may become "big men" by gathering and giving gifts. Some of these societies have female traders and "big women," but key positions of political and economic power are often monopolized by men.

Still, the horticultural surplus produced by a small village is likely to be limited and perishable, so there is little hoarding and little intergenerational accumulation. The big man's power is based largely on his own charisma and influence; he has little coercive power. His privileges are based on his ability to redistribute goods to everyone's satisfaction and to his own advantage. And his prestige is based largely on his ability to reward supporters generously.

Even if the top leaders are most often men, women often play a key role in the social organization and governance of horticultural societies because they are the gardeners. Men may do the "ax work" of clearing the land, but it is often the women who do the "hoe work" of tending the gardens. Men may supplement the food supply by hunting and fishing, but it is the women's gardens that sustain the village. The rule that he or she who controls the economy also controls much of the governance again seems to apply. Economic power leads to social and political power—a lesson not lost on feminist activists in modern industrial societies as well. Among the Iroquois, an Eastern Woodland group of horticulturalists who lived in what is now the U.S. Northeast, the men were the tribal leaders who met in council but the women were the electors who chose those leaders. Further, it was elder women who controlled the longhouses and the clan structure; when a man married, he went to live with his wife's relatives. Horticultural societies in the Pacific often trace descent and lineage through the women's rather than the men's ancestry, and behind every male leader there are well-connected and often influential women.

Age often brings respect in horticultural societies: The village elder, the wise "medicine woman," the vision seeker, the canny clan leader—all had places of particular prestige. These societies' attitudes toward sexuality were often what caught the attention of the first European traders in the Americas, the Pacific, and parts of Africa. Restrictions on female sexuality were sometimes less stringent than in European society, perhaps because inheritance did not necessarily follow a male line in which men had a prime concern in establishing the "legitimacy" of their offspring. In this atmosphere, alternative sexual orientations were also more likely to be tolerated—in some cases, they were even highly regarded. The Native American *berdache*, a person who crossed traditional gender lines, taking on the attributes, and sometimes the marital options, of the opposite sex, is one example of this.

The domestication of plants is one way to cope with scarce food resources; the domestication of animals is another. Rather than just hunting animals, people began to control some animals' movements and eliminate the competition of predators, becoming herders, or pastoralists. Herding societies emerged alongside horticultural societies in arid and semiarid regions that were too dry for horticulture but had grasses on which large herd animals could graze.

Herding societies are often marked by distinct social inequalities. Individuals in such societies can accumulate wealth in the form of herds. Further, they must defend their property. It's not an easy task to swoop down and make off with a horticulturalist's sweet potatoes, but cattle and horses must be guarded. This is generally the job of armed men (women and children are more likely to be given herding responsibilities for smaller animals, such poultry, sheep, and goats). Given that tending and guarding the herds are male responsibilities, men tend to dominate these societies. In fact, in some herding societies wealthy men acquire harems of women just as they do herds of horses, cattle, or camels. Men without herds are left to serve as hired hands, often with little prospect for advancement. In some herding societies this servanthood develops into hereditary slavery.

Age and gender work together as the route to privilege in herding societies: A senior male can become a patriarch, accumulating a vast herd along with the servants to help tend the animals and a large family lineage dependent on him. Still, in a society that produces few luxuries and is constantly on the move, even a patriarch has to be hardy and able to live with hardship and rigors—that is, unless he and his followers take to raiding the luxuries of settled rulers. In a society with strict gender divisions and great importance placed on inheritance of livestock, there also tend to be stricter punishments for those who blur the clear lines of gender or sexuality.

For the past 7,000 years, horticultural and herding societies have been disappearing, giving way, like hunting and gathering societies, to larger and more powerful societal forms. This is a process that began in the Middle East with the rise of agrarian societies at what we have come to call the dawn of civilization.

Agrarian Societies

Agrarian societies, like horticultural societies, are based on cultivation, but they practice intense and continuous cultivation of the land rather than rely on shifting gardens. The term *horticulture* comes from the Latin word for garden; *agriculture* comes from the Latin word for field. The shift from horticulture to agriculture is largely one of scale. As people domesticated both plants and animals, and as they faced increasing demand to bring more land under cultivation, they realized that oxen could turn up more ground than they could. The keys to agriculture are irrigation and the plow, which allow much more intense cultivation of a given plot of ground. This shift first occurred in the Middle East, which had good candidates for domestication in the horse and the wild ox, good candidates for seed cultivation in the wild grasses, the oats and wheat that blew in the highland, and good opportunities for irrigation in several major river systems. Agriculture provided enough food to support not just villages but whole cities filled with people who were not cultivators themselves. The city, and that urban-based form of social control we call civilization, was born. Again, the order in which things occurred is not entirely clear. Some have suggested that growing food surpluses *allowed* the establishment of cities of priests, artisans, rulers, and their soldiers. A darker view suggests that once ruling groups established themselves, perhaps at key trade and ritual centers, they *demanded* more supplies from the countryside and forced the intensification of work and production that became agriculture.

In either case, agrarian societies became vastly more stratified than their predecessors. As the cultivation process shifted from groups of women to men, or to families led by men who worked with the animals, women's prominence in society declined. Even more striking was the division of societal members into true and largely fixed classes. Land that was continually under cultivation could be owned, and private property now became very important. A food surplus could support metalworkers and artisans who fashioned lasting items of great value that could be accumulated. The surplus could also support standing armies that gave rulers great coercive power to enforce their demands. Continuously cultivated private property could be passed from father to son, and so inheritance became important in maintaining the class structure. As rulers expanded their domains, simple redistribution became difficult, and valuable metals were made into coins to support the first money-based market economies. Land was still the main source of wealth, however, and the way to increase one's privilege, prestige, and power was to bring ever more land under one's control. Rulers could do this with standing armies using metal-edged weapons, and great empires were built.

Increased centralization of power further increased the wealth and power of a few relative to the bare subsistence-level existence of the many. Those who worked the fields, the new class of peasants, may have produced a surplus, but as the lords who owned or controlled the land laid claim to it, the peasantry were often left with only enough to survive so that they could produce more. The obligations of peasants to their lords were often as high as 50% of their total production. Local landlords and petty rulers could use their command of a region to gain impressive privileges. The rulers of the great empires extracted enough surplus from both the land and subjugated cities to live in fabulous luxury. The kings and emperors of agrarian empires often commanded absolute power (although often fearing palace coups), ultimate prestige as god-men, and all the privileges that their societies could supply. The Egyptian pharaoh came to be considered a god-man who rightfully owned all the land and all who lived on it to serve him (or, in one case, her) as he pleased. In twelfth-century England, the average noble's income was 200 times that of a field hand, and the king's income was 24,000 times that of a field hand. The artisans who worked on the manors and in the cities were often no better off than the peasants, and many agrarian cities were crowded with beggars; these destitute "expendables" may have constituted one-tenth to one-third of their populations (Lenski 1966). Yet limited land and a high birthrate meant a steady supply of unskilled labor. When labor was in short supply, warfare could provide not only new land, but also slaves or servants.

Agrarian societies also provided opportunities for one other group that was usually kept to the margins. Merchants could travel between cities, using coined money and expensive goods as the medium of exchange. The nobility generally looked down on the merchants, and yet there was always the possibility that merchants could amass so much wealth that the largely idle nobles would need them as creditors. As Marx realized, in this odd relationship was one of the contradictions that would destroy the old agrarian systems. Agrarian societies spread from the Middle East across Asia, Europe, and North Africa. For almost 5,000 years they were the dominant form of human organization: in ancient Babylon and Egypt, in ancient Greece and Rome, and in medieval Europe, as well as in the great Chinese empire and imperial Japan.

Agrarian empires were carried to the Americas by European colonizers. But ever larger trade routes and expanding monetary systems, fueled in part by American gold and silver, gave ever greater power to traders and merchants relative to the landowners. Capitalism began to replace feudalism as the economic core of agrarian societies. The early great agrarian rulers, such as the Egyptian pharaoh of the Old Testament who has Joseph gather surplus grain for times of famine, were largely just very mighty redistributors. But with the expansion of a money economy, redistribution gave way, in part, to **markets**. Markets, where goods are bought and sold with common currency and prices are set by a balance of supply and demand, were the domain of merchants. As markets grew in power, merchants grew in wealth, amassing money that they could reinvest to create still greater profits.

Fueled by new interests in science, along with this new science of money, the Industrial Revolution began to transform agrarian societies. Bolstered by the science of war, expanding industrial societies began to displace the agrarian order. Still today, many of the world's so-called developing nations in Asia and Latin America are industrializing agrarian societies. These areas are also home to some of the world's most economically unequal societies.

Along with the dawn of civilization, the rise of cities, the invention of writing, and the widespread use of the wheel, the agricultural revolution brought the beginnings of the world's great religions and philosophies: The period around 3000 B.C. (somewhat later in the Americas) witnessed one of the greatest flourishings of human creativity ever. Agrarian societies brought other things as well: widespread slavery and serfdom, chronic warfare, forced taxation, devastating plagues and famines, and malnourished poor who labored in the shadow of luxury and indulgence. These societies were based on hierarchies of power, seen clearly in both European and Asian **feudalism,** in which peasants served lords who served greater lords who served the king or emperor. They were "deference societies" (Stephens 1963), in which individuals at each level showed great respect for those on the level above them: Peasants groveled in the dust before lords, who fell on their knees before kings. This pattern of deference, or showing great respect, carried down to the local level and the household. A cautious peasant who controlled a bit of land through the favor of a lord bowed (literally and figuratively) to the lord's every wish. Yet returning home, this man would expect the same gestures of deference from his wife and children. Age and masculinity might bring control of land, and land was power and the source of prestige. In such a system, wives, daughters, and younger sons were often at the mercy of senior men, who in turn answered to "noblemen." As in most **patriarchal societies**, controlled by senior men, any actions that bent the bounds of traditional gender roles or strict sexual norms were viewed as threats to the system of power and inheritance and were most often severely restricted (Skolnick 1996).

Agrarian empires brought under one rule many diverse peoples, typically giving prominence to the conquerors and subordinating other ethnic groups. Some ancient empires incorporated diversity without much regard to color or ethnicity. It appears that Alexander the Great, with whom we began this chapter, irked some of his followers by not just spreading Greek culture, but marrying and promoting Persians and adopting some of the ways of the diverse, multiethnic Persian Empire. For rulers such as Alexander, personal loyalty often mattered more than race or ethnicity. Within medieval Mediterranean empires, religion—Christianity or Islam—was often the key divider and determiner of privilege. For their Asian counterparts, such as the great Mongol Empire, religion seemed to matter little. Finally, as European states colonized vast regions of the Americas, Africa, and Asia, they formed empires in which privilege was often based on European-ness, or "whiteness," and racial divides of color became common. Even as the world industrializes, the patterns of 5,000 years of agrarian society remain built into many traditions and social structures: male privilege, white privilege, class hierarchies. Some elements have been undermined somewhat, such as respect for age, but others have remained quite persistent, such as suspicion of gay and lesbian sexuality.

Life on the Edge: Frontiers and Ports

Before they succumbed to industrial pressures, agrarian societies dominated the planet but they also shared the world stage with several other societal types, each with its own patterns of inequality. Alongside agrarian societies were a handful of maritime societies in places like Phoenicia, Venice, and, ultimately, the Netherlands. These societies, which depended almost entirely on sea trade, tended to be merchant dominated. Many were republics rather than monarchies, with groups of wealthy and influential traders forming their governments.

In places where native populations were displaced by newcomers there was also the possibility of forming frontier societies. Frontier societies survived through farming and herding, but they differed from agrarian societies in that they were populated by newcomers. In general, the landed elites stayed home, so frontier societies were more equal and often, like maritime societies, more republican-minded. Labor was often scarce in these societies, so laborers were able to command higher wages and more influence. In time, as elites emerged or were transplanted from elsewhere, frontier societies came to look more like their agrarian counterparts—unless other events intervened. The colonial United States, Canada, Australia, and New Zealand were essentially frontier societies. Although they were agrarian in their livelihood, they developed social patterns that differed from those in their home countries of France and Great Britain. Newcomers had to develop new forms of social organization to work the land they had taken from Native American horticulturalists (or Australian hunter-gatherers or Polynesian horticulturalists in New Zealand). By the middle of the 1800s, the American and Canadian frontiers had moved westward, creating a fleeting frontier herding society that lives on in the popular imagination, encouraged by Hollywood westerns. By this time, an economic pattern very similar to older agrarian societies had taken root in the U.S. South: plantation agriculture supported by African slave labor. The great divides that had separated medieval European landowners and their peasants or serfs (peasants tied to the land as virtual slaves) were replicated and supported by a racial divide and a racist ideology. In the northeastern United States, frontier society never had time to establish older agrarian patterns. New England first became a largely maritime society of traders and merchants and then was fully gripped by industrialization. These patterns of the past two centuries have left their mark on the social structure and stratification patterns of the United States, as we will see.

Industrial Societies

The first social change that accompanies industrialization is often increasingly obvious social inequality. The rich may not garner any larger share of a society's goods, but with industrialization there are more goods to amass. The industrial poor may not be any poorer than the agrarian poor, but they now labor alongside the symbols of urban wealth and prosperity. The first country to industrialize was Great Britain in the middle of the 1700s. Industry made Britain the wealthiest and most powerful nation on the earth. It also led to the horrific conditions and gross inequalities that Charles Dickens made famous in his novels, such as *Oliver Twist,* with its begging and stealing orphans, and *A Christmas Carol,* with its greedy Ebenezer Scrooge and his struggling employee, Bob Cratchit. By the middle of the 1800s, France, Germany, and the United States were rapidly industrializing in a race to catch up with Great Britain. These nations faced the same experience: greater production, greater power, and increasingly blatant inequalities. In the United States, this was the era of desperately poor immigrant slums in industrial cities and of the fabulously rich "robber barons," whose wealth surpassed anything previously imagined. In the latter portion of the 1800s, European workers revolted and battled with soldiers and police, U.S. workers staged sit-down strikes and fought with state militia, and Karl Marx strode to the British museum to work on books and pamphlets denouncing the evil of it all.

Then an odd thing happened: Inequality declined. Middle classes filled some of the gap between rich and poor, sharing some of the privilege of the rich. Power was still largely class based, but united workers found they could win political concessions. Prestige came to be based on multiple criteria, with political leaders, business leaders, and even entertainers sharing the elite circles. The relation between economy and society is more complex than observers such as Marx could have initially realized. In a famous statement, Marx contended that history does repeat itself, the second time as a parody, or mockery, of the first. One reason we need to take the time to look at disappearing societies is that, in some remarkable ways, they resemble our own. In some aspects of social stratification, it seems, prehistory repeats itself.

Social inequality is at its lowest in hunting and gathering societies. It widens in horticultural and herding societies as noble or privileged family lines are established. It reaches its greatest extremes in agrarian societies, where a rigid class structure often divides the landowning nobility from the peasantry and poor artisans. Inequality is still extreme in early industrial societies, where handfuls of individuals amass great fortunes while workers confront long hours, miserable conditions, low wages, child labor, and cramped, wretched living conditions. As societies move into an advanced industrial stage, inequality again declines as a more complex class structure with more middle positions emerges (see Exhibit 3.1).

This is also a general picture of gender inequality over time. Hunting and gathering societies are marked by considerable gender equality, especially if gathering provides a large portion of the food supply. Women's position remains strong in many horticultural societies unless chronic fighting elevates the role of the male warrior. Women's position is considerably diminished in most pastoral and agrarian societies. Women still contribute greatly to daily survival, but they are more likely to be cloistered in domestic settings, veiled or otherwise hidden from view, and limited in their ability to own property. In some aspects they may be treated as the property of fathers and husbands. Early industrial societies do little to elevate the role of women as poor women are brought into the lowest wage sectors of the new economy in textile mills and sweatshops, doing "piecework" for factories in their homes, and doing domestic work in the homes of the wealthy. Advanced industrial societies, however, create new opportunities for women, allowing them to excel in education and industrial skills on par with men. Industrial

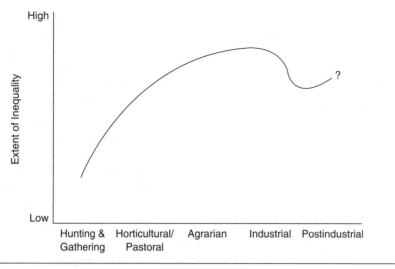

Exhibit 3.1   Inequality by Societal Type

societies also create new dangers and problems for women, which we will explore in more depth, but often industrialization is followed by women's demands for new roles and improved access to power and privilege. As Rosen (1982) notes:

> Under the impact of industrialization . . . The mystique of male dominance that had for generations kept the female in a subordinate position becomes tarnished, and women, supported by an ideology of sexual equality, challenge their husbands' omnipotence, often with success. The scepter of patriarchal authority does not exactly fall from nerveless male hands; sometimes the wife, emboldened by her new freedom of power, snatches it brusquely from her husband's grasp. (P. 3)

More flexible roles for women often mean gradually greater acceptance of variation in family forms and sexual orientation, although this is a slow process. Youth also tends to become less of a reason to exclude individuals completely from power and rights; for instance, voting is no longer restricted to landowning males. Yet, as young people once had to wait to inherit land to gain access to power and privilege, in industrial societies they often have to wait to acquire educational credentials, or to inherit the family enterprise. Life for the elderly can also be precarious in industrial societies, as they may lose their assurance of family support at the same time they often get little social support. Retirement can mean poverty. Only in advanced industrial societies does the idea of social support for the elderly become common. Even this can remain precarious, as the continued anxiety in the United States over the future of Social Security illustrates.

An optimistic interpretation of the trend illustrated in Exhibit 3.1 has come to be called the **Kuznets curve,** after economist Simon Kuznets (1955), who first called attention to this trend in national development. The inequality within a society increases until the society reaches a certain point in industrialization at which it declines. Kuznets argued that this describes the experiences of Great Britain, Germany, the United States, and many other advanced industrialized nations. He was less sure it would apply to later-developing countries in Asia, Africa, and Latin America. The optimistic view is that they will also follow this pattern and that inequality within the poor nations of the world will decline as these nations further their drive to industrialization. Other theorists maintain that poor nations will be prevented from following this pattern because they will be relegated to a subservient role in the world economy. At this point we can only note the strong influence of past patterns of economy and society on the world's nations.

The wealthiest nations in the world are the advanced industrial—some would now say *postindustrial*—societies of Western Europe, North America, and the Asian rim. The poorest nations of the world fit Lenski and Nolan's (1984) description of "industrializing horticultural societies." These are nations in sub-Saharan Africa and isolated portions of South Asia where disparate horticultural and herding societies were united under European colonial rule into single administrations and are now independent nations with populations too large and land too degraded for the inhabitants to continue as small-scale horticulturalists and pastoralists, but with limited background in centralized government and national-scale economy. These are the poorest of the poor—Niger, Mali, Mozambique, Somalia, Nepal, Afghanistan—which some have started to refer to as the Fourth World.

The nations in the world with the highest levels of social equality have been the Eastern European remnants of the Soviet "Second World." Hungary and the Czech Republic emerged from years of enforced socialization with inefficient industries and environmental degradation, but also with societies far more equal than those to the west. Their transformation has brought new freedoms, new business opportunities, and prosperity for some, but also unemployment and hard times for others. Their economies are growing, but inequality is also growing. The leader in this divide is Russia itself. With a handful of entrepreneurs and speculators gaining control of most of the national wealth while the rest of the economy withers under the feet of workers with declining incomes, the first communist nation is now more unequal than the United States.

The nations of the world with the lowest levels of social equality are the "industrializing agrarian societies" (Lenski and Nolan 1984) of Latin America and, to a somewhat lesser extent, South Asia. Inequality at its most raw extremes is found in Brazil, with its old plantations and new industry; in Guatemala, with its years of struggle between wealthy landowners and desperate highland *campesinos,* or poor peasants; in El Salvador, where 14 elite families own most of the country; in Panama, where a small handful control the profits of the country's strategic location; and in Bolivia, where 6% of the population own 90% of the land. Pakistan and India also struggle as the inequalities of an ancient caste system are superimposed on the modern inequalities of unequal development.

Of the advanced industrial countries, the most unequal are no longer those with rigid class divides inherited from an agrarian past. Great Britain remains socially class-conscious, but 250 years of industrialization have accustomed the British to a society in which cash-strapped dukes and earls rent out their great manors while upstart industrialists grow rich through investments in global manufacturing. France is also still socially class-conscious and a bit more unequal than Great Britain, but all across Western Europe in the twentieth century inequalities were gradually lessened both by demand for technical and skilled labor and by welfare state policies that tax the rich to support small farmers, unemployed workers, the elderly, and children. Decades of social welfare policies in Western European nations have helped to reduce the divide between rich and poor. Most equal are the Scandinavian countries of Sweden and Norway. Another advanced industrial nation with a fairly small gap between rich and poor is Japan, where defeat in war destroyed old elites and postwar policies have combined with social pressures to limit huge incomes.

The most unequal advanced industrial nations are the former frontier societies. These are often also the least class-conscious, with strong ethics concerning equal standing and equal opportunity, and general suspicion of social snobbery. Yet the frontier approach to limited federal government and largely unrestrained markets, coupled with extensive land and resources, has led to vigorous but very unequal income growth for these now industrialized societies. They include the United States, Canada, Australia, and New Zealand, with the United States leading the list as the most unequal (see Exhibit 3.2).

When countries have a frontier heritage and are still in the midst of industrializing, the inequalities are often vast: Brazil and South Africa vie for the title of the most unequal large nation in the world. Former frontier societies also have another element in common: The most entrenched inequality is usually not along old class lines—there is no nobility—but along racial lines. The United States, Canada, Australia, South Africa, and, to a certain extent, Brazil all have native populations confined to reserves, reservations, homelands, or other isolated, set-aside areas. New Zealand, the country in this group with the largest indigenous population (about 20% of New Zealand's inhabitants are Maori), has done perhaps the most to integrate that population into the immigrant majority society. South Africa, Brazil, and the United States all have histories of turning to African laborers to sustain their economies, imported across the Atlantic in the case of the latter two. Race relations are very different among these three societies, and the nature of race relations is changing rapidly in all of them, but the correlation between color and class remains strong in all three.

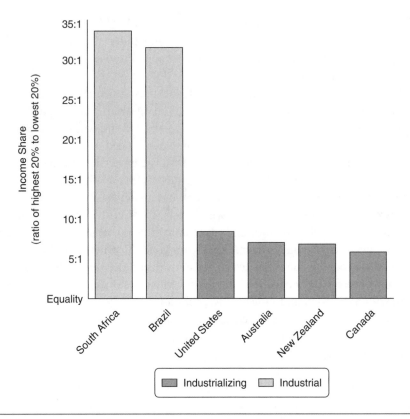

Exhibit 3.2   Inequality in Industrial and Industrializing Former Frontier Societies
Source: Data from United Nations, Human Development Report (2004).

The Coming of Postindustrial Society

Have we already begun the shift to another social and economic form? If so, what will its consequences be? Beginning around the mid-1960s and well under way by the 1970s, there has been a shift within advanced industrial economies away from a manufacturing base and toward a service base, a shift we will examine in more detail in Chapter 4. In the early 1970s, Daniel Bell (1973) referred to this shift as "the coming of post-industrial society." Bell was quite optimistic that this change would create new opportunities for many, as the possession of knowledge rather than the control of physical capital would become the key asset. The widening of the ranks of the middle and upper-middle classes would continue as it had with industrial society, as new managers, professional service providers, and skilled technicians would be needed. Because women could provide these skills as well as men, gender inequality would diminish. As skills would become more important than old social divides, racial inequality might also decrease. This optimism is countered by the pessimistic analysis of "deindustrialization" (Bluestone and Harrison 1982), which asserts that the loss of industrial work undermines the gains of labor unions and the working class, creating unemployment and a "race to the bottom" as workers in advanced industrial economies must compete with poorly paid workers in newly industrializing countries. Hardest hit are the least protected workers—older, female, and nonwhite workers in particular. Inequalities could therefore increase along lines of race, class, and gender. A few profit enormously while many others lose job security and see falling wages, leading to a shrinking middle class.

The American case has been cited as proof that the Kuznets curve can be inverted: The United States has seen two decades of rapid growth accompanied by widening inequality. The decline in inequality stalled in the 1970s, and by the Reagan years of the 1980s inequality was growing rapidly, with the United States ahead of Britain, then ahead of France, and far ahead of Scandinavian countries and Japan. Given that the 1980s and the 1990s were periods of strong growth in the U.S. economy, it was not so much that the poor got poorer but that they got nowhere. It took the entire unprecedented growth of the Clinton era for the poorest groups just to recover what they lost in income in the recession of 1990 to 1992. Even that small gain was lost entirely in the recession of 2000–2004. Meanwhile, the wealthiest one-fifth of Americans watched their incomes soar. The greatest gains went to those who owned a piece of the national wealth. Upper-middle-class Americans with significant holdings in popular mutual funds gained some, and those few who controlled the vast majority of the stocks and assets of the country saw amazing windfalls.

We will look more closely at the dimensions of inequality—class, race, gender, status, and power—in the chapters that follow. Although we can't completely untangle the Gordian knot, we should at least be able to see clearly that the knot was twisted, strand by strand, by human hands. The dimensions of inequality are **social constructions**; that is, they're not facts of nature but the results of societal forms and patterns of power. The demands of the economy and the desires of ruling groups have divided societies along various lines, and cultural constructions follow to justify and explain societal patterns. What the members of one type of society find "obvious" about the nature and place of women may sound absurd to the members of another. Differences of color, language, ethnicity, or geography may be crucial determinants of an individual's place in one society and completely irrelevant in another. Yet once these differences become intertwined with class, prestige, and power, it can take years of hard effort to untangle the knot.

KEY POINTS

- Inequality occurs along various dimensions. Max Weber noted three: class, status, and party. Gerhard Lenski termed the same basic divisions privilege, prestige, and power, respectively.
- Sociologists have grown increasingly interested in the intersecting dimensions of class, race, and gender, along with other dimensions such as ethnicity, religion, sexual orientation, and age.

- Inequality grows as societies become larger and more powerful. Advanced industrial societies have seen a decrease in inequality with the growth of a large middle class. The effects of global, postindustrial economies are still uncertain, but many postindustrial societies are again seeing rises in inequality.
- Different societies vary in the emphasis they place on their constructions of class, race, and gender. Each society creates its own explanations for inequality along these dimensions.
- Attitudes toward race, ethnicity, and gender have shifted in the United States, yet these dimensions still divide the workplace, as seen in the food industry.

FOR REVIEW AND DISCUSSION

1. How has social inequality varied over human experience and history? What has characterized the divides of different forms of societies?
2. Is inequality likely to grow or diminish in the future? Explain.

MAKING CONNECTIONS

United Nations

Several organizations based in the United Nations gather and disseminate reliable information on race, class, and gender around the world. For example:

- The United Nations Educational, Scientific, and Cultural Organization (UNESCO; see www.unesco.org) makes available very good information on racial, ethnic, and gender issues around the world.
- The United Nations Children's Fund (UNICEF; see www.unicef.org) offers good information on the status of children, including separate data on boys and girls, reports on the Year of the Child, and other materials on children.

4 Class Privilege

Money is like muck, not good except that it be spread.

—Francis Bacon, English philosopher (1561–1626)

Our inequality materializes our upper class, vulgarizes our middle class, brutalizes our lower class.

—Matthew Arnold, English essayist (1822–88)

If you can count your money, you don't have a billion dollars.

—John D. Rockefeller, American oilman (1839–1937)

Americans don't like to talk about class. The topic smacks of snobbery and elitism and seems sure to offend. To many Americans, talk of class is crass. Besides, most of us claim to be middle-class, implying that we're all in the same mix together despite evidence to the contrary. The elite medical specialist may wave hello to the security guard on arriving at work, and both may describe themselves as middle-class, but the medical specialist may easily earn 10 times as much income as the guard, and their respective worlds may hardly ever intersect except at work. That is one reason Americans are often reluctant to talk about money, at least about our own relative to that of others. "What do you earn?" is as common a question in China as "What do you do?" is in the United States, but most North Americans would be startled and even embarrassed by the first question from a relative stranger. Honest answers would quickly point to the reality that "all of us in this together" are really living in very separate financial worlds.

Some social scientists have also grown leery of class analysis. Some have suggested that if there is considerable mobility among various financial conditions and many gradations along the way, perhaps we should give up the language of class and just focus on a continuous variable such as "socioeconomic status," or perhaps just income. Yet others insist that class is as much a key to understanding our social world as it has ever been (Wright 1997; Sorensen and Grusky 1999).

According to Marx, the primary class divider is wealth: who owns the means of production. Erik Olin Wright has demonstrated that Marx's view of class still predicts a great deal of income inequality, especially if we follow Dahrendorf's idea about authority being as important as property. To Marx's categories of owners and workers, Wright adds the third category of managers (Wright and Perrone 1977). These three are the essential aspects of contemporary social relations of production: owners, managers and professional-level employees who don't own but still command, and the third category of workers who labor under the power of the first two (Wright 1985). Weber contended that a social class comprises those who share similar life chances in the marketplace. This may be based on wealth, but it may also include education, occupation, and income.

Class matters. It may be difficult at times to establish clearly the boundary between one social class and another; however, once given the labels, most still recognize the core features of life in the working class, the upper-middle class, and the elite capitalist class. We seem to recognize crucial class markers readily, even if as a society we often deny they exist. In this chapter, we will examine the objective dimensions of social class: wealth, occupation, and income. Later we will explore the more subjective dimensions of class.

Wealth and Property

Wealth is the accumulated property owned by a person or family. A person may have a high income but little wealth if he or she is spending as fast as he or she is earning. Likewise, a person with a low income may have inherited great family wealth. In recent years, some titled British nobles have had to endure the indignity of allowing bus tours of paying tourists to trek through their fabulous family homes because that is the only way they can afford to pay the taxes on those homes, as well as pay the high costs of upkeep and their astronomical utility bills. These British "peers" have inherited great wealth yet have limited incomes, and so must find new ways of generating income from their wealth.

For Karl Marx, the key variables in determining social class are an individual's accumulated private property and his or her relationship to that property. The recent past had been dominated by two groups: a landowning elite who passed what they owned on to their children and a peasantry who worked the land but often did not own it. In such a system, land is wealth. Marx was more interested in the new wealthy, however—the capitalists. They owned the new means of production, factories and great productive enterprises, and earned large returns on this wealth. The laboring class had no wealth and so could receive income only by selling their labor to the wealthy capitalists.

As the case of the "wealthy" but "cash-strapped" British peers shows, the relationships among wealth, income, and well-being have become increasingly complicated. Yet, in some ways, little has changed. Members of the wealthy capitalist class are distinguished by the large returns they receive on their accumulated assets: land, corporate stock, and maybe "intellectual property," such as patents and copyrights. Microsoft chairman Bill Gates earns a fairly "modest" salary as "chief software architect" but is worth many billions because of his stock holdings. In any profitable year his stock holdings will increase by many times the value of his salary. This provides him with an "income" of several billion dollars per year, although it is neither taken nor reported as "income" but merely accumulated as ever greater "personal worth," an archaic term for wealth. To a humanitarian the "personal worth" of Mother Teresa may have been inestimable, but to an economist her wealth was one sari and her "personal worth" a few rupees!

When we think of "the rich," we may envision famous sports and entertainment stars with multimillion-dollar contracts and fabulous incomes. Yet in any given year, the top earner in the United States is never a sports or entertainment star but rather a "corporate star" whose compensation package, often mostly in stock options rather than salary, is in excess of $50 million. A few of these individuals may be well-known if not exactly famous: Disney's CEO Michael Eisner, for example, whose annual compensation has ranged between $40 and $750 million (*Forbes* 1999 estimates; remember, most of this amount is in stocks and benefits rather than salary), or Nike's Phil Knight, who by some estimates regularly receives more compensation than all of his Indonesian workers' wages combined. Others may be largely unknown, such as Anthony O'Reilly, who recently earned (or at least received) $67 million as CEO of Heinz. These figures boggle the mind and dwarf even the finest sports contract. What these CEOs are receiving are returns on accumulated wealth: a worldwide network of production facilities, as in the cases of Nike and Heinz, or globally recognized and protected copyrights, as in the cases of Disney and Microsoft.

Even the richest entertainers don't move into the billionaire category until they accumulate wealth. Over his lifetime, Michael Jordan will earn far more from his business interests and sponsorships than even he ever earned on the basketball court. The wealthiest black woman in the United States, and perhaps one of the wealthiest women not to have received a large inheritance or survivorship, is Oprah Winfrey. The largest share of her income, however, no longer comes from the six-figure earnings she may receive to do a single show, but rather from her $400 million in accumulated wealth, much of it holdings in major media corporations (*Forbes* 1996). Yet her accumulated wealth may never reach the levels of other lesser-known figures, such as the late Joan Kroc, heir to the McDonald's fortune, or any one of the family heirs to Sam Walton's vast Wal-Mart fortune.

The pattern is familiar to any player of the game Monopoly. It is nice to "pass Go" and receive the $200 "salary," and it's always nice to get a favorable Chance card. But the big money comes from acquiring monopolies and amassing wealth in the form of properties, houses, and hotels. The "rents" or returns on these assets invariably determine the winner of the game.

Wealth is somewhat more fluid today than in the past, when fortunes could be lost overnight but often had to be accumulated over one or more lifetimes. Yet wealth continues to reside primarily in families that have benefited from years or even generations of accumulation. For years, several of the wealthiest people in the United States were members of the Rockefeller and Du Pont families, whose wealth had its origins in old oil and chemical monopolies. The world's current wealthiest people include two Microsoft founders: Bill Gates and Paul Allen. As noted above, Gates receives a comparatively modest salary as chief software engineer for Microsoft, but he controls billions of dollars of stock in a company that makes the software that operates 95% of the world's computers, making him the world's wealthiest individual. Warren Buffet, one of Gates's bridge-playing buddies (one wonders what the stakes are at that table) also made his fortune through stock holdings. Ties to state-supported monopolies also help: oil in the case of Prince Alwaleed Bin Talal Alsaud of Saudi Arabia and Mikhail Khodorkovsky of Russia (recently imprisoned), telephone monopolies in the case of Carlos Slim Helu of Mexico. Retail has become increasingly important, as seen in the case of the German Albrecht brothers, owners of Aldi stores, and most dramatically, in the case of the Walton family. Of the 10 richest people in the world as of 2004, 5 are named Walton, all of them heirs to Sam Walton's Wal-Mart fortune. If Sam Walton were still alive, his $100 billion would make him the world's richest man; since his death, that wealth has been divided among five of his family members, each of whom is now among the top 10 wealthiest (*Forbes* 2004). (See Exhibit 4.1.)

In the United States, wealth is far more concentrated than is income. The wealthiest 1% of Americans control more of the total wealth of the nation than the bottom 90% combined (see Exhibit 4.2). Although the figures vary slightly from year to year depending on such factors as the state of the stock market, wealth in the United States falls into a fairly neat three-part division: slightly more than one-third goes to the wealthiest 1%, the next wealthiest 9% own about one-third, and less than a third is left for the remaining 90% of the population. That remaining 90% can be further divided into halves, with the top half controlling almost all of the remaining wealth and the least wealthy group, almost 45% of the U.S. population, with almost no net assets whatsoever. In fact, those in this last group are ever more likely to be in debt, and, given that they have few fully owned assets, they have, in fact, "negative wealth." They may be working hard, they may even have moderate incomes, but in the language of "personal worth," they are worse than worthless.

How can this be in the age of "penny capitalism" and the rapid expansion of mutual funds? A recent investment advertisement shows "Grandma" looking over her retirement portfolio of mutual funds, essentially shared stock investments, and notes that "the face of capitalism in America has

Rank	Name	Age	Worth ($ billions)	Country of Citizenship	Residence
1	William Gates III	48	46.6	United States	United States, WA, Medina
2	Warren Buffett	73	42.9	United States	United States, NE, Omaha
3	Karl Albrecht	84	23.0	Germany	Germany, Donaueschingen
4	Prince Alwaleed Bin Talal Alsaud	47	21.5	Saudi Arabia	Saudi Arabia, Riyadh
5	Paul Allen	51	21.0	United States	United States, WA, Mercer Island
6	Alice Walton	55	20.0	United States	United States, TX, Fort Worth
6	Helen Walton	84	20.0	United States	United States, AR, Bentonville
6	Jim Walton	56	20.0	United States	United States, AR, Bentonville
6	John Walton	58	20.0	United States	United States, AR, Bentonville
6	S. Robson Walton	60	20.0	United States	United States, AR, Bentonville
11	Liliane Bettencourt	81	18.8	France	France, Paris
12	Lawrence Ellison	59	18.7	United States	United States, CA, Redwood Shores
13	Ingvar Kamprad	77	18.5	Sweden	Switzerland, Lausanne
14	Theo Albrecht	81	18.1	Germany	Germany, Foehr
15	Kenneth Thomson & family	80	17.2	Canada	Canada, Toronto
16	Mikhail Khodorkovsky	40	15.0	Russia	Russia, Moscow
17	Carlos Slim Helu	64	13.9	Mexico	Mexico, Mexico City
18	Michael Dell	39	13.0	United States	United States, TX, Round Rock

Exhibit 4.1   World's Largest Fortunes

Source: Forbes (2004). Reprinted with permission from Forbes.com.

changed." In reality, it has changed very little. In fact, after declining for a while, the share of total net wealth held by the top 1% has increased steadily since the late 1970s. "Grandma" may indeed now have a stake in the stock market, but unless she is a surviving Du Pont heir or the like, she has hardly joined the capitalist class. With low interest rates that make bank deposits unattractive compared with stocks over the long term, as well as the relative ease of investing through means such as mutual funds, many more Americans are likely to hold shares in corporate assets (stocks and stock funds) today than ever before, but they are also less likely to have large bank accounts, and the share of total assets they hold has not increased.

The bottom 40–45% of American households have accumulated almost nothing. If they "own" their homes, that property is in fact heavily mortgaged, and the same is true if they have cars. If they sold all their possessions and paid all their debts, most would have less than $3,000 left over. Many would actually not be able to even pay off their debts. Their solvency is based on their ability to continue to generate enough income to make payments on homes, cars, and credit cards. Those who have no accumulated wealth must depend on regular paychecks. In Marx's terms, they have nothing to sell but

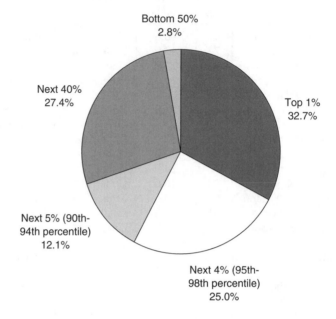

Distribution of U.S. Wealth Ownership, 2001

Bottom 50%
2.8%

Next 40%
27.4%

Top 1%
32.7%

Next 5% (90th-
94th percentile)
12.1%

Next 4% (95th-
98th percentile)
25.0%

Total Net Worth in U.S.: $42.3892 trillion ($42,389,200,000,000)

Exhibit 4.2   Pieces of the American Pie: Wealth Distribution
Source: Data from Kennickell (2003).

their labor. Recent studies have found that most Americans are less than three months of paychecks away from desperation and homelessness. Many in this category are only one month's income ahead of evictions, foreclosures, and repossession. Those in this group have far more possessions, from microwaves to VCRs, than did the so-called popular classes of the past, but this is in large measure a result of the easy availability of consumer credit.

The next 40–45% are the savers. They are more likely to own their homes and to have accumulated some equity in that property. In fact, the accumulated equity in their homes is likely to be their largest asset. They are also quite likely to have some retirement savings. They may even have some personal property of significant value, such as debt-free cars or campers. These may be older working-class families that have slowly accumulated savings and equity and have not faced major financial crises such as expensive illness or sudden unemployment. This group also includes many middle-class savers who have been able to save a portion of their discretionary income as a nest egg. Some are saving toward retirement, and many others are saving in anticipation of their children's college expenses.

The upper 10% are the investors. They typically have planned portfolios of investments that include stocks, bonds, and other investments selected to provide likely growth, protection from taxes, and stable accumulation over time. Many of these are upper-middle-class earners who still depend on high salaries but are eager to protect these salaries from taxes and to plan for a comfortable future for themselves and their children. Of this group, only the top 1% have the truly large investments that allow them to control corporate activities, hold large quantities of investment real estate, and invest in major business ventures. This latter group depends on returns on assets—growth, dividends, profits, rentals, and profitable sales and mergers—for the largest part of incomes that often exceed $1 million (see Gilbert and Kahl 1982; Gilbert 2003).

Wealth begets wealth. The wealth of the investor class may buy elite college educations that will help ensure that the children of this class enjoy a similar position. The wealth of the top 1% of the capitalist class may generate returns for heirs for many generations. Income-generating talent may show up in surprising places, but wealth holds to elite families. In this way, wealth is by its very nature conservative: It helps to ensure that through the reinvestment of assets, which will in turn bring new assets, the rich will indeed become richer. It ensures that those near the bottom will be called on to spend almost all of their incomes and that any wealth they may acquire, such as an aging automobile or an aging house in a vulnerable neighborhood, will more likely depreciate than increase in value, and the poor will get nowhere.

The conservative nature of wealth can also undermine progressive social changes. Many African Americans, for example, have made substantial gains in both occupation and income since the beginning of the 1960s. Because they are the first generation to "make it" to middle-class positions, however, they have little accumulated wealth. Oprah Winfrey and Michael Jordan notwithstanding, a huge gap remains between "white wealth" and "black wealth" (Oliver and Shapiro 1995). This means that the members of the new black middle classes are very vulnerable. If they are "downsized" or face some other unexpected downturn in fortunes, they have little personal or family wealth to fall back on. They also have less to pass on to their children to continue and expand on the progress they have made. Whatever the personal talents, vulnerability to losses and lack of cumulative gains can make progress very slow and uncertain.

Occupation

Most of us obtain our incomes not through returns on assets but through **occupation**, the work we do for compensation. Our occupations also have a big effect on the people we associate with, our lifestyles, our standing in our communities, and maybe even our values and outlooks.

Around the world, the occupational distribution of the labor force is continuing to change at a remarkable rate. Before 1700, the vast majority of people in the great agrarian societies of Europe and Asia were peasant farmers. The cities and towns held a few officials and merchants and various craftsmen and artisans, but these were never more than 10–20% of the total population, sometimes less. As cities and towns grew and diversified in the late Middle Ages, more people left the hardships of the land for the uncertainties of the cities. Cities and towns helped absorb the growing population that would eventually outstrip what the land could support. They offered a place of escape from ravaging armies and from brutal landlords. They also collected the landless and the displaced, who came to find whatever livelihood they could.

With industrialization, the trickle from the land to the cities became a movement, both voluntary and forced. As Great Britain surged into the industrial era in the mid-1700s, new mills needed more workers to spin, weave, sew, haul, and ultimately export textiles. As water power gave way to steam, people were needed to work the mines to fuel the coal fires of London, Liverpool, and Birmingham. By and large, these workers came from the land. Conveniently, ever more land was needed to raise the sheep that would provide the vast quantities of wool that could now be processed. Clearances drove small farmers from their land to make room for sheep. The newly displaced families could emigrate abroad, often first to Ireland and later to the American colonies, or they could work in the mills. Many did the latter. Whole families that had worked together were displaced, and whole families often went to work in industry, although not always together. The small stature of boys was especially valued in the crowded mines, and many followed their fathers into this profession or replaced lung-diseased fathers who could no longer work. Women's nimble fingers were valued in the delicate work of sewing, and many women, especially young women, found this as available employment, setting a trend of young women in textile work that still continues around the world, from Nike plants in Indonesia to immigrant-filled stitching shops in New York (Waldinger 1986; Waldinger and Lichter 2003).

Not all of the new workers crowded into cities at first. Cottage industries—which were just that, sometimes located in peasant cottages—provided some employment, especially where little technology was needed. The use of water power kept early industry dispersed as well, for it had to remain close to the key waterways, which provided both power and transport. With the advent of steam power, however, bigger factories were possible, as was rail transport, which worked most efficiently from key rail centers and hubs. Cities grew rapidly. By the early to mid-1800s, workers were leaving agrarian trades across the United States to find work in more urban areas, first in the mill towns of New England, then in the great new industrial cities, such as Baltimore and Pittsburgh and, somewhat later, Memphis and Birmingham. They paralleled the movement of French and German workers about the same time. By century's end, the Japanese were quickly following the pattern, pulling people from rice fields to work in coal mines and in great industrial centers served by European-style railroads. The Industrial Revolution had become a worldwide phenomenon.

The twentieth century saw the high tide of heavy smokestack industry and its promise: more products and faster production than anyone ever could have imagined. It also saw the evils of the industrial age: a legacy of workers too poor to share in the abundance, horrific industrial pollution that blackened the skies and poisoned the soil and water, the particular vulnerability of urban industrial populations during economic depressions, and the horrors of industrial war fought on a global scale. The exodus from the land continued, but by midcentury new occupations were being created in large numbers. Someone had to keep track of all this production, and clerical work expanded from the desk clerks of Bob Cratchit's time to great accounting firms and "stenographic pools." Someone else had to manage and coordinate the efforts of all these new workers, fix their machines, and design better ones. As urban industrial workers were far less self-sufficient than their rural grandparents, new service workers were needed. All these new products had to be marketed and delivered, and retail sales positions expanded. At the same time, governments grew as they tried to encourage prosperity, fend off economic depression, and grapple with the ills that accompany an industrial world. Service-related work, both professional and menial, grew rapidly in the advanced industrial countries.

During the second half of the twentieth century, another trend was well under way: **deindustrialization**. The shifts from coal to electric power, from rail to air and truck, from clumsy cargo ship to giant container ship, from products forged from steel to those based on plastics and microchips, all made industry more mobile. Plants could locate where labor was the cheapest. The logical strategy became apparent to these new managers and accountants and to the asset-holding capitalists they served: a new international division of labor. Design, engineering, sales, service, and marketing would be done in the home country close to the consumers and sources of financing and expertise, while the labor-intensive work of stitching and soldering, and especially of assembly, would be transferred to accessible locations where labor costs were low. Many of these locations are now experiencing a continued shift from agricultural to industrial occupations as their workers leave the land for the new industry. As before, this is sometimes voluntary and sometimes forced, and many groups are left vulnerable, especially older men, who may not be considered desirable workers, and young women and even children, who may be highly desired workers yet are poorly paid. Meanwhile, in the now "postindustrial" countries, the industrial workforce fears plant closings and may be forced to shift to either service-related work or new light industries, both of which may pay much less than the unionized work of the peak industrial boom.

The United States has seen a dramatic decline in farm labor. This trend continues although it can no longer be described as a dramatic shift in the workforce force as a whole, given that there are already so few farmers and farmworkers. Less than 3% of the U.S. labor force is now working in agriculture (although a much higher percentage of workers are employed in

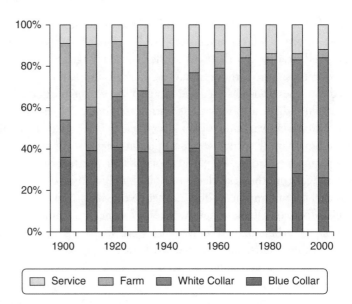

Exhibit 4.3   Occupational Changes by Sectors
Source: U.S. Census Bureau (1976); U.S. Department of Labor (2001); Gilbert (2003).

industries that transform agricultural products: potatoes into French fries, corn into ethanol, sunflowers into vegetable oil, and so forth). This is in large measure a result of the combination of the forces of capitalism and industrialization.

Agribusiness uses large amounts of investment capital coupled with large expanses of land to mass-produce agricultural products for a carefully media-cultivated consumer market. This production is supported by a number of technological interventions: mechanical interventions, such as large, special-ized machinery; chemical interventions, such as fertilizers, pesticides, and herbicides; and biological interventions, including new hybrid seed varieties and now even genetically altered varieties. The high expenses of agribusiness operations are repaid by huge **economies of scale.** To produce vast amounts of goods to be shipped to countries all around the world, companies must make enormous investments, and this gives the advantage to large corporate producers over small producers. Around the world, small farmers are being displaced by American and European agribusiness corporations that rely on economies of scale to enable them to export vast amounts of desired prod-ucts: peanut oil from West Africa; bananas, sugar, and coffee from Central America and the Caribbean; pineapple and tropical fruit from East Africa and the Pacific. Often these countries then in turn need to import food to feed their growing urban populations, food that is likely to include grains grown by agribusiness in North America, Australia, and other temperate locales.

Economists find an overall efficiency in this system. Indeed, world hunger is declining, although slowly, and great famines such as those that occurred in the agrarian past are fewer (Food and Agriculture Organization of the United Nations 2000). At the same time, many farmers have been displaced. Again, economic benefits can result from such change: One reason so many goods and services are available in the United States is that so little of the U.S. population is engaged in basic food production. The social costs of this trend, however, have also been considerable, as family farms and family homes have been lost, and the rural communities they supported are likely to fall into decline unless they can become new bedroom communities to sprawling metropolitan regions.

The steep decline in agricultural jobs in the United States over the twentieth century was paralleled by an equally steep increase in white-collar occupations (see Exhibit 4.3). My own family's experience is typical: from farm to factory to white-collar (well, maybe turtleneck in my case) profession in three generations. The growth of white-collar employment has been one of the hallmarks of the changing American class structure over the past century (Mills 1951). The nature of white-collar workers has also changed. In their classic study of Middletown (Muncie, Indiana), Lynd and Lynd (1929) noted that most white-collar workers early in the century were self-employed professionals and businesspeople (with rare exceptions, business*men*): dentists, physicians, private attorneys, small shopkeepers. By midcentury this had changed. The growing white-collar workforce was still largely male, but it increasingly included examples of the kind of worker Whyte (1956) labeled "the organization man"—employees of large businesses and sometimes huge corporations.

It is interesting to note that by century's end, a trend in the other direction had begun. Large corporations were downsizing their managerial ranks, and more people were becoming free agents, doing consulting or contract work from home or moving from firm to firm, looking for new opportunities or just in desperation to secure professional employment. It remains to be seen whether this trend will continue, but the definition of white-collar work does seem to be changing to accommodate a new **social contract** (Rubin 1996) in which long-term commitments between employer and employee are replaced by temporary and flexible arrangements. Some white-collar workers, like some free agents in the world of professional sports, have profited greatly from such arrangements, but others have lost jobs that they will never be able to replace in terms of income or quality (Newman 1988). Likewise, the new arrangements have been positive for some families, allowing them greater flexibility as they try to meet the demands of both work and home life, but negative for others, creating new stresses as they try to balance several competing occupations and work commitments to maintain what they have come to see as the white-collar lifestyle.

Employment arrangements have also shifted as the composition of the white-collar workforce has shifted, with women making up an ever larger share of this group. All professions have seen increasing proportions of female workers, but this shift has taken place more quickly in professions that require verbal and managerial skills than in those requiring technical and quantitative skills. Almost half of all new law school graduates are now women, for example, but only a small proportion of engineering graduates are female. Old patterns also persist among new trends: There are many more female physicians today than in the past, but they tend to be concentrated in family-related specialties, such as pediatrics and obstetrics, whereas some of the highest-paying specialties, such as general surgery and cardiology, are still largely male. Substantial growth has also taken place in the proportion of white-collar professionals who are nonwhite. Asian Americans, Hispanic Americans, and African Americans are all represented in this population in increasing numbers. Gains for African Americans since the 1960s have been particularly dramatic, although this group is still underrepresented, and some African Americans who do achieve professional success must face alienation from their home communities as well as suspicion and isolation from their professional colleagues (Cose 1993).

The trend line for blue-collar employment is less clear. After rising during the early decades of the twentieth century, it dipped during the Depression in the 1930s and then recovered during World War II, peaked about 1950 (although at levels roughly proportionate to 1920), then began a decline that has become steeper with each decade. This is a graphic depiction of American deindustrialization. A similar trend can be plotted for Europe, although the devastations of war and its aftermath began the decline there even earlier. It is interesting to note that the trend is similar for some middle-income "industrializing" countries. One might suppose that the flight of industry to south of the U.S.-Mexico border would mean a great increase in Mexican industrial employment to accompany the U.S. losses, but this is not the case. Mexican industrial employment also increased to 1920 and then leveled off before some recent declines. Mexican gains in border industry employment, mostly involving young workers, especially young women (Fernandez-Kelly 1983; Peña 1997), as well as employment in a handful of plants established by Japanese, American, and European companies (such as Nissan, Ford, and Volkswagen) have been matched by losses in local heavy industry (Sernau 1994). The total Mexican blue-collar workforce has not increased, it has merely shifted: The muscle-bound laborer of Diego Rivera's murals has given way to a tired-eyed young woman under the watchful eye of a clipboard-carrying engineer.

U.S. industry has also shifted. Significant production still takes place within the boundaries of the United States, but it is much less likely now to be found in a large brick building in a midwestern central city with a unionized workforce. Instead, it is likely to take place in a single-floor warehouse-type structure constructed of sheet metal (maybe with a brick facade for the front door) along a major highway on the edge of a growing city in the South or the West. The products are likely to be lightweight, frequently changed or adjusted to meet consumer demand or shifting orders, and probably shipped by truck or air. The typical workforce is almost half female, half black and Latino, and largely nonunionized.

A blue-collar worker in my home city of South Bend, Indiana, will find nothing but hollow brick shells in the old industrial corridor on the west side. Neighborhoods once filled with Polish immigrant industrial workers have given way to low-income African American renters and new Latino immigrants. The Studebaker plant closed in 1963, and Oliver Plow Works merged with larger agribusiness firms and left. If the new residents want automotive industrial employment, they will need to travel to a suburban location where AM General builds Humvees for the U.S. Army (a contract that would be politically unpopular to export) and is now leveling houses to build a bigger one-story plant to produce a new Humvee sport utility vehicle (don't try to fight it for a parking space at the PTA meeting!) that is popular with the more prosperous. New labor migrants now travel to neighboring Elkhart to work in the recreational vehicle industry in nonunion jobs that pay about $9 an hour; this industry currently employs a workforce that is 50% Mexican American. Further on, in a neighboring small city, Biomet builds hips and kneecaps and spinal rods for the custom orders of orthopedic surgeons. Biomet is also a nonunion employer, but it promises to advance workers who are skilled and patient. Repeated across the country, this is the new face of blue-collar employment.

The decline in blue-collar employment is matched by a corresponding increase in nonprofessional service-sector work. This includes low-wage and often female-dominated employment in clerical work and personal services. These have been dubbed pink-collar jobs: The people who hold them are secretaries and data entry clerks, nurses' assistants, home health aides, and retail workers in big-box one-story suburban marts. The largest employers in South Bend are now the hospitals and the universities, and I often meet my students when I shop in the stores where they work on an ever-growing strip that starts with K-Mart and leads on to Wal-Mart, Home Depot, and many others. Along with these jobs, and including more men, are the "green-collar" jobs. These workers clean offices, medical facilities, and stores; change the oil in our cars in 10 minutes or less; and do many of the other manual service tasks of modern urban America. It may be unfair to term any of these jobs "unskilled"—anyone who can read my physician's handwriting or do anything to help maintain my car in less than 10 minutes is not unskilled—but these are quickly learned, low-investment, low-wage jobs. Coupled with these are the "burgundy-collar" fast-food jobs that provide first employment for many young people and, increasingly, supplemental or last-resort employment for the elderly, part-time homemakers, and others. Analysts project continued increases in the numbers of such jobs (see Exhibit 4.4).

Current projections concerning job growth offer reason for both optimism and pessimism. Overall, the U.S. economy has created new jobs to replace those that are disappearing. The quality of these jobs has varied greatly, however, with substantial growth in the no-benefits, low-wage portion of the service sector. The good news, especially for college students working on graduate and undergraduate degrees, is the anticipated growth in the white-collar professions.

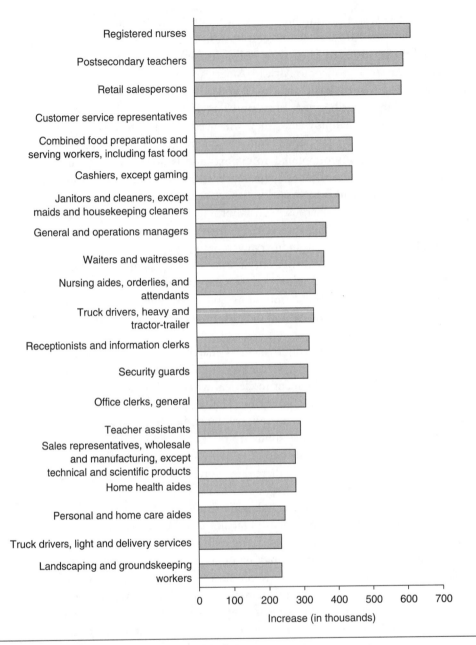

Exhibit 4.4   Occupations with Largest Growth

Source: Data from U.S. Department of Labor, 2004.

Continuing technological development and expansion, coupled with an aging population that will be retiring and then needing health care, means continued job growth for the skilled and the degreed. Yet we must not overlook the underside to this growth. There will also be continued and growing need for routinized workers: more nurses' assistants and health care aides than physicians, more data entry clerks than systems analysts, as well as people to clean and

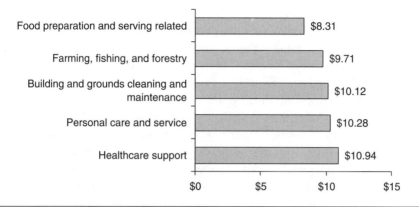

Exhibit 4.5   Occupations with the Lowest Wages

Source: U.S. Bureau of Labor Statistics, 2003.

guard the offices of the managerial class. Two key questions for the beginning of this century are how we will fill these positions and how we will properly compensate the people who do the work. The cry has gone out for more skilled workers, yet we must also face the question, Who will do all the low-skill, low-wage work left over? The uncomfortable fact is that as a society we are dependent on people who have few options—immigrants (both legal and illegal), the poorly educated, and the ill prepared—to do the service work that cannot be automated (see Exhibit 4.5).

Income

People in other times and places may have looked to family background or even inherited title as the key class divider, but Americans today are likely to think first of income. **Personal earnings** consist of the money an individual receives in wages, commissions, and tips for work performed. Personal earnings are a good measure to use in considering returns on work, but a focus on personal earnings misses people who may be out of the workforce. **Family income** is the total amount of money coming into a family unit. Because the U.S. Bureau of the Census collects data by households, which are based on place of residence, this figure is often referred to as **household income.** The choice of which measure is best depends on the questions being asked. For example, an examination of differences in wages between men and women needs to focus on personal earnings, because many men and women share households and mingle their incomes. On the other hand, an examination of the well-being of children needs to focus on family income rather than on the children's earnings. Some differences are more subtle; for instance, gaps between blacks and whites are larger for family income than for personal earnings (because there are fewer wage earners in the average black household), whereas Asian Americans do better than white Americans on family income measures (in large measure because there tend to be more wage earners in Asian American households) but lag in personal earnings. Household income is the measure most often used, because data on household income are readily available from the Census Bureau, which provides periodic updates.

What does the distribution of household income in the United States look like? Some who describe the country as a "middle-class society" might expect to see very few rich, very few poor, and many in the middle—a diamond-shaped distribution. Others who are worried about the shrinking or "disappearing" middle class might imagine that the distribution is becoming hourglass shaped, with a bulge at the top, another at the bottom, and a narrow place in between. Over history, the distribution of income in many societies has approximated a pyramid: an absolute ruler, a few wealthy nobility, a somewhat larger middle or merchant class, and many near the bottom. No society really has the hourglass form, but many still have some form of the pyramid, with a few wealthy families (sometimes topped by a dictatorial figure who has enormous wealth, such as Indonesia under Suharto, the Philippines under Marcos, and Congo under Mobutu); a small middle class of businesspeople, professionals, and technocratic experts; and a large group of rural and urban poor and near poor (see Exhibit 4.6).

The distribution in the United States perhaps most closely resembles the outline of a potbellied stove. It begins at the bottom with a significant group of very poor with incomes well below the poverty line, a larger group of near poor with incomes near the poverty line, and a bulge of lower-income households. The households of many elderly and many nonprofessional service workers help fill this bulge. Above the bulge, the distribution gradually narrows upward through middle and upper-middle incomes, then is capped by a very tall but narrow "pipe" of high-income households. To proportionately capture every income level, the pipe for this stove would need to soar several stories high.

The placement of households in this distribution corresponds partly with high- and low-wage occupations. Many occupational categories span a wide range of incomes, however. For example, small business owners may run single-person operations from their kitchen tables and generate incomes that fall below the poverty line, or they may operate extremely profitable enterprises that generate million-dollar incomes. Certain professions, such as many in the arts or in the music industry, may have a similar spread. As noted above, household income also depends on the numbers of persons in a household who are working. Many of the households in the upper-middle part of the distribution are there only because they combine the earnings of two or more wage earners. Single-parent-headed households often do not have this option, and these families are much more likely to be in the lower income brackets (see Exhibit 4.7). Families of color are found in all income brackets, but African American, Hispanic American, and Native American households are disproportionately represented in the lower income levels. Households of

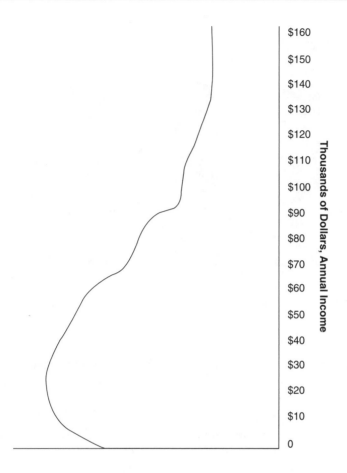

Exhibit 4.6   Income Distributions

Source: Data from Rose (2000).

elderly couples and individuals are no more likely to be poor than the overall population (the exception being very elderly women) but many of these "gray households" are in lower income brackets, including many near poor. (For an excellent visual representation of these patterns, see Rose 2000.)

The **central tendencies** of a distribution such as this may be described numerically in several ways. The bulge in the distribution, the point that holds the most people, is the **mode**. The statistical **mean** is simply the average—that is, the total divided by the number of cases. The **median** is the point at which half the cases are greater and half are less, regardless of how far above or below. In a balanced or bell-shaped distribution, these measures are the same. Income, however, is a skewed distribution. There is a clear limit to how poor one can be ("flat broke"), but there is no upper limit to income. As a result, a few extraordinarily high incomes can pull the average upward. For this reason, discussions of income levels often refer to the median—half of the incomes fall above this point, and half fall below.

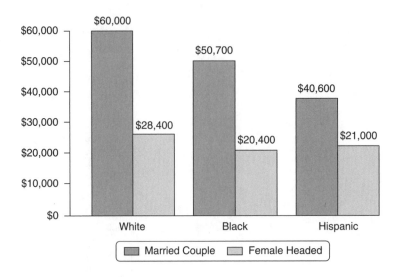

Exhibit 4.7   Median Income by Family Characteristics, 2002

Source: U.S. Census Bureau, 2003.

Although the United States is not in danger of becoming an "hourglass society" in the near future, income inequality is increasing, and markedly so. The economic expansion of the 1980s favored upper-income groups far more than the lowest income groups, some of which lost ground because of a minimum wage that did not rise with inflation and reductions in government services and payments. The recession of 1990–92 hurt everyone, but people in the lower income groups were hurt particularly badly, especially as manufacturing jobs were eliminated. The long economic growth period that started in 1992 eventually helped those in the bottom brackets slightly, but only enough so that they regained what they had lost at the beginning of the decade. When recession again struck in 2000, the gains that lower-income groups had made over the entire 1990s were again eliminated. The recession accelerated the job losses caused by deindustrialization as fragile industries collapsed or moved to remain competitive, but recovery has not brought "reindustrialization," and most of the current job growth is at the lowest end of the income scale. For the wealthy, the recession's effects were cushioned by a large tax cut and a strong real estate market, and they were well poised to benefit as the economy began to rebound in 2004. The scenario begins to sound like the old adage that the rich get richer and the poor get poorer. It might be more accurate to say that, in the United States at least, the rich are getting richer while the poor are getting nowhere. The height of the "stovepipe" in the income distribution has soared several more stories while the bulge at the bottom has remained. One way to chart this movement is to return to the measure of percentage of income going to each **quintile,** or fifth of the population. The growing spread becomes quite evident, as Exhibits 4.8 and 4.9 illustrate.

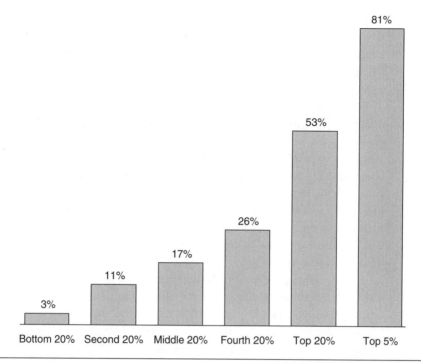

Exhibit 4.8   Change in Family Income by Quintile

Source: U.S. Census Bureau, 2005, Tables F-1 and F-3.

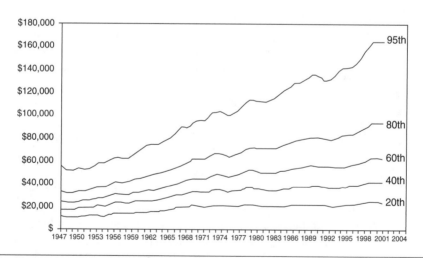

Exhibit 4.9   The Growing Spread in Family Income

Source: U.S. Census Bureau, 2005, Table 5.3.

The middle class in this picture is stressed but not disappearing. The fortunate have moved up into upper-middle-class positions with rising incomes. The unfortunate may still see themselves as "middle-class" in their outlooks and perspectives, but they find that they are no longer able to maintain middle-class lifestyles with their declining incomes (Newman 1988). One reason for the income spread is an integrating global economy based on new technologies. Global markets and new technologies create new opportunities, and new sources of profit, for investors. At the same time, workers' wages face downward pressure from international competition and from displacement by automation.

Similar trends are at work around the world. During the 1980s, the old adage worked internationally: Rich nations grew richer while poor nations grew poorer. It was a good decade for the advanced industrial countries: Reagan's United States and Margaret Thatcher's United Kingdom, with their conservative governments, but also Mitterrand's France, with its "socialist" government. It was a disastrous decade for poor nations in Latin America and Africa. Yet, with a growing economy, the wealthy in the United States could grow much richer without the U.S. poor growing much poorer. Many had a sense of being "better off" until the downturn during the first Bush administration.

Government policies do have some effect on income distribution, although less than we may be led to believe. Tax changes during the Reagan administration, enacted in the hope of fueling further economic growth, allowed the wealthiest groups to keep more income and did spur greater inequality. Inequality also grew in European countries and Japan during this time, but most did not see the same wide income divides.

The Clinton administration, especially under labor secretary Robert Reich, made helping those who were left out of the benefits of the global economy a national priority. The minimum wage again began to increase, although it is still below real dollar amounts of the 1970s. Finding a way to provide health care for those workers without benefits was made a priority, although the ranks of the uninsured have actually increased. Some government services to poor families were extended, paid for in part by a slight increase in taxes on upper-income earners. Yet inequality continued to increase.

In Europe, the demands of staying competitive with expanding American and Asian economies have led to a reduction in government spending and services. The most dramatic cases are the Eastern European countries that left the old Soviet sphere of control with hopes of joining the European Union. In cases such as Poland, where the economy has grown, so has inequality. In particular, with fewer national guarantees and rapidly shifting economies, worker insecurity has increased markedly. Many developing countries have long claimed to strive for "growth with equity," yet "growth with growing inequality" has become the norm for the industrialized nations. Currently, the United States is the economic leader but also the leader in growth in income gaps and worker insecurity.

A historical perspective may also help to frame this picture. A world of intense international competition with stagnant wages, job insecurity, and weak unions is not new. It was the world of our grandfathers and great-grandfathers. A few grew rich (people called them robber barons) and many stayed poor or near poor. Only during the period of unquestioned U.S. global economic dominance, roughly 1945 to 1965, did Americans come to expect ever-rising wages, stronger organized labor, expanding benefits, and greater job security as the norm. U.S. dominance has by no means vanished, but a new global economy has meant a return to the "bad old days" for some workers. So far, government efforts have been of some use in alleviating the pain and cushioning some of the dangers of this economy, but they have not reversed the trend toward great divides in income.

Class Structure

By placing together wealth, occupation, education, and family income, we can get a picture of the **American class structure**. In a book by that title, Gilbert and Kahl (1982; see also Gilbert 2003) suggest a six-part division of American society. In many ways their "synthesis" updates the work of previous class researchers (Coleman and Rainwater 1978). Gilbert and Kahl's labels are not without controversy, but they provide a more complete picture than do the terms *upper, middle,* and *lower,* and they may be more revealing than *upper-upper* and so on:

- **Capitalist class:** investors, heirs, and executives, typically with prestige university education and annual family incomes over $2 million, mostly from assets
- **Upper-middle class:** high-level managers, professionals, midsize business owners with college education, most often with advanced degrees, with family incomes of $120,000 or more
- **Middle class:** low-level managers, semiprofessionals, some persons in sales and skilled crafts, foremen and supervisors with at least high school education, usually some college, technical training, or apprenticeship and family incomes of about $55,000
- **Working class:** high school–educated operatives, clerical workers, most retail salesclerks, routinized assembly and factory workers, and related "blue-collar" employees with family incomes of about $35,000
- **Working poor:** poorly paid service workers and laborers, operatives, and clerical workers in low-wage sectors, usually with some high school and family incomes of around $22,000
- **Underclass:** persons with erratic job histories and weak attachment to the formal labor force, unemployed or able to find only seasonal or part-time work, dependent on temporary or informal employment or some form of social assistance, and with typical family income of $12,000

This scheme emphasizes several important elements of the American class hierarchy. At the top are those with more than just high salaries, who most often gain the largest part of their income from returns on investments and assets. That broad-band all-American middle class divides into highly paid professionals and upper management, a "white-collar" core, and a largely "blue-collar" working class. Among the working poor, collar color often gives way to the smocks of poorly paid clerks and service workers; whether they actually experience poverty often depends on whether they must support dependents on their incomes. The term *underclass* is sometimes decried as demeaning, implying that members of this group have poor values or poor work habits (Gans 1990, 1995), but here the category simply encompasses those with poor job histories, who are unable for one reason or another to maintain steady, formal employment.

As always, it is easier to describe the "typical" member of each class than to define the boundaries between the classes. Yet movement between classes is not fluid, and many people remain in their classes of origin. The underclass is distinguished from the working poor by the lack of regular, enduring, formal employment. The working poor are distinguished from the working class by incomes that do not provide for what Americans consider to be basic necessities without outside assistance. The working class is distinguished from the middle class by jobs that may require skills but limited higher education. The middle class is distinguished from those above by occupations and incomes that keep them tottering in the middle with many comforts but few luxuries and ever fewer assurances. The upper-middle class has the sought-after education and experience to command high-wage positions but lacks large quantities of assets, so members of this class must "still work for their money rather than allowing their money to work for them."

Although this overall class structure seems quite stable and enduring, the spread between levels is increasing. In particular, in accord with Marxist analysis, the gap between capitalist owners and everyone else is growing. Vast global consumer markets coupled with fearful unions, states, and nations, each willing to offer investors more to keep jobs in the competitive global labor market, present tremendous opportunities for wealth accumulation to a well-placed few. In accord with Weberian analysis, there also appears to be a growing divide between those in the top three classes, whose skills and backgrounds afford them some market power, and those in the bottom three classes, whose power in the labor market continues to erode, forcing them to accept whatever they can find to survive. The labor provided by the working class and the working poor remains vital, but in an era of the global movement of people and jobs, those in these classes are always at risk because someone new may perform the same work for less pay.

The organization of the class structure described above is based on the interactions among wealth, education, occupation, and income. In large measure, these factors determine the life chances of individuals and families. Class position, however, interacts with both race and gender, as well as the political structure, and it is these interactions we consider next.

KEY POINTS

- Although Americans may be reluctant to think in terms of class divisions, class position affects many aspects of our lives and our future life chances.
- Wealth is more concentrated than income in most societies; in the United States, 10% of the population controls two-thirds of the wealth.
- The occupational structure of the United States and other advanced industrial countries has shifted with the decline in farm and manufacturing positions coupled with the rise in both white-collar and low-wage service professions.
- Income inequalities continue to grow in the United States, with the greatest gains going to the highest-income groups. The middle class is not disappearing, but the numbers of well-off upper-middle-class families are increasing while others face declining incomes and the prospect of joining the working poor.

FOR REVIEW AND DISCUSSION

1. What economic and social forces tend to make wealth as concentrated as it is in the United States? What are the consequences of this concentration of wealth?
2. What major changes have occurred in sectors of the U.S. economy over the past 100 years? How have these affected occupations and class structure? What are the current trends in occupation growth, and what effects will these likely have on the distribution of income?
3. What have been the major trends in the distribution of income over the past several decades? How do median incomes differ by race and ethnicity, by age, and by family type?

MAKING CONNECTIONS

U.S. Bureau of the Census

The U.S. Bureau of the Census gathers detailed data on wealth, income, occupations, and national trends and makes both current and historical data available in numerous publications and on the bureau's Web site at www.census.gov. To examine data on housing and household economics, you may wish to go directly to the Web site's "Housing and Household Economic Statistics" page, at www.census.gov/ftp/pub/hhes/www/index.html. This will take you to useful summary tables providing information on poverty, income, wealth, wages, and occupations. You can also zoom in to explore your local region: On the site's home page, click on "State and County Quickfinder" to look at conditions in your local state and county. You can find more detailed information, down to the city and neighborhood level, at American Factfinder (www.factfinder.census.gov), which is also linked to the Census Bureau home page. Explore one aspect for the nation as a whole or focus on a range of attributes of your local community. Prepare a brief profile of what you find.

5

Race and Class

Nature still obstinately refuses to co-operate by making the rich people innately superior to the poor people.

—Sidney and Beatrice Webb, English social critics (1923)

Becoming White

How do you know when you have truly arrived, when you have succeeded, when you are welcomed and "at home"? For many immigrant groups in the United States, it was when they became white. "Becoming white" may seem like a strange notion—after all, isn't one born into a race? But many of the various and diverse individuals who make up the group we think of as white Americans have little in common in terms of heritage: a pale blond Finn in Minneapolis whose ancestral language can be traced to central Asia, a black-haired and medium-complected Iranian in Los Angeles with Farsi-speaking grandparents, and a freckled, red-haired Irish child in Boston of mixed Celtic heritage may all be "white." Think for a moment about whiteness not as some biological category but as a category of privilege, and this starts to make sense. The first immigrants to the United States weren't "white," they were of the "British races," the "Irish race," the "German races," and so forth. The term *white* was apparently first employed in the colonial United States to distinguish between persons of European ancestry and those of African ancestry in early marriage laws. This was no small distinction. In time, poor European-born indentured servants would become citizens, and their children would inherit the rights and privileges that go with citizenship. Poor African-born indentured servants, however, were "black" and thus subject to what became 200 years of intergenerational slavery. Whiteness mattered.

In time, other immigrants came in large numbers. They weren't white either. The largest group was made up of members of the Irish "race," or "races," as some distinguished between the "black Irish" and the "red Irish; height and hair color rather than skin color were the key markers. The Irish "races" were said to be inferior: intellectually slow, given to drunkenness and irrational emotion, easily manipulated by unscrupulous priests and politicians. The anti-immigrant Know-Nothing Party and the Protestant Ku Klux Klan understood this and were glad to share their opinions about it. Over time, however, the Irish "became white." While holding on to certain markers of ethnicity, such as religion, they began the long struggle to reach working-class and middle-class status, and, eventually, even though many stereotypes persisted, it seemed silly to keep them out of the club—at least they were from the United Kingdom (although not by choice).

New immigrant groups were coming, and they were even less white than the Irish. Sicilians, southern Italians, and Greeks with olive-toned complexions and black hair, along with unusual clothing and languages, were arriving in large numbers. These southern European races were said to be diluting the "pure" white race of America. They were "colored." Also arriving were Russian Jews, who looked white but were part of the "Jewish race." These groups also variously hung on to and relinquished aspects of their ethnicity but were eager to be accepted as white. Although stereotypes and prejudice persisted, eventually they were grudgingly included (see Brodkin 1998).

Some new groups stretched the boundaries of racial divisions even further. When Chinese immigrants were brought to segregated Mississippi late in the nineteenth century, they were supposed to be "colored" and to work in the fields, maybe with more diligence than the increasingly disgruntled "colored" African Americans. Instead, they opened small businesses such as grocery stores and started to become middle-class (Loewen 1988). This posed a real threat to a system based on a clear two-category division. Eventually the tension was resolved: As long as the Chinese didn't push so far as to intermarry with southern whites, they could be white and live in white rather than black neighborhoods. If they were to marry outside their group, however, it would have to be into black families, and they would then again become "colored." Like Roman citizenship in an earlier era, whiteness is not so much about origins as it is about privilege (see Tuan 1998).

On the subject of race, sociologists, anthropologists, historians, and other scholars are coming to a point of agreement: Race is a social construction, not a biological fact. Societies have created racial divisions in attempts to categorize the range of human physical diversity. Arrange a sampling of the world's population in a line from lightest to darkest, and the result is a continuous gradation. Further, there are interesting combinations that don't easily fit established "racial" categories: Australian Aborigines with dark brown skin but green eyes and reddish hair, Japanese Ainu with "Caucasian" features, Melanesians with ties to Southeast Asia that are far more recent than their ties to Africa but with skin and hair that would make them "black" in the United States. Further, color, the favorite racial marker in the United States, doesn't correlate with much else. We're used to speaking of "blood," as in "He has Indian blood" or "one drop of African blood," but blood types do not correspond to our color divisions.

Other societies make much finer distinctions regarding color: For example, in South America, a person may be white, or black, or *café con leche* (cream colored), or *café sin leche* (darker brown), and so forth. One African American reporter has written about how astonished he was to realize that a "black" Brazilian woman with whom he was talking didn't even consider herself "black." In Brazil, racial identities are tied as much to social class, education, and cultural identity as to color. For a moment, he was dizzied by the realization that his color, one of the most basic identifiers of his status in U.S. society, might not be important in Brazil, and in fact he might even be able to "claim" a new racial category!

> I've always been black. The surprising news that there was a place where I wasn't, necessarily, or at least didn't have to be, was imparted to me one hot summer's afternoon in Rio de Janeiro on the beach at Ipanema. It was a special moment in my life, a time when suddenly I felt free as a bird and open to all sorts of new possibilities. (Robinson 1999; quoted in *New York Times,* September 17, 1999)

Some racial categories bring together enormously diverse groups. "Asian American," for example, can include light-complected North Indians with language ties to Europe, South Indians with coloring as dark as many African Americans but with more "European" features, and East Asians with what Americans have come to consider "Oriental" features and coloring. "Hispanic" makes no sense as a category for a biological race, as it brings together descendants of Mayan and Incan Amerindians, "white" Cubans of Spanish heritage, and black Caribbeans descended from African slaves, yet "Hispanic" is often treated as a racial category in U.S. society. Hispanics constitute a "race" only in that they may have a common language and may occasionally share a common agenda.

The possibilities for confused claims and misunderstood identities are boundless. I take my students to a community agency, La Casa de Amistad, which serves our area's growing "Hispanic" population. We are greeted by Angel, a "black" Cuban from New York, who grew up immersed in English. His black friends reject his claim to be "Hispanic" (wrong color, no accent) and insist he must be a "homeboy who got confused about who he is." He works with Adriana, who is Mexican American with mixed Spanish and indigenous ancestry. She grew up amid Anglo-Americans and has had to learn Spanish as a second language. A university student, her skin is dark enough for her to be accepted readily by Latino groups on campus, but some members of these groups claim she is acting "white" when she brings along her longtime Anglo friends; in fact, they claim she is "rejecting her people" when she brings her blue-eyed brother! Angel and Adriana work with Carlos, an immigrant from Guatemala. With his light complexion, he gets little sympathy as he struggles to learn English: He is supposed to be "white"! He has also had to learn about racial categories that are entirely different from those he has known in his home country.

The complexity of race division is also seen in the growing numbers of multiracial individuals. This is nothing new. In early colonial times, terms such as *mulatto* were used to describe such people. Mixed categories pose a problem for clear black/white segregation systems, however, and so in time the United States came to be dominated by what became known as the "one drop rule": Any amount of African heritage made one "black." Yet many African Americans have some European ancestry, and many "whites" who are not of recent immigrant background have some African heritage. Both groups often find they have some Native American ancestry. Again, what matters is not biology but perception. When having Native American roots became socially desirable, many "whites" suddenly discovered and proclaimed their Indian great-grandmothers. Interestingly, few "whites" seem eager to discover black grandmothers, even though black-white liaisons may have been as common as white-Indian relationships. Race matters because of the importance people give to it.

The perception of shared race can unite divergent groups for common political action, as candidates seek to court the "black vote" or to take advantage of growing Hispanic political clout, for example. Issues of race can also be used to divide. Ronald Takaki (1993) tells how the "giddy multitude" of early Virginia, consisting of poor whites and blacks, all of whom were indentured servants, came to be divided by race. Those higher up in the social hierarchy benefited as this division diluted the power of the indentured servants' threatening numbers. At other times, racial and ethnic divisions have kept workers and citizens' groups from coming together on a common agenda.

Americans have often treated race as an attribute of certain individuals rather than as a boundary or category of privilege (Lucal 1996). From this viewpoint, race is of concern only to "minorities" who are not part of the white club. My students of color often tell me that they've thought hard about the American racial system and their place in it; my white students, in contrast, generally say they've thought little about race, except as a "black issue" or a "minority issue." Whiteness, however, has been rediscovered, and not just by white supremacists. In 1997, a *Washington Post* article noted with some amusement the gathering of the Second National Conference on Whiteness and quoted the conference organizer: "One thing about white people, we tend to either be proud or ashamed of being white. Proud in a supremacist way or guilty in a liberal way. Very seldom do you find the balance." Finding such a balance will require confronting the roles that race and whiteness have played in defining American identity and shaping American history.

A Debt Unpaid: Internal Colonialism

In 1903, black sociologist W. E. B Du Bois described racial exploitation as the "debt unpaid" and predicted that the "color line" would continue to haunt the twentieth century. Gunnar Myrdal (1944) watched the irony of black Americans fighting in segregated units to liberate Europe from Nazi racism and spoke of "the American dilemma." More recently, Wahneema

Lubiano (1997) titled a collection of essays on race in American history *The House That Race Built.* The United States was "built" by many groups from many shores, but some of the builders are still waiting to be truly welcomed into the house. These groups are often termed *racial minorities,* but by the middle of the twenty-first century they will constitute close to half of the U.S. population. A newer term gaining wide currency is *people of color,* for they are the groups that never became white.

Perhaps the situation these groups have faced has been entirely different from that experienced by the white immigrant groups. In the early 1970s, Robert Blauner (1972) adopted the term "Third World peoples" for these others, given that many have roots in parts of the world that comprise the so-called Third World, and that these groups seem to form a "Third World" of poverty within the First World. Blauner argued that in fact these groups are also victims of colonial oppression, only for them the oppression has occurred within their own country rather than being imposed by a foreign power. The United States has held only a few external colonies (most notably the Philippines) but has often relied on the low-wage labor of many groups within its borders. Blauner termed this process **internal colonialism** and outlined its four basic elements:

1. *Control over a group's governance:* The colonial group is allowed neither full autonomy nor full participation in the national government.

2. *Restriction of freedom of movement:* Colonial peoples are not willing immigrants but are involuntarily incorporated into the national society. Often their ability to choose where they live and work is severely restricted.

3. *Colonial-style labor exploitation:* A "cultural division of labor" exists in which the colonial peoples are assigned to the most menial or dangerous work and given the least compensation for that work.

4. *Belief in a group's inferiority:* In this model, prejudice follows from, rather than causes, discrimination. The exploitation of the colonial group must be justified, and this is done through an ideology that asserts the group's moral, intellectual, and cultural inferiority. Thus the domination of the group's members by others is "for their own good."

This description fits, although in differing ways, the experiences of several groups of color in the United States.

Native Americans

European and European American views of Native Americans have vacillated between "the noble savage" and "the savage killer." Both images have proven very dangerous to Native peoples themselves. When Columbus first encountered the Arawak peoples of the Caribbean, he saw in them children of nature: Mostly naked yet innocent and even childlike, they were perfect candidates for subjugation as subjects of the Spanish king. When the Spanish then encountered the more aggressive Caribs, whose name gives us both *Caribbean* and *cannibal,* they came to a different conclusion: These were savage killers to be exterminated. The pattern was set for the next several hundred years. Followers of romantic thinkers such as Rousseau were looking for the noble savage and thought they found their children of nature in the unspoiled peoples of the New World. But children need caretakers to look after and educate them, and this view often led to paternalism. Some thought they could assimilate the Native peoples if only they could educate them properly. Benjamin Franklin recounted an offer by the commissioners of Virginia to educated Iroquois youth in "All the Learning of the White People." The Iroquois declined, noting that in their previous experience of this type of education, their young men returned "totally good for nothing." The Iroquois, however, returned the offer: "If the Gentlemen of Virginia will send us a Dozen of their Sons, we will take great Care of their Education, instruct them in all we know, and make Men of them" (quoted in James Madison Center 2004). Needless to say, the gentlemen of Virginia declined the offer.

Along with attempts to assimilate were attempts to annihilate. When hostilities broke out between European powers (as well as the emerging European American power of the United States), all sides were very ready to enlist Native forces. Native peoples were by no means united in one great peace-loving brotherhood when Europeans arrived (Bordewich 1996), but they fought their bloodiest campaigns in the service of France, England, and the United States. Yet the image of the savage killer grew to justify the elimination of entire peoples. Note how "nobly" Andrew Jackson, known as the "Indian Fighter" and the "Frontier President," could speak of what amounted to genocide:

> The fiends of the Talapoosa will no longer murder our women and children, or disturb the quiet of our borders. . . . How lamentable it is that the path to peace should lead through blood, and over the carcasses of the slain!! But it is in the dispensation of that providence, which inflicts partial evil to produce general good. (1817, quoted in Takaki 1994:64)

Many Native peoples did attempt to assimilate. By the time of Jackson's presidency, many of the Cherokee lived in villages with blacksmith shops and white steeple-topped churches that could have fit in a Currier & Ives lithograph of a New England town. They were farmers. Ironically, it was Native farmers who taught the Europeans how to grow corn, cotton, and tobacco. Yet Jackson was determined that the Indians must go, and they were marched on the Trail of Tears to Indian Territory in what is now Oklahoma. With few exceptions, the remaining Native populations of the eastern United States were to be moved west of the Mississippi. A few remained in rugged pockets in New York and the Carolinas. A few resisted, such as the Seminoles, multiracial descendants of the Creek and escaped black slaves who held out in the Florida Everglades. But most were removed. Before Columbus, the eastern United States was far more densely populated than the harsher plains and dry West, but soon Indians were associated largely with the West, their relegated home.

As Americans moved westward in ever larger numbers, the Native peoples were again in the way and the same two solutions were proposed: assimilation and removal. In the 1860s and 1870s, the fight was to move mobile Native populations onto small and remote reservations. Some, such as Crazy Horse and the Lakota, resisted; some, like Joseph and the Nez Percé, tried to flee to Canada. Most, however, had no choice but to accept the harsh terms.

> The little children are freezing to death. My people, some of them, have run away to the hills and have no blankets, no food; no one knows where they are—perhaps freezing to death. . . . Hear me, my chiefs. I am tired; my heart is sick and sad. From where the sun now stands I will fight no more forever. (Chief Joseph, quoted in Foner and Garraty 1991)

The cruelty directed toward the Native Americans is best seen not as a uniquely American ruthlessness but as the American version of the empire building of the time. While the U.S. army moved west with Gatling guns and Hotchkiss guns, the British were loading their Maxim machine guns onto gunboats to decimate indigenous resistance in Africa, Asia, and the Pacific. The Russian czarist army was moving east against tribal peoples and incorporating them as internal colonial people of the Russian Empire. Meanwhile, the British, French, and German armies looked south and, in a matter of just over a decade, turned the tribal peoples of Africa into external colonies (Bodley 1990). Australia moved its aboriginal peoples onto reservations to make room for grazing and mining, just as the United States was doing.

Then, in the late 1880s, with nowhere left to move the Native peoples to, the mood in the United States again swung toward demanding assimilation. The Dawes Allotment Act of 1887 provided that the reservations would be divided into 40-acre plots and sold. The Plains Indians were to become small farmers on the Great Plains, where farming has been an impossible undertaking for anyone without a riverside location or modern irrigation. The Plains people failed, of course, and lost their land. Many Natives refused "allotments," and, in the 1898 Curtis Act, Congress declared that those who refused allotments would be "terminated," losing reservation status. The press for assimilation continued through the 1920s with more terminations as well as the belated granting of U.S. citizenship to all American Indians.

Native Americans' hopes for partial autonomy returned with Franklin Roosevelt's **New Deal**. John Collier, then U.S. commissioner of Indian affairs, promised an "Indian New Deal." The 1934 Indian Reorganization Act ended allotment, extended credit, promoted the revival of Native cultures and crafts, and encouraged tribal self-government. The changes were welcomed on the reservations, but with the country in the midst of the Great Depression and the plains gripped by the drought of the Dust Bowl, progress was slow and hard. The country's attention was diverted by World War II, and many Native Americans served in the armed forces. They returned to a country that was eager for unity and often uniformity during the 1950s. In 1952, a new program was introduced to encourage Native peoples' relocation off of reservations to the growing cities. Termination returned in 1953 with the Termination Act, which was designed to eliminate reservation status. Once again, Native-controlled land was lost and poverty persisted. By the 1970s, these policies were again reversed: The 1973 Restoration Act restored lost reservation status, and the 1975 Self-Determination Act again promoted autonomy.

A mood similar to that of the 1950s returned to the country in the 1980s. Ronald Reagan's secretary of the interior, James Watt, claimed that reservations showed "the failures of socialism," and so he promoted assimilation and Indian "enterprise zones" like those in inner cities along with mining and resource extraction on Native lands. With deep budget cuts and limited success in entrepreneurship, poverty only grew deeper among Native Americans, especially on isolated reservations. Once again, by the time of the first Bush administration and the Clinton years, the pendulum had shifted back to self-determination: After years of suppression, Indian languages were again encouraged, and tribes began to experiment with self-determination apart from control of the Bureau of Indian Affairs. No law had greater impact, however, than the 1989 Indian Gaming Act, which stated that any state that allows some form of gambling within its borders must also negotiate to allow Indian gaming. As states across the country sought to replace declining federal revenues with gambling revenues, Native tribes also began to open casinos. This was entrepreneurship that worked: With special status, many Indian casinos became hugely profitable and provided jobs for thousands. Yet many, Indian and non-Indian alike, have wondered about the effects of this single "industry" on Native cultures, traditions, and social structures. Memories of the battle over the Little Bighorn have given way to battles over bingo, some fiercely fought. The controversy and the casino openings continue.

Along with these other changes, there has been a shift in the perception of Native Americans. The savage killer has galloped out of the scripts of Hollywood producers to be replaced by the noble savage. In the films of the 1950s, John Wayne and the cavalry bravely fought off swarms of attacking savage killers. In the films of the 1990s, the cavalry are the savage killers, and outnumbered bands of noble savages try to fight them off to save their way of life. Meanwhile, interest in Native spirituality, Native ecology, and Native crafts has mushroomed. The American Indian population is growing rapidly,

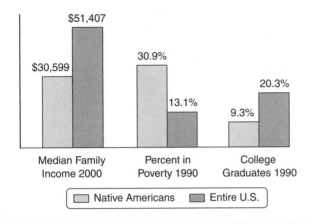

Exhibit 5.1   Socioeconomic Standing of Native Americans

Source: U.S. Census Bureau, *Statistical Abstract of the United States,* 1999 and 2003.

less because of high birthrates than because "being Indian" is now a mark of distinction, and many "whites" are reclaiming their partial Native heritage. Descendants of interracial couples who once fought to be accepted as "white" are now eager to emphasize the "red" and its assumed nobility and wisdom.

Most American Indians are now urban—a full two-thirds live in major metropolitan areas. Los Angeles is currently home to the single largest concentration of Native peoples in the United States. Those who remain on reservations, especially isolated rural reservations, are the most likely to be poor. In the 2000 census, the single poorest county in the United States was not one containing urban slums but Buffalo County, South Dakota, home to the Crow Creek Reservation. Unemployment among Native Americans on reservations has topped 30%, and poverty rates are over 60%. For those too isolated to benefit from a thriving casino and who have few exploitable resources, the choices are familiar: They can risk poverty for the autonomy and community of the reservation, or they can take their chances on "assimilation" in major urban areas and risk isolation, dislocation, and the vagaries of the urban economy (see Exhibit 5.1).

In discussing the experiences of Native American nations, Matthew Snipp (1986) contends that they have moved from "captive nations" to "internal colonies." Early on, they were treated as independent nations but never as equals. Later this model was abandoned in favor of more direct "colonial" control over these nations' governance and more emphatic restriction of their freedom of movement, always bolstered by beliefs that the Native peoples were morally or intellectually inferior to whites. Labor exploitation was used heavily in Latin America, were concentrated populations had previously been subjected to Aztec and Inca masters who assigned their labor and extracted tribute. Without this imperial tradition in what would become the eastern United States and Canada, Europeans found the Native peoples to be poor subjects, servants, and slaves, and soon turned to other populations for laborers. They sought to exploit Native land and resources, however, a process that continues to this day in hydropower projects on Cree land in Canada and mining projects on Hopi and Navajo land in Arizona. The visual essay that appears following this chapter provides a glimpse into the land and lives of the Navajo today.

African Americans

Perhaps the classic internal colonial group in the United States has been African Americans. Brought to this continent involuntarily starting with a Dutch trader in 1619, Africans were conscripted into what may have begun as indentured servitude but was soon intergenerational slavery. The slaves' movements were closely controlled, and even those who had escaped across borders to "free" states were to be returned. Black slave labor filled the gap left by the destruction and removal of the Native population. In Latin America, great haciendas and plantations grew up, served by the labor of Amerindians and their mixed-race descendants. In coastal areas of the Caribbean, from Cuba to Brazil, where the labor demands of producing sugar were the greatest and where disease and oppression had taken the greatest toll on the Native population, the Native labor force was supplemented by African slaves. This "African alternative" quickly became the preferred choice of cotton and tobacco plantation owners as well. To this day, African rhythms of life, work, and diversion still color the coasts of the Americas from Maryland to the southern reaches of Brazil.

Growing European populations meant ever greater demand for American products, and inventions that increased production, such as the cotton gin, only increased the amounts of raw materials that were needed. As with the Native population, the actions taken toward the African slaves were justified by stereotypes. The stereotypes of the two groups were similar, but with new twists for new situations: Africans were foolish children who needed wise overlords for their own protection, but, at the same time, Africans had poor control over their sexual and aggressive urges and so the strict "discipline" of overlords was necessary for everyone's protection.

A growing social conscience in Europe and the United States in the late 1700s brought changes, but stereotypes and colonial relations were slow to crumble. The horrors of the slave trade, in which people were taken by force in Africa and then packed shoulder to shoulder in the fetid, dark holds of ships to languish in their own waste until released to the slave block on the other shore, provoked belated remorse. Britain, in the midst of a religious "reawakening" and humanitarian reforms, went from using its naval power to promote the slave trade to opposing it. The Netherlands soon followed. Gradually, Portugal and its Arab middlemen abandoned the trade. Christians in Europe were coming to believe that trafficking in human beings violated Christian principles, and Muslim leaders in North Africa argued that it was against the Koran to sell fellow Muslims into slavery. Opinions on the subject were less clear in the Americas, which by that time had grown dependent on slave labor.

Thomas Jefferson included the promotion of slavery among the "crimes" of King George III in his original draft of the Declaration of Independence, but he was later persuaded to remove it. Jefferson knew slavery was immoral but was uncertain what to do about it. "We have taken the wolf by the ears," he said. Convinced that blacks and whites could not form a single society, he thought that the children of slaves should be returned to Africa until the institution disappeared (what these unaccompanied children were to do in Africa was less than clear). Jefferson promised to free his own slaves as soon as he was out of debt; unfortunately for them, he died still deeply in debt. Benjamin Franklin was president of an abolitionist society in Pennsylvania, but he also thought that Africans would need to be returned to Africa, to leave the Americas for the "lovely white and red." His assimilationist sentiments extended to Native peoples, but not to Africans. As provided in the Constitution, the United States was closed to the slave trade in 1817. Oddly, this had a double effect: Because existing slaves were now more valuable, they may have been better treated, for working or beating a slave to death meant a big financial loss; however, slave owners were also less likely to reward loyal slaves by freeing them or their children and were more likely to pursue those who escaped.

In the northern United States, which had no plantations, other ways of providing cheap labor prevailed. The needs of growing mills were filled by poor Welsh and Irish immigrants and by increasing use of child and female industrial labor. The eloquent abolitionist poet John Greenleaf Whittier sang the praises of the "Yankee Girl" who was poor but free and glad to work in the North rather than benefit from slavery in the South. Yet working conditions in the North were often only marginally better than in the South. Black freemen were limited in the jobs they could hold and the positions to which they could aspire. Their situation showed that freedom from slavery would be only the first step in a long road toward real opportunity for African Americans.

The United States ended slavery in 1865, and Brazil followed in 1889. But long-established institutions changed slowly. In the southern United States, rural slaves simply became sharecroppers. They still worked the land for landowners but were now bound by debt obligations rather than formal slavery. In the North, employers used newly freed slaves as strikebreakers, threatening to replace striking workers with blacks desperate for any work. The pattern of internal colonialism remained in place: Segregation controlled African Americans' movements, wage and job discrimination continued their labor exploitation, voting restrictions limited their self-government, and stereotypes about them persisted. The nature of these factors has changed as the U.S. economy has changed. In a widely debated book titled *The Declining Significance of Race,* William Julius Wilson (1978) outlines three major periods in U.S. history in relation to African Americans. The period of *plantation economy and racial-caste oppression* was one in which the life

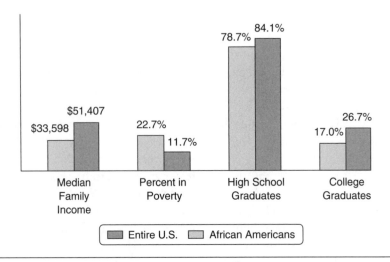

Exhibit 5.2   Socioeconomic Standing of African Americans

Source: U.S. Census Bureau, *Statistical Abstract of the United States,* 2003.

experiences and chances of African Americans were largely dictated by their racial designation. (The term *caste* refers to an intergenerational system of racial or ethnic divisions that determine an individual's place in society. In a caste system, there is almost no room for upward mobility.) The second period, which Wilson terms the *period of industrial expansion, class conflict, and racial oppression,* includes the last quarter of the nineteenth century and roughly up until the New Deal era of the 1930s. During that time, competitive labor relations combined with racial exclusion to leave African Americans the most vulnerable of workers seeking a foothold at the bottom of the occupational hierarchy. In the years since World War II, Wilson contends, there has been a systematic shift from racial inequalities to class inequalities. Racism remains in this third period, but Wilson asserts that the life chances of black Americans are more strongly tied to the disadvantaged social class position of many poor and working-class blacks in a changing economy than to their race (see Exhibit 5.2).

Others have challenged or extended Wilson's thesis, and he himself has modified it, noting how race and class continue to intersect in U.S. society. Although much attention—starting with Wilson's book—has been given to the rising black middle class and the falling black underclass, a large portion of African Americans remain in the blue-, pink-, and green-collar sectors of the working class, often in danger of becoming working poor (Duneier 1992). Persistent segregation means that many blacks are concentrated in central-city areas where there are few jobs available to them (Wilson 1996; Massey and Denton, 1993). Travel to growing suburbs for work can be limited by lack of public transportation as well as by the racial suspicion that still guards the suburban frontier. African Americans' attempts to claim a greater role in governance have often led to racially charged politics. And old racial stereotypes are frequently reincarnated to explain continued poverty: From "childlike irresponsibility" to an "inability to control sexuality and aggression," these stereotypes have long-standing roots.

Hispanic Americans

The experiences of those groups often placed together as Hispanics or Latinos are also closely bound up with colonial-type relations between the United States and Latin America. In 1836, Anglo settlers from the south-eastern United States fought alongside disgruntled *Tejanos,* or Spanish-speaking Texans, to wrest Texas from newly independent Mexico. The Anglos dominated, and 10 years later Texas joined the United States as a slave state. A border dispute resulted that led to the Mexican-American War. The 1848 Treaty of Guadalupe Hidalgo ending that war gave fully half of Mexican territory to the United States, land that is now the states of California, Nevada, Arizona, New Mexico, Utah, and much of Colorado. As had happened in Texas, residents of Spanish heritage, some of whose families had had lived in this territory since the 1500s, lost land and political influence, and many became poor laborers (Takaki 1993).

The new Mexican Americans contributed a great deal to the culture and economy of this region, yet they came to be viewed with suspicion as "outsiders" in an Anglo-dominated Southwest. Workers with mining experience from north-central Mexico brought important skills that bolstered the California gold rush and later mining in Nevada and Colorado. Mexican *vaqueros* brought the skills, clothes, and techniques that would become the symbols of the American cowboy. While black workers built the Southeast, Mexican American workers provided much of the labor that built the Southwest. In time they were joined by Chinese workers (especially in the mines, dockyards, and railroads) and by westward-migrating blacks (especially on ranches, orchards, and farms), but the Mexican Americans were the strong backbone of the labor force.

Poorly paid, they provided the "colonial labor." They were quickly excluded from governing coalitions. The Lone Star Republic of Texas had an Anglo president, Sam Houston, and a *Tejano* vice president, Lorenzo de Zavala, but soon state governments in the Southwest came to be thoroughly dominated by Anglos. Regionally, Mexican American miners, cowboys, and laborers were a people on the move, but locally their movements were often severely restricted to "the Mexican side of town." Towns and cities across the Southwest were often divided, usually by rivers or other physical features, into two sides, one Anglo, the other Chicano. Latino-dominated East Los Angeles is a remnant of this pattern. The original Mexican city is preserved along Olvera Street right next to the closest thing Los Angeles has to a downtown, but the Mexican American population was moved, and is largely confined, to East Los Angeles across the bridges that span the Los Angeles River. The river itself is no longer obvious, but the abrupt change in the residential character of the neighborhoods remains. Mexican Americans were also stereotyped: They were said to be less capable, less industrious, less clever than whites—they were seen as best suited to working under the watchful eye of Anglo supervisors.

The 10-year turmoil of the Mexican Revolution, from 1911 to 1920, brought many more people of Mexican heritage into the Southwest, now as refugees from violence and war. U.S. general John Pershing made forays into northern Mexico during this time to try to catch the revolutionary "Pancho" Villa, but no attempt was made to guard the border to prevent the influx of immigrants and refugees, especially as the United States went to war in 1917 and needed agricultural laborers. Finally, in 1924, as anti-immigrant sentiments prevailed, the U.S. Border Patrol was established. Even with the Border Patrol in place, however, the permeability of the U.S.-Mexico border has varied greatly depending on shifts in the U.S. economy. When the economy has been growing, Mexican laborers have been welcome, but when the United States has faced economic downturns or shifts in politics, the welcome mat has been withdrawn. During the Great Depression of the 1930s, many Mexican American immigrants were sent back to Mexico, including some longtime U.S. residents and citizens who looked Mexican. The tide shifted during World War II, when laborers were again needed, and the Bracero Program (named for the Spanish term for a manual laborer, from *brazo,* the Spanish word for arm) recruited Mexicans to work in the United States. During a period of economic slowing and political suspicion of all things and people "un-American" in the mid-1950s, Operation Wetback returned to Mexico many Mexican laborers who had stayed in the United States. Often the very people who had been welcomed as "braceros" were later thrown out as "wetbacks."

The Bracero Program finally ended in 1964, the same year that the Border Industrialization Program made it easier for U.S. firms to locate across the border in northern Mexico. Many believed it would be better to move the plants south than to bring the workers north of the border, where they might become a permanent presence. Meanwhile, in the mid-1960s Cesar Chavez's newly formed farmworkers' union staged strikes against the owners of California's vineyards to protest poor wages and working conditions. The economic recession of the 1970s and new conservatism of the 1980s saw increasing attempts to control the growing movement of people across the border. One claim made by supporters of the 1993 North American Free Trade Agreement (NAFTA) was that, like the Border Industrialization Program before it, it would slow the northward tide of Mexican workers by creating more jobs in Mexico.

In recent years, strong anti-immigrant sentiment has again prevailed in California, where voters have passed propositions intended to limit the rights of immigrants and their children. Yet, at the same time, a growing economy with a tight labor market has led both landowners and industrial employers in California and the Southwest to call for a new version of the Bracero Program that would again allow Mexican laborers to come north to work—but perhaps not to stay.

The creation of two other Latino groups in the United States, Cuban Americans and Puerto Ricans, is directly tied to U.S. colonial expansion at the end of the nineteenth century. The Spanish-American War of 1898 gave the United States control of the Philippines and of the island of Puerto Rico, as well as protection of the newly independent country of Cuba. The Philippines gained independence after World War II, but Puerto Rico remains a U.S. territory. In 1917, Puerto Ricans became citizens of the United States (and eligible for the draft during World War I), but the territory's government was appointed from Washington. During the 1950s and 1960s, while many colonies around the world were getting their independence, a Puerto Rican independence movement also emerged and even engaged in armed actions, such as a shooting in the U.S. House of Representatives. In nonbinding referenda, however, Puerto Ricans have shown themselves to be divided regarding the options of statehood, independence, and continued territorial status. They now elect their own island government but have no official representation in the U.S. Congress.

During much of this time, many Puerto Ricans have moved from the island to urban areas in the United States in hopes of employment, most significantly to the New York area but also to Chicago and other major metropolitan regions. Because Puerto Ricans are U.S. citizens, they do not face immigration limits in coming to the U.S. mainland. Many have experienced poverty and dislocation, however, as they have shuttled between the island and the mainland. Puerto Ricans are better off than some of their Caribbean neighbors, but both on the island and on the mainland they remain the poorest Latino group. Their rates of poverty and unemployment, as well as rates of family disruption, are closer to those for African Americans and Native Americans than they are to most other Latino groups. This may partly reflect the ongoing disruption of a group caught between two worlds. Many Puerto Ricans have considerable African heritage and so face a particular burden of stigma and discrimination on the mainland; on the island they consider themselves "tan," but in the urban United States they "become" black. With this, they also have the greatest degree of segregation of any Latino group.

The Cuban American experience is far more mixed. Given the close proximity and close ties of the two countries, a handful of Cubans moved regularly between the United States and Cuba before Fidel Castro's rise to power in 1959. Once Castro made his break with U.S. ties and declared himself a communist, many middle-class Cubans fled for the United States. They were refugees, but refugees who arrived with more money and education than many. They also found in the United States a sympathetic government that

Median Family Income 2001
Entire U.S.: $51,407
Hispanic Americans: $34,490

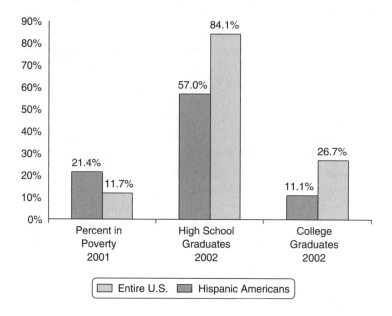

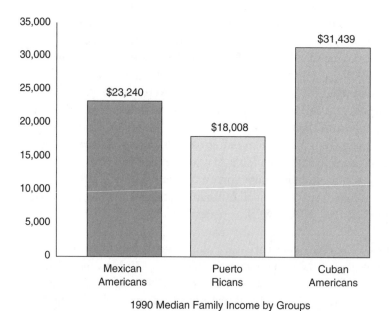

1990 Median Family Income by Groups

Exhibit 5.3   Socioeconomic Standing of Hispanic Americans

Source: U.S. Census Bureau, *Statistical Abstract of the United States*, 1999 and 2003.

shared their anti-Castro sentiments. Although U.S. immigration policy was designed to disperse this incoming group, many found their way to the Miami area, which was poised to take advantage of a growing Spanish-speaking population of workers and consumers and yet only 90 miles from the home island to which they hoped to return. Travel from Cuba became difficult until a brief period in 1980 when Castro opened the port of Mariel. The group that fled Cuba during that time were more likely to come by boat than by plane and were a poorer and less elite mix of refugees. The rumor in the United States was that Castro was emptying his prisons and asylums, yet these newcomers seemed to have no more criminal tendencies than any other. They were, however, more likely to be working-class and less likely to be "white" than the first arrivals.

Other Latino newcomers have faced similarly mixed receptions. Haitians and Dominicans, also often "black," have found some opportunities but also have been stereotyped as "bizarre, exotic, and unruly." Refugees from wars in Central America have included middle-class exiles from Nicaragua, who in the 1980s were fleeing the socialist Sandinista government and, like the early Cubans, had the advantage of middle-class status, "white" features, and opposition to a government that was out of favor with the U.S. government. Less welcome were Salvadorans and Guatemalans, sometimes refugees from peasant groups targeted by right-wing governments and death squads using U.S.-supplied arms and tactics.

Irish Americans and New European Groups

In describing internal colonialism, Blauner (1972) intended to distinguish the experiences of groups of color in the United States from those of white immigrant groups who decided to make the journey and were more likely to be accepted. Indeed, the experiences of African, Latino, and Native Americans may be quite different from those of European immigrant groups. The boundaries, however, are often blurred. The Irish, in particular, had been on the poor fringes of the United Kingdom for a long time and were often treated as an internal colony of Britain (Hechter 1975). When they fled the famine and disruption on their island to find opportunity in the United States, they often found a similar reception. They were treated as colonial labor: They worked the hardest, most dangerous jobs at the very lowest wages. Advertisements from the late 1800s often show blatant wage discrimination: "Laborers two dollars per day, Irish one dollar." They were discouraged from entering the political process for fear that their Catholic faith would make them, and hence the United States, subservient to the wishes of the pope. They were segregated into bleak slums and shantytowns of tacked-together shacks. And they were heavily stereotyped. As Andrew Greeley (1972) notes, much that has been said of African Americans was also said of Irish Americans:

Practically every accusation that has been made against the American blacks was also made against the Irish: Their family life was inferior, they had no ambition, they did not keep up their homes, they drank too much, they were not responsible, they had no morals, it was not safe to walk through their neighborhoods at night, they voted the way crooked politicians told them to vote, they were not willing to pull themselves up their bootstraps, they were not capable of education, they could not think for themselves, and they would always remain social problems for the rest of the country. (Pp. 119–20)

How did Irish Americans overcome their disadvantaged position? They became white. People stopped talking about the "Irish race." Instead, a class distinction was made between the middle-class "lace-curtain Irish"—so called because they could afford to pay attention to decoration in their homes—and the poor working-class "shanty Irish" who still populated the slums. The Irish had advantages, including familiarity with English and a northern European appearance that blended in once they abandoned their old-country clothes and brogue. They used ethnic solidarity to help them advance, avoiding discrimination in the public schools by opening Catholic schools and electing their own political leaders, who could then offer patronage jobs until the Irish policeman (instead of the Irish gangster) became so commonplace as to become its own stereotype. At the same time, their ever greater assimilation meant that the Irish could limit their ethnic solidarity to church, holidays, and associations as they increasingly entered the mainstream.

New groups arrived to take the most menial positions vacated by the Irish. New immigrant laborers from Italy and Poland often found that they took orders from Irish foremen, were served communion by Irish priests, and maybe were even courted by Irish politicians. Each of these groups also experienced labor exploitation. They faced stereotypes as well—even today, we still have lingering jokes about the foolishness of Poles and the criminality of Italians. They initially faced segregation: In the 1920s and 1930s, all of the major cities in the United States had distinct ethnic neighborhoods that were often not particularly welcoming to other groups. The residents of these areas competed with and told jokes about one another. Yet draftees from all of these new European American groups fought together during World War II (blacks, and to some extent Latinos and Native Americans, were still in segregated units). When they returned from the war, the GI Bill and the Veterans Administration invited them to attend the same colleges and to borrow for homes in the same growing suburbs. European Americans had all become white (see Exhibit 5.4). Left behind in the urban areas and emphatically not welcome in the suburbs were the remaining groups, most particularly blacks and Chicanos. To a large extent, these groups still share the same schools and compete for the same dwindling urban resources.

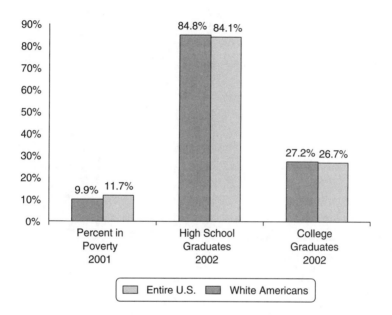

Median Family Income 2001
Entire U.S.: $51,407
White Americans: $54,067

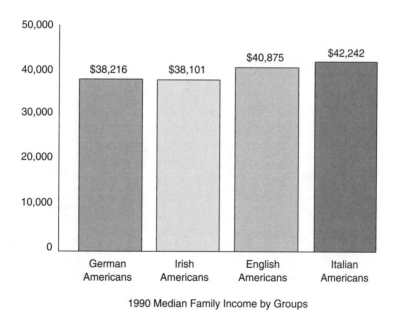

1990 Median Family Income by Groups

Exhibit 5.4   Socioeconomic Standing of White Ethnic Americans

Source: U.S. Census Bureau, *Statistical Abstract of the United States,* 1999 and 2003.

Median Family Income 2001
Entire U.S.: $51,407
Asian Americans: $60,158

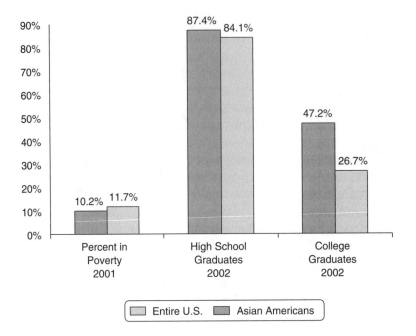

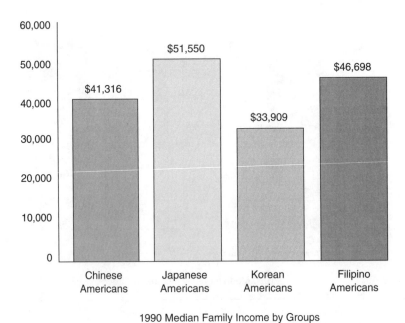

1990 Median Family Income by Groups

Exhibit 5.5   Socioeconomic Standing of Asian Americans

Source: U.S. Census Bureau, *Statistical Abstract of the United States,* 1999 and 2003.

Middleman Minorities
and Ethnic Solidarity

Alejandro Portes and Ruben Rumbaut (1990) describe four types of immigrant incorporation. Workers with limited credentials and marketable skills who arrive in large numbers and are dispersed across the country are likely to work in the **secondary labor market** of low-wage jobs with few benefits and little security. In the United States, this group includes many recent Latin American and Caribbean immigrants as well as labor migrants and "internal colonial" groups such as African Americans and Mexican Americans. Workers who arrive with select skills and credentials that are in great demand may move directly into the **primary labor market** of professional and technical employment. This group includes Asian Indian physicians and other professionals, Filipino and Hong Kong professionals educated in English, and select immigrants from East and South Asia and from Europe and elsewhere. In a phenomenon sometimes called a brain drain, because in their flight such immigrants may drain their home countries of needed talent, they often come to seek professional opportunities that may not be readily available in their home countries. Few in number and widely dispersed, they may face initial hurdles to success but often quickly move up the economic ladder (see Exhibit 5.5).

Being integrated into a society as "colonial labor" is much less appealing. Some scholars have suggested that one way immigrants may avoid this is by mobilizing ethnic solidarity within close-knit groups. The members of such groups work closely together to operate small businesses in otherwise poor or neglected areas (often the neighborhoods of Blauner's colonial peoples), functioning as **middleman minorities** (Blalock 1967). Jewish merchants have often served this role in both Europe and the United States. In recent years, Lebanese, Korean, and Chinese immigrants have moved rapidly into small business in the United States as owners of small groceries, liquor stores, laundries, small motels, and variety shops. These businesses are typically located in areas that are not well served by larger competitors, giving the middleman minorities a chance to gain an early foothold. Small business is always risky, and many fail, but others find economic success—perhaps on their second, third, or fourth try (Nee, Sanders, and Sernau 1994).

Middleman groups, with their persistent upward mobility, may be accepted by dominant groups, even seen as "model minorities," but they often incur the hostility of the people they serve. Family-operated businesses may rarely hire people from the neighborhoods in which they are located, and such businesses can be seen as exploiting or profiting from the hardships and limited opportunities of the neighborhoods they serve. Cultural misunderstandings can occur between business owners and customers. Sometimes, these businesses simply provide a convenient focus for the ill-defined resentments of people who are unable to move ahead economically. Asian-owned stores in black and Latino neighborhoods in New York have been the targets of boycotts and demonstrations, and during the rioting in Los Angeles after the police officers accused of beating Rodney King were acquitted at trial, many residents in black neighborhoods targeted Korean-owned businesses. A dangerous cycle can be created: Hostility may encourage group solidarity, and then this solidarity may breed new hostility if middleman minorities are perceived as clannish or greedy, or as exploiters of poor neighborhoods (Bonacich 1973).

An alternate strategy some immigrants use is to cluster into an **ethnic enclave** (Wilson and Portes 1980), an area in which ethnic entrepreneurs hire and serve the needs of coethnics, creating their own ethnic labor market. This is possible only where large numbers of coethnics share a common metropolitan area, such as Cuban Americans in the Miami area. It also requires a mix of backgrounds among the immigrants themselves: some with the human and financial capital to begin businesses, others with fewer resources who are willing to work in those businesses. Poorer recent refugees from Cuba can find employment in the businesses of middle-class Cuban Americans who came first, and recent workers from the People's Republic of China can find work for Taiwanese business owners. Whether such employment allows the workers to avoid the secondary labor market or merely leaves them to struggle in an enclave version of low-wage labor is a matter of debate (Sanders and Nee 1987). In time, the desire of employers to expand their businesses and the desire of workers to move up into better-paying jobs can serve to diversify an ethnic enclave (Nee et al. 1994). In Los Angeles, Chinese restaurant owners now often need to turn to Latinos for kitchen workers, and in Miami, Cuban business owners may need to hire Dominican or other Caribbean workers.

The Analytic Debate: Cultures and Structures of Poverty

Why do some new immigrant groups experience relatively rapid upward mobility while others struggle with low wages and little mobility? As Blauner (1972) reminds us, different groups came to the United States under widely varying conditions. And, as Portes and Rumbaut (1990) note, the size, distribution, timing, and mix of any given group may affect the strategies available to its members.

One key factor is timing: Groups that have arrived in the United States when labor has been in great demand have been more readily accepted and easily incorporated into mainstream society than have groups that have been perceived as competing for limited jobs (Lieberson 1980). Yet some groups that have arrived at the same time have had quite different experiences. Rodney Stark (1996) discusses the sharp contrast between the experiences of Jewish immigrants and Italian immigrants at the start of the twentieth century. Jewish immigrants rapidly moved into commercial and then professional employment with rising wage levels, whereas many Italian immigrants struggled in low-wage positions in the secondary labor market. Why the difference? One possibility is cultural differences. Jewish families had long emphasized literacy and education (Perlmann 1988), even if only to read and interpret the Torah, as essential. In the United States, Jewish parents strongly encouraged their children to pursue higher education, and this propelled their upward mobility. Italian families, in contrast, often distrusted the American school system and encouraged their children to leave school early and begin working. Many Italian children went from jobs as newspaper boys and errand boys into other unskilled jobs that offered little advancement.

Closer inspection reveals that this is too simplistic an answer, however. Jewish immigrant families also struggled with getting all of their children through high school, and only rarely were they able to get their children into higher education (Steinberg 1981). They prospered first, and then used their higher incomes and higher status to overcome the barriers to higher education. In contrast to the cultural argument, Steinberg (1981) proposes a more structural argument. Jewish immigrants were not the top-hat-clad financiers of the cartoons of the time, but neither were they poor, illiterate peasants. Most had been urban middle-class tradespeople in Europe, and they arrived in the United States with readily marketable skills that were in great demand in the growing cities of the northeastern United States.

In many ways the Jewish "immigrants" were actually middle-class refugees, fleeing anti-Semitic governments and oppression throughout Eastern Europe. In this they were more like the middle-class Cuban refugees of the 1960s. They had one added incentive to succeed: Most knew they could never return to their home countries. They would have to prosper in the United States and pass this success on to their children. Many Italians immigrants, in contrast, did hope to return to Europe someday; these "immigrants" began as something closer to labor migrants, seeking to work and then to return to their sunny welcoming homeland with new prosperity. Of course, many never found enough prosperity to return, but significant numbers did go back. The ones who stayed and worked had few marketable skills beyond a willingness to work hard, and they become concentrated in the secondary labor market. Sometimes they would save their earnings and invest in businesses such as restaurants, but just as often they sent the money to family back home as "remittances." In this, they were more like recent Mexican American immigrants and labor migrants. From these different structural positions in U.S. society come differing cultural stereotypes: "Jews are smart and ambitious but greedy and shrewd" and "Italians love to enjoy life and are committed to their families but are unreliable and unlikely to get ahead." Ironically, Jewish and Italian American businesspeople in Boston and New York are now likely to repeat the same comments—the first about their Asian American competitors and the second about their Puerto Rican and other Latino and black employees.

The most compelling and influential statement to date regarding cultural position comes from Oscar Lewis (1961, 1968), who coined the term **culture of poverty** to describe common attributes he found among the poor in Mexico, Puerto Rico, and the United States: fatalism, a present rather than future orientation, mother-centered families, and suspicion of outside institutions. Lewis (1968) himself believed that these cultural traits are rooted in the overall structure of a society: "The culture of poverty is both an adaptation and a reaction of the poor to their marginal position in a class stratified, highly individuated, capitalistic society" (p. 188). He found a culture of poverty in U.S.-dominated Puerto Rico, but not in socialist Cuba. Yet he believed that once in place, such a culture could perpetuate poverty through its effects on children:

> The culture of poverty, however, is not only an adaptation to a set of objective conditions of the larger society. Once it comes into existence, it tends to perpetuate itself from generation to generation because of its effect on the children. By the time slum children are age six or seven, they have usually absorbed the basic values and attitudes of their sub-culture and are not psychologically geared to take full advantage of the changing conditions or increased opportunities that may occur in their lifetime. (P. 188)

Others who picked up on this idea of "culture as villain" (Zinn 1989) were less subtle in expressing themselves. Cultural arguments became a convenient excuse for avoiding social change. In a book titled *The Unheavenly City* (1970), Edward C. Banfield stated this position strongly:

> So long as the city contains a sizable lower class, nothing basic can be done about its most serious problems. Good jobs may be offered to all, but some will remain chronically unemployed. Slums may be demolished, but if the housing that replaces them is occupied by the lower class it will shortly be turned into new slums. . . . New schools may be built, new curricula devised, and the teacher-pupil ratio cut in half but if the children who attend these schools come from lower-class homes, they will be turned into blackboard jungles. . . . If, however, the lower class were to disappear—if, say, its members were overnight to acquire the attitudes, motivations, and habits of the working class—the most serious and intractable problems of the city would all disappear with it. (Pp. 210–11)

In a book by the same title, William Ryan (1971) dubbed the cultural argument "blaming the victim." He noted that the old racist argument popular at the beginning of the twentieth century—that one racial group is inherently, biologically superior to another—had fallen out of favor, so in its place conservatives created a new scapegoat: culture, or, more properly, subculture. Instead of arguing that "Negroes" themselves were inferior, one could argue that they were the unfortunate victims of an inferior set of cultural norms and practices. By taking this perspective, Ryan contended, conservatives could subtly blame the victims of poverty and discrimination for their own misery. Cultural arguments shifted the focus from poor schools to poor educational values, from the problems of poor families to problems of poor family values, and so forth. The structural counterarguments of Ryan (1971), Valentine (1968), Stack (1974), and others during the early 1970s were persuasive and propelled interest in opening the opportunity structure of U.S. society as a whole. During that time of economic stagnation and deindustrialization, however, government programs met with little success. By the 1980s, cultural arguments were again favored, and there was much talk of restoring the "work ethic" and proper "family values." Many working-class whites read statements such as Banfield's and said, "Yes, the man's right," and became newly conservative "Reagan Democrats." By the beginning of the 1990s, however, recession and deindustrialization were squeezing ever larger portions of the working class into declining job opportunities and declining neighborhoods with declining schools. People with the "right" values were having the "wrong" outcomes, and Bill Clinton was able to briefly build a winning coalition around his call for basic structural changes in the economy as well as in health, education, and welfare.

At the beginning of the twenty-first century, these differing perspectives are still often hotly debated. Politically conservative commentators such as Rush Limbaugh and Alan Keyes often advance cultural arguments, noting declines in "family values" and work values. Politically liberal commentators such as Jesse Jackson as well as many labor leaders continue to call for reforms in the national and global economic structure to protect workers and open opportunities for others. A few populist commentators such as Patrick Buchanan and Ross Perot manage to denounce both declining family and work values as well as greedy corporations that abandon their workers. We have even seen a return to long-discredited racist assertions that race-based differences in intelligence are at the root of income differences, as Herrnstein and Murray argued in their 1994 book *The Bell Curve*. That book drew widespread press attention even though the vast majority of sociologists, psychologists, and anthropologists have challenged the authors' conclusions, noting that their assertions about race-based intelligence founder on the fact that both racial identity and measured IQ are social constructions. It should be no surprise that their "most intelligent races" are the middleman minority groups who strongly stress education, Jewish and Asian Americans, while their "least intelligent races" are the internal colonial groups, most particularly African Americans, who have long been denied access to consistent and high-quality education.

Social scientists emphatically reject the new race-based arguments about individual ability just as they did the older version of these arguments that held sway in the 1920s and 1930s (Cohen 1997), yet cultural and structural arguments may be more intertwined than political divides would suggest.

Poor or hostile schools may cause some groups to hold education in great suspicion. This happened with southern Italians and Sicilians who were ridiculed and rejected in northern-dominated schools in Italy and then found a similar experience in Anglo-Protestant-dominated schools in the United States. Newly freed slaves in the United States turned to education and literacy with passion following the American Civil War, yet years of being subjected to poor schools and finding few opportunities on graduation, coupled with Jim Crow segregation in the South and urban segregation in the North, left many African Americans distrustful of the education system and skeptical about what it has to offer them. Likewise, those who have faced intergenerational discrimination in the labor market may develop hostility and suspicion toward the idea that hard work will always pay off.

Gunnar Myrdal (1944), a Swedish economist who was a keen observer of American life, called the great discrepancy between this country's claims to ideals of freedom and equality and the reality that the nation was built on racial oppression and inequality "the American dilemma." He noted how many people tried to justify this discrepancy by using perceptions drawn from this cycle of structure and subculture, a vicious cycle of "cumulative causation" in which discrimination and prejudice mutually reinforce one another. The poor attended poor schools and were then faulted for failing to value education, they were segregated in dilapidated housing in poor neighborhoods and then faulted for failing to keep up their property, they were denied promotions and job opportunities and then faulted for failing to value hard work, and they were denied the resources for self-improvement and then faulted for dressing, speaking, and acting "poorly."

This cycle continues today and perhaps can be broken only if addressed from both directions. In school programs among poor blacks in Chicago, poor Mexican Americans in San Antonio, and poor Italian Americans in south Boston, educators and church and community leaders seek to instill in young people a new sense of possibility, an understanding that education, accomplishment, and advancement can and should be a part of their heritage, self-identity, and community pride. At the same time, these young people will accept this possibility as real only if the opportunity structure open to them proves that education and hard work do indeed pay off, including for groups long excluded from advancement. We will explore these efforts in more detail in the chapters that follow. They reflect issues of inequality and opportunity that extend beyond race and ethnicity. Yet in many ways, race, class, and perception are intricately interwoven and continue to define acceptability and opportunity. A good example of this is seen in language.

Colorful Language

More than a hundred years ago, W. E. B. Du Bois (1903) stated that the central problem of the twentieth century would be the "color line." For a long time, Americans have been obsessed with color. Two colors, black and white, have dominated as poles, pulling everyone in between in one direction or another. Color and class became intertwined. Lighter groups fared better than darker ones. Even within the black community, color often mattered: Having a light complexion was an advantage, whereas dark coloring carried a stigma. The most advantageous of all was to be light enough to "pass": to be included in the white category. Many early black leaders and intellectuals, including Du Bois himself, gained greater acceptance if they weren't "too black." This peculiar emphasis on color seems to be fading now, but only as new markers take its place.

The racial line of the twenty-first century may be the linguistic line. A nonwhite person's acceptability to members of mainstream society is often contingent on the language he or she uses. The most accepted black leaders are those who neither look particularly "black" nor speak in a particularly "black" way. Colin Powell, for example, has a large number of political supporters among whites, many of whom do not think of him as "black." In particular, his speeches have none of the cadence and expression generally associated with black activists. Language matters. The new criterion seems to be whether a black person can "pass" on the telephone.

Perceptions of the acceptability of language are bound up with issues of both ethnicity and class. Early immigrants worked hard not only to learn English but to remove traces of their "old country" dialects. The Irish arrived speaking English, but they still needed to lose their distinctive "brogue" in order to move up the socioeconomic ladder.

Some ethnic accents, such as French and British English, may be considered marks of distinction, whereas others, such as many Eastern European accents, are accorded much less prestige. Regional accents are also received differently: A New England accent, particularly if it is upper-class Boston Brahmin, may be accorded much more prestige than a highland-southern Appalachian accent. Those accents associated with wealth, education, and long-standing privilege tend to be accorded more favor. Try envisioning an introduction to an episode of *Masterpiece Theatre* delivered not with Alistair Cooke's upper-class British accent but with a dialect from the hills of Tennessee (a dialect drawn, in fact, from eighteenth-century British English). For most, such an introduction would carry far less weight.

Language marks class as well as ethnicity. If one has a British accent, it is far better to sound like a member of the royal family than to have a working-class cockney accent, an accent often used by ridiculous characters in Monty Python sketches and that of the flower girl in *My Fair Lady*. French should be Parisian and not rural Cajun. If one has an American southern accent, it is better to sound like a southern gentleman or lady than to have a "hillbilly" accent. Some accents have a narrower window of acceptability. A native Spanish speaker's accent is rarely a mark of distinction unless the speaker can master the peninsular diction of Ricardo Montalban as he extolled Chrysler's "rich Corinthian leather" and stay far from the urban

Chicano diction of Cheech and Chong humor. Black English, the accent and dialect of African Americans, may have the narrowest window of acceptability of all. The rich baritone of James Earl Jones is acceptable, but in general the "whiter" the English, the greater the acceptance. Urban black "street" English is continually associated with low education, low social status, and often even low intelligence or morality.

As is true of other such social distinctions, no one accent can be demonstrated to be superior to any other. What matters are the associations—class, status, ethnicity, race, and region—that accompany an accent. A black American with white diction may be assumed to be more educated and more assimilated and so may be more acceptable to the white community. At the most basic level is the issue of communication. Those with accents from languages other than English or from non-Standard English dialects may need to work to attain more broadly understandable speech to gain greater acceptability. "Standard" speech in the United States carries something of an urban midwestern-Californian accent, yet someone with an upper-class Bostonian accent will rarely feel the need to switch accents (unless he or she has hopes of becoming a sportscaster). For less favored groups, the challenge is greater.

A few years ago, a debate raged in California over the idea of schools' acceptance of Black English, or "Ebonics," as a separate dialect rather than as a "deficient" form of English. Many African Americans readily shift the dialects they speak between their home communities and white-dominated workplaces, just as Mexican Americans often shift between Spanish and English. Yet linguistic prejudice may mean that both groups have to do more than make their language intelligible to others; they may need to remove as many traces of their ethnicity from their speech as possible in order to make it more acceptable to others.

In one sense, linguistic prejudice may be a less absolute obstacle to acceptability than color prejudice. People (with the possible exception of Michael Jackson) do not generally change their skin color! Yet people in the United States have long worked to change their speech to match mainstream acceptability. Language is adaptable. However, if some find their home speech broadly accepted whereas others find that both academic and professional success are dependent on their erasing any traces of their home speech, those in the former group will begin with a clear advantage. Language reflects class, ethnic, and racial advantage and can serve to perpetuate such advantage.

If I were to be so presumptuous as to restate Du Bois's prediction for this century, I would contend that the problem of the twenty-first century will be the class line. Ironically, just as some racial divides are diminishing, class divides are growing. A well-off black (or Latino, for that matter) professional may still have a difficult time hailing a taxi or convincing the security guard that he is a partner in the law firm and not an accused client (Cose 1993). Such a professional may still face snubs and slights but nonetheless have a life more like that of a white professional than like that of the custodian—black, white, or brown—who cleans the building. Yet the elements that constitute the social construction of race in the United States and the world—such as discrimination in housing, employment, and marketing; segregation in housing and in personal networks; and stereotyping of culture and language—continue to interact with class privilege to confine many people of color to the underside of the class line.

KEY POINTS

- Race, although hard to define objectively or biologically, has been and continues to be a key social boundary in the United States and around the world. In the United States, the key to economic success and social acceptance for many groups has often been to come to be considered "white."
- Internal colonialism theory posits that oppressed groups are used as colonial labor within the dominant society. To achieve this, the dominant society often suppresses the freedom of movement and rights of self-expression and full political participation of internal groups and creates an ideology that tries to justify oppressive structures by asserting these groups' basic inferiority.
- The elements of internal colonialism—exploited labor, suppression of political rights and freedoms, and assertions of inferiority—can be seen in differing combinations in the experiences of many nonwhite groups in the United States, including Native Americans, African Americans, Mexican Americans, and Puerto Ricans as well as other groups, such as Irish Americans, at some point in their histories.
- Immigrant groups have had very mixed experiences of struggle and success in the United States depending on the timing of their arrival and their class backgrounds. Cuban, Chinese, and Filipino Americans have been subjected to colonial and internal colonial experiences but have also had high rates of entrepreneurship and professional employment. Many immigrant businesspeople become middleman minorities, serving the needs of other ethnic groups. Immigrants may seek the support of their coethnics through ethnic solidarity and by concentrating in ethnic enclaves.
- "Culture of poverty" arguments contend that the attributes of certain cultures and subcultures may be passed on to poor children and limit their aspirations and success. Critics of cultural arguments contend that different levels of acceptance and structures of opportunity are what really explain the successes and struggles of different ethnic groups.
- Some argue that class rather than race is becoming the primary divider in U.S. society. Yet in many ways race and class are closely interrelated in people's experiences, particularly in residential segregation. Stereotypes about language and accents are one good example of how racial and ethnic differences together with class background can affect people's opportunities and status.

FOR REVIEW AND DISCUSSION

1. In what ways has whiteness been a category of privilege in U.S. history? In what ways is this still true, and in what ways is it changing? Are there new categories of privilege?

2. The concept of internal colonialism best fits the experiences of which ethnic and racial groups?
Provide examples. Are there currently internal colonial peoples in the United States or elsewhere? Who might these be?
3. Give examples of the ways in which language is used as a marker of class position as well as racial and ethnic identity. Are any other such markers predominant in social interaction?

MAKING CONNECTIONS

In Your Family

Prepare a chart of several generations of your family to examine your own racial and ethnic heritage. See how much you can find out about the ethnic backgrounds of the family members represented. If the information is available to you, you might also note the wealth, education, occupations, and class positions of your family members. Were any immigrants or labor migrants? What were their origins and experiences? Compare their racial/ethnic backgrounds with their occupations: Were any working in positions that look like internal colonial labor? Were any entrepreneurs or middleman minorities? What were their strategies for getting ahead? Did they use ethnic solidarity by means of membership in ethnic churches, clubs, schools, or other organizations? Did any of them marry interracially or interethnically? If so, were these marriages signs of assimilation or sources of tension? Did they experience forms of discrimination?

In Your Family or Neighborhood

Conduct a brief oral history with a family member, neighbor, or acquaintance who is an immigrant (recent or otherwise), is or was a Mexican American or Puerto Rican labor migrant, or was an African American transplant to the urban North from the rural South. Where did this person come from? Why did he or she choose to move? How did this person select his or her destination (work, family, or other)? Did this person move to an ethnic neighborhood or community? If so, what were that community's key organizations and meeting places: churches, clubs, bars, parks, civic or political groups, community centers, or something else? How was the person received by the community as a whole: Was there hostility or discrimination? What did the person do for work? What social and cultural changes did he or she experience? How has the person's community/neighborhood/life changed over time?

Ethnic Organizations

Look into the work of a national organization that seeks to improve race relations or to advance the economic standing or preserve the culture of a racial or ethnic group. Such groups include the NAACP (www.naacp.org), the Urban League (www.nul.org), the National Council of La Raza (www.nclr.org), and the American Indian Movement (www.aimovement.org), among many others. You might also look at any of a wide range of European American ethnic group organizations (such as the Polish National Alliance; www.pna-znp.org), refugee resettlement agencies, and other organizations concerned with particular racial or ethnic groups. What is the target population of the organization you have chosen? How was it founded, and how has it developed? What is the organization's current mission? What are its concerns, and what activities does it conduct? Is there a local chapter of the organization near you? If so, consider visiting the chapter to find out more about the organization's local initiatives.

Photo Essay: Navajoland

Catherine S. Alley

The Navajo reservation is the largest in the United States. The harsh but majestic land has sustained the Navajo for generations in changing ways. Farming and ranching, particularly sheep, have been mainstays. More recently, mining and resource extraction, along with tourism, have become important. These uses have brought the Navajo new revenues, but they have also led to conflicts over the uses of the land. Such conflicts have included the U.S. government, large mining companies, and their Hopi neighbors.

First-time visitors to the Navajo Reservation in northeastern Arizona are often surprised to find that such conditions exist within the borders of the United States. While often familiar with stories of wrong-doings that have been wrought against the various Native American tribes, few realize the extent of continuing poverty in groups such as the Navajo Nation.

Caught between traditionalism and attempts to find a place in modern U.S. culture, the Navajo (or Dine) struggle with issues of unemployment, alcoholism, diabetes, and other issues of poverty and well-being. In Ganado, Arizona, neither running water nor electricity is available to all.

Nor can everyone understand the written and spoken English used by the U.S. government and the B.I.A. (Bureau of Indian Affairs) to make important decisions about these people's lives.

Younger Navajo are faced with the realization that life on the reservation is hard, and they often leave. While Navajo culture is taught at the local high school, some of the elderly worry that members of the next generation are forgetting their Dine roots.

In this traditionally rural, sheepherding society, the universals which used to describe Navajo families are changing, as people move to Albuquerque or Phoenix in search of work. While some live in trailers, others still live in the traditional seven- or eight-sided hogan.

Hogans intermingle with more recently built homes, some with and some without water and electricity.

In a society where many fall below the federal poverty line, status is based upon wisdom, respect, and community involvement. The Navajo hope that these values will also carry them through the changes to come.

6 Gender and Class

When Men Were Men

Masculinity as Privilege

Discussions of **gender** in books such as this one typically begin by addressing issues related to women. Let me shift the focus for a moment by looking first at men. Gender issues are not women's issues only, although textbooks in particular might give that impression. Categories of inequality are not just about disadvantaged groups; rather, disadvantage in any group implies that someone else must be privileged, holding advantages. One way to think of masculinity is as a category of privilege, like whiteness. Historically, men have been accorded multiple dimensions of advantage: in power (they could vote), in privilege (they could own land), and in prestige (they were accorded special deference and respect for their gender as well as for their accomplishments). In most of the United States until early in the twentieth century, none of these were also true of women. It is true that to be a *respectable* man, a *good* man, a *gentleman,* a man was supposed to use these advantages to protect, guide, and support his family. This role as head of household and family provider also served to increase the power and prestige that went with successful masculinity. As with racial distinctions, these privileges have become more subtle in recent decades, but they have not disappeared.

Men still have greater access to power than do women. After years of slow gains for women, men still hold more than 85% of all legislative positions and governorships in the United States and more than 80% of governmental positions around the world (United Nations Development Program 2004). In the corporate world, men still hold well over 90% of executive positions. Men still have greater control of wealth and property than do women: Of the world's wealthiest people, all but a handful are men or the heirs of wealthy men (*Forbes* 2004). And men still command privileges. They do less housework than their female partners, regardless of their marital status or whether or not both the man and woman work outside the

145

home (Stapinski 1998). The size of the gap between men's and women's contributions to housework varies by country, but men do less housework around the world (Scarr, Phillips, and McCartney 1989). Although the amount of time that men spend with their children is increasing, most of this time is spent in play and other interaction; women still do the majority of daily child-care tasks. Again, this suggests that men get to pick and choose the types of tasks they take on, whereas women often do not (Hochschild 1989). And men still hold remnants of the greater prestige they have always enjoyed. Male professors, male physicians, and male attorneys are still more likely to be considered "highly knowledgeable" than are their female counterparts. Even our language tends to confer greater deference to men. The term *chairman* has been slow to disappear; it is often used generically unless it is known that the specific person being referred to is a woman (how often have you heard anyone mention the "chairperson of the board"?). *Sir* refers to men and knights; its counterpart, *madam*, refers to women and the owners of brothels. *Master* clearly implies power; its counterpart, *mistress*, implies a woman in an extramarital relationship.

Differences in the language and titles used in reference to men and women may be subtle and unconscious, but that is just the point. The privileges of masculinity are rarely overtly expressed, and so men may be largely unaware of them. Rarely does anyone ever say, "I'll give you this job [or this promotion] because you are a man." Nor is a student adviser likely to state, "Because you are a man you should consider professional school." Because the advantages of being male are embedded in the social structure and not articulated, men are likely to attribute their successes to hard work and ability rather than to their gender. They are probably partly right, in that they may have had to work hard to advance. It is only when compared with the disadvantages of women—who may also work hard but be overlooked or discounted, or discouraged from advancement—that the privilege of masculinity becomes apparent. In a widely cited article on "white privilege and male privilege" Peggy McIntosh (1995) contends that men are trained over time not to notice the privilege that accompanies their gender, just as whites may not notice racial privilege. Privileged groups take for granted that they may go where they like and do what they want without arousing suspicions and animosities or being told they are "out of their place." Privileged groups take for granted that much of the world is tailored to their tastes, interests, and desires. Because, like most people, they tend to work hard and to have a variety of talents, they assume that their advancements and accomplishments, along with the respect they receive for these, are wholly earned.

In explaining men's resistance to changing gender roles, family sociologist William Goode (1992) proposes a "sociology of superordinates." He suggests that relations between men and women share aspects that are likely true of those between any other superordinate (or privileged) group and any other subordinate (or disadvantaged) group:

1. Many groups may misunderstand one another, but men need to know less about the world of women than the other way around, because the world in which men operate is not female dominated, just as whites often know less about the lives of blacks than the other way around.

2. The current generation of men did not create the social system that exists today, and so can claim innocence of any "conspiracy" to oppress women. Note that whites, especially younger whites, often make the same argument in reference to oppressed racial groups.

3. Men, like all superordinates, take for granted the system that gives them their status and so are unaware of how the social structure may give them many small but cumulative advantages. People at the top almost always assume they have earned their accomplishments.

4. When men look at their lives, they are more likely to note the burdens and responsibilities of their gender roles than the advantages.

5. Superiors, in this case successful men, do not readily notice the talents and abilities of subordinates, such as female colleagues. Given that people who have few opportunities to take risks rarely accomplish much, this situation becomes self-validating.

6. Men are likely to view small losses of position and deference toward them as significant changes (even if, and maybe especially if, they willingly give up some power or privilege). On the other hand, they hardly notice activities and events that maintain the status quo.

This last point also explains why some men may react angrily or dismissively to attempts to change the social structure. Likewise, whites may be unaware of their racial advantages, may be overly optimistic about the changes that have taken place in race relations, and may be completely unaware of the frequent but often subtle racial slights that black members of their communities experience. In their daily experiences at work and in their homes, men may be less aware of privilege than they are of vulnerability and loss.

Masculinity as Vulnerability: The Harder They Fall

In one sense, male privilege has always been precarious. That is, respect, prestige, privilege, and power have gone to successful men, by the current social standard, but not necessarily to all men. In an article titled "The Good Provider Role: Its Rise and Fall," Jessie Bernard (1981), one of the first prominent female sociologists in the United States, notes that successful men have long been expected to use their advantages to be good providers to their families. This is not just an ancient idea; the importance of this role for men actually increased in the 1800s as the partnership of men and women in agrarian societies gave way to a market economy dominated by business*men*. But not all these businessmen were successful. Some never achieved their dreams, some experienced economic disasters. Whether for economic or personal reasons, many men found themselves unable to fulfill the good provider role and withdrew from the competition. Frequently, in a time when divorce was rare, they simply abandoned their families. The good provider role seems to have had many deserters. Although these deserters were clearly vulnerable to downturns and wounded by social expectations, in one sense they were still privileged relative to their women. The men could move west, or move north, or go to the big city, and start over; the women they deserted were often left to raise families with few means of support.

Other men were always vulnerable because of their race, ethnicity, or class. Black sharecroppers and Irish American stockyard workers could never support families in the proper style and had to either forgo family life or admit to being less than good providers. Men still held more power, and could claim more privileges, in poor families and communities than did women, but these men's vulnerability is a good reminder of the ongoing interaction of gender with other aspects of social structure, such as race and class. This interaction continues, making some men particularly vulnerable.

Within the African American community, men are more likely than women to not finish school and to be unemployed. They are also far more likely to be incarcerated. As black women struggle to maintain their families and communities without large numbers of contributing men, it is not clear that they are therefore privileged, but it is clearly that poor black men are particularly vulnerable to loss and humiliation. In societies that accord other men added respect, advantage, and power, they find only disrespect, disadvantage, and disempowerment. Note that *societies* here is plural, in that this is increasingly true not just of poor black men in the United States, but also of poor black men in Brazil, the Caribbean, Europe, and across Africa.

The vulnerability of manhood is also realized by Latino men in both North America and Latin America, where changing economies limit their ability to be good providers. It is increasingly realized by working-class white men as well, both in North America and now across Europe, who are far more aware of their vulnerability than of their privilege. In an era of heavy industry, men from low-income backgrounds often found their best opportunities in manufacturing. If these were unionized jobs with relatively high wages, they contributed to male advantage. Men in these jobs earned considerably more than women who worked in routinized clerical, service, and domestic positions. Yet as these unionized heavy-industry positions have become increasingly scarce, the men and their industrial skills have become extremely vulnerable to termination, protracted unemployment, or new employment at far lower wages. Women in industry are also very vulnerable, but those with skills applicable in the service economy may have more secure employment than the men of their families and communities. The men may still have certain privileges, such as greater freedom from family responsibilities than women have, but they are most acutely aware of their vulnerability. Their severe class disadvantage "trumps" their gender privileges. This is not a purely new development—in the 1800s, the paneled offices and boardrooms were dominated by men, but so were the decrepit boardinghouses, flophouses, and rescue missions.

The place of women in society is also changing and also in complex ways. Women are more likely today than in the past to be seen in the boardrooms, but they are also more likely to be found in the rescue missions and homeless shelters. To complete the picture we now turn to women's roles.

From Glass Slippers to Glass Ceilings

You've Come a Long Way, Maybe

The Victorian ideal of womanhood, which paralleled the rise of the good provider role in the 1800s, emphasized that a woman's loftiest aspiration should be motherhood. She should have many children and devote herself to their nurture. Her husband should dutifully devote his time to his family's economic comfort and sustenance. A man's home was his castle, and his wife's time went into maintaining that castle for her prince and his children, with ample help from servants and day laborers. Of course, only the upper classes and the small, well-off middle classes of the time could attain this cultural ideal. The reality for most women was more like the life of Cinderella before the fairy godmother showed up. Completing all the tasks necessary to maintain a household was long and arduous work without the help of servants.

It is also misleading to note simply that many women of the time did not work outside the home. Many businesses were family owned and operated, and farms were family farms; women contributed a great deal of the labor and expertise needed for economic survival, even if they were not recognized as landowners or business owners. Typically, women of all classes but the most elite were involved in economically productive activity, even if it was not wage labor. Only the wealthiest families could afford to have women in hoopskirts in grand Victorian homes or plantation houses supervising the work of others while doting on their children, even if this was the cultural ideal.

Women began the movement into wage labor early in the U.S. and European industrial periods. Many women, especially young, unmarried women from lower-income backgrounds, went to work in the mills that were becoming common in the United States and Western Europe from about 1840. Textile mills in particular were major employers of women as industrial extensions of the textile work that women had often done in their homes. By the turn of the twentieth century, a full fifth of U.S. women were in the paid labor force, many of these immigrant women trying to help their families survive and become established in growing U.S. cities. Women from middle-class backgrounds were expected to leave the labor force when they married. Some professions, such as teaching, even made this a requirement (male teachers could continue teaching after marriage as long as they could support their families on the meager wages). If a middle-class woman lost her husband to death or desertion, she found herself in a very difficult situation; many such women turned to "quiet employment" such as using their homes as boardinghouses, to gain needed income.

The surge in immigration to the United States declined sharply after 1920, but the rise in women's labor force participation did not. Women in the 1920s worked sometimes as an expression of newfound independence (as they did again in the 1970s) and sometimes as a way of trying to share in the highly touted prosperity of the "Roaring Twenties," which was not reaching everyone (as they did again in the 1980s and 1990s). Women in the 1930s often worked out of necessity as hard economic times forced many to postpone marriage or to replace or supplement the lost income of unemployed and underemployed husbands (a pattern that also returned in the 1970s). Female labor force participation reached its peak during the war years of the early 1940s. It was considered to be women's patriotic duty to go to work to keep the industrial "arsenal of democracy" churning while the men were at war. Women worked in heavy industry (as "Rosie the Riveters") and in meatpacking plants, as well as taking jobs as bus drivers and "milkmen," fulfilling traditionally male-dominated service roles.

When World War II ended, however, the women were told to go home. European countries that had lost huge portions of their male populations to war casualties, Germany and the Soviet Union in particular, continued to need women in industry and encouraged them to continue working. But in the United States, millions of servicemen were returning home, looking for college educations and jobs. Women were encouraged to make room for them. Female university enrollments declined and female labor market participation plummeted. Correspondingly, women married at younger ages than ever before, and, after years of declining fertility rates, they started to have more children and at younger ages. The Victorian ideal of the domestic woman whose sole devotion is to home and family was revived, but now a growing middle-class meant that more families could live this ideal. Women were accepted in the paid labor force, especially in female-dominated occupations, but mostly with the understanding that they were working while waiting to get married or to buy a few extras, and as long as their employment did not involve taking "men's work." Suspicion of career women was often fierce; this attitude is illustrated by Merle Miller's rebuke in *Esquire* magazine in 1954 to "that increasing and strident minority of women who are doing their damnedest to wreck marriage and home life in America, those who insist on having both husband and career. They are a menace and they have to be stopped" (quoted in Miller and Nowak 1977:164).

In 1956, *Look* magazine extolled the virtues of the newly domestic woman, although in language that acknowledged she was different from the women of the preceding decades:

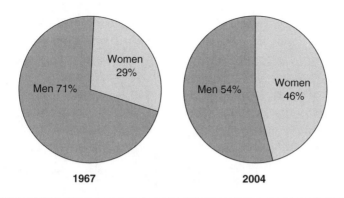

Exhibit 6.1   Women's Labor Force Participation
Source: U.S. Department of Labor (1994, June).

The American woman is winning the battle of the sexes. Like a teenager, she is growing up and confounding her critics. . . . No longer a psychological immigrant to man's world, she works, rather casually, as a third of the U.S. labor force, less towards a "big career" than as a way of filling a hope chest or buying a new home freezer. She gracefully concedes the top jobs to men. This wondrous creature also marries younger than ever, bears more babies and looks and acts far more feminine than the "emancipated" girl of the 1920's or even 30's. Steelworker's wife and Junior Leaguer alike do their own housework. . . . Today, if she makes an old fashioned choice and lovingly tends a garden and a bumper crop of children, she rates louder hosannas than ever before. (Quoted in Friedan 1963:52–53)

Even in the 1950s, a full third of the U.S. labor force was female (and most employed women were not working "casually"!). During that time, growing numbers of other women, many with significant education and work experience, began to find complete devotion to home less than satisfying (despite the "hosannas"). Yet this cultural vision of the domestic woman was so strong that many people were shocked when young women, as well as less-than-content older women, rejected the "old fashioned choice" and followed in the footsteps of the "emancipated girl" (often their own grandmothers) of the 1920s and 1930s. They returned to higher education in large numbers in the 1960s and to the professional workforce, including "the big career," in the 1970s (see Exhibit 6.1).

In one sense, the return of large numbers of women to the paid labor force in the 1970s was merely a continuation of trends that began earlier in the century, trends that had their roots in the expansion of industrialization. Yet there was something new about these working women: An increasing proportion of them were seeking true careers rather than temporary or low-wage employment. In 1900, wealthy women often received good educations, but their schooling was intended to help them be more elegant, more refined,

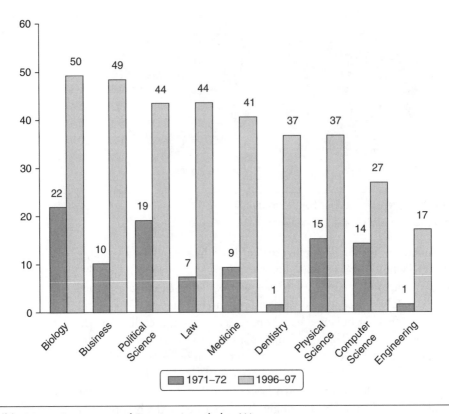

Exhibit 6.2   Percentages of Degrees Awarded to Women

Source: General Accounting Office Report 01–128; U.S. National Center for Educational Statistics 1999.

and, sometimes, more pious domestic women. There were exceptions: women who traveled to other countries as missionaries, for example, and the extraordinary group of wealthy, highly educated women who advanced both social work and public health practices and progressive social theory at Jane Addams's Hull House in Chicago. But these cases were remarkable in part because they were so few. Poor women, in contrast, often worked for wages but received little education. In the 1970s, the idea that large numbers of women wanted to pursue higher education and then use that education in professional employment was new, and even shocking. This "new idea" has since transformed many occupations. Law was once a male-only domain, but law schools now enroll about as many women as men, and significant numbers of these women are older, returning students. With their emphasis on reading, writing, and communication skills, law schools have been more permeable to women than have engineering schools, where gaining entrance often depends on a person's having received early encouragement to pursue science and higher mathematics. Medical schools, with similar entry requirements, have been slower than law schools to enroll large numbers of women, but by 1996 more than 40% of the medical doctorates awarded in the United States went to women.

Closing the Gaps

Despite the changes described above, women's earnings continue to lag well behind those of men (see Exhibit 6.3). Women's earnings for full-time work are currently about 80% of those for men, a fact often stated in dollar terms: Women earn 80 cents for every dollar earned by men. The gap between women's earnings and those of men has narrowed over time, but very slowly. In the 1980s, it showed some signs of closing, but the reason for this change is not so encouraging: Deindustrialization meant that many working-class men were losing unionized industrial employment and having to take lower-wage work or face unemployment in such large numbers that men's earnings were actually declining. Women also lost industrial employment, but given that they were more likely to lose clerical and service positions, as a whole they felt the impact of deindustrialization less. Thus the wage gap started to close only because of the declining wages of a major segment of the male labor force—not because women were making the progress they had hoped for. When men's earnings showed a slight improvement in 1993, the gap again widened. It has begun to close again in this century, but, once again, the major factor appears to be declining men's wages due to the loss of unionized industrial work.

Another reason that women's earnings lag behind those of men is that women are still concentrated in some of the most undercompensated occupations (see Exhibit 6.4). Of the 10 lowest-paying occupations in the United States, only one—farm laborer—is male dominated. One of the fastest-growing occupations, child care (both domestic and in commercial facilities), is still overwhelming female and also offers some of the lowest wages. This concentration of women in low-paying, feminized jobs has been dubbed the "pink-collar ghetto," and pink-collar jobs remain some of the lowest paying of all.

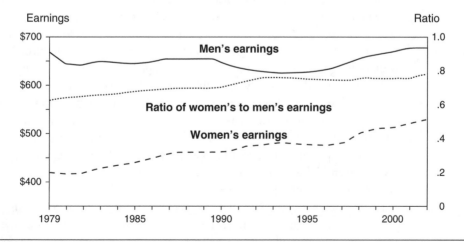

Exhibit 6.3   Earnings of Men and Women
Source: U.S. Department of Labor (2003, September).

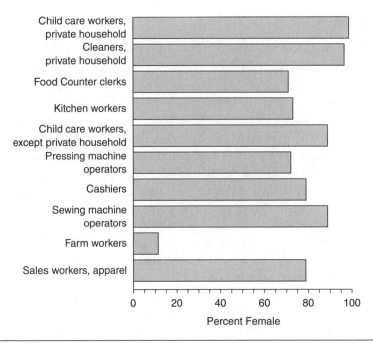

Exhibit 6.4   The Pink-Collar Ghetto: Lowest-Paying Occupations
Source: U.S. Department of Labor, 1988.

It is interesting to note that even when women enter higher-wage, previously male-dominated professions, they still often earn less than men—in many cases, substantially so. This can be explained in part by the fact that women are more likely to be newcomers to these professions and so have less experience and seniority. Women in the legal profession are more likely to be junior associates than senior partners in law firms, women in academia are more likely to be assistant professors than full professors, and women in business are more likely to be junior managers than senior executives. Some of this gap will presumably close as more women gain experience and seniority in professional positions. Yet some professions seem to be changing as they become "feminized." The incomes in real estate sales, for example, dropped markedly as large numbers of women found this an easily accessible profession and even second career. The earnings of senior partners in major law firms remain very high, but the average incomes of new attorneys have been dropping as more women have been entering the profession. This may simply be the result of greater competition for limited clients, or it may be that both employers and clients are used to paying women, even highly skilled women, less than men, and these assumptions are carrying over into newly feminized occupations.

Differences between men and women in earnings are also due in part to the fact that men and women tend to pursue different specialties within professions. For example, although there are now many more female physicians than in the past, high proportions of women doctors cluster in the specialties of family practice and pediatrics, which have lower average earnings than specialties such as surgery and cardiology, which are still heavily male dominated. Again, the reasons for this are less than clear and the subject of some controversy. Do women "naturally" tend toward the specialties that are, let's say, more holistic, or does the medical establishment channel women into specialties that focus on family and children? Do these specialties offer less income because they are less demanding than others or because they are more feminized? As with all gender differences, it is easier to present the differences than to explain them.

Among the various challenges working women face, one of the most difficult to explain fully is the phenomenon that has come to be known as the **glass ceiling**. Women have been moving into managerial positions in increasing numbers for several decades now, but they are still scarce in top executive positions. They seem to be able to move up into middle management fairly readily, but then their careers "stall out" as though they have run into an invisible ceiling that blocks their advancement. When this problem was first noted in the 1970s, two large East Coast companies decided to investigate the reasons. Both companies employed many women as middle managers, but very few women were moving up into senior management. These companies were interested enough in finding out why to hire social science researchers to look at the problem.

One of the companies hired a team of Harvard social psychologists, Margaret Hennig and Anne Jardim, who interviewed a wide range of employees, both male and female, junior and senior, and came to a series of conclusions based on the tenets of social psychology. The researchers reported that **differential socialization** was the problem; that is, men and women are socialized to behave differently in the workplace and this has divergent consequences. Hennig and Jardim (1977) found that the men in the company developed career plans early on, were willing and even eager to take on risks, liked to work in groups where they were noticed, were sensitive to company politics, and knew how to cultivate relationships that would help them get ahead. In contrast, the women tended to come to think of their work as a lifelong career only later, generally avoided risk, preferred to work independently rather than in committees or groups, and tended to help subordinates rather than mingling with superiors who could help them advance. These women believed that their hard work and dedication to the company would be rewarded. It wasn't. Most were fine employees—in fact, it may be better for a company to have managers who are dedicated to doing a good job rather than playing company politics—but they were not the ones singled out for promotion.

When Hennig and Jardim interviewed the company's small handful of female senior managers, they found a different pattern. These women behaved, if you will, like men. They planned their careers early, they worked the informal networks adeptly, they liked to take on risks, and they enjoyed being in, and especially leading, groups. These few women rose to near the top. Hennig and Jardim had an explanation for this: These women had been socialized to behave more like the men than like their female counterparts. Almost without exception, these women were only children or the eldest in families where all the siblings were girls. The researchers speculated that all the attention that usually goes to sons therefore went to them: They were encouraged to climb trees, to play competitive sports, to take risks, and to take charge. Hennig and Jardim suggested that although the company couldn't change employees' family backgrounds, it could, through management seminars, train female middle managers how to think more like the company's men and the executive women.

About the same time, another company in a nearby city turned to a Yale sociologist, Rosabeth Moss Kanter, with a similar question. Kanter (1977) also interviewed a wide range of employees (and their spouses), but as a structural sociologist she was inclined toward conclusions that differed somewhat from Hennig and Jardim's. Kanter saw the root of the problem as lying not in the socialization and behavior of individuals within the corporation, but in the structure of the corporation itself. She noted that a corporation (and almost any other kind of organization) has three hidden structures. The first of these is its *demographic structure:* how many people hold differing types of positions and what the ages, races, genders, and so forth of these people are. In this company she found that in many key departments women were still "tokens." That is, they were present but in very small numbers. Kanter asserted that tokens, in whatever setting, behave cautiously. They know that they are under scrutiny and that any mistakes they make may reflect on their group as a whole; they may be intimidated by majority opinions, left out of key inner circles, and burdened with extra responsibilities just because they are so few. These tendencies may be observed in many settings—the only woman in an otherwise all-male lab, the only black student in an otherwise all-white class, the only Asian American in an otherwise all-white law firm. Kanter believed that the women in the company behaved in ways that were different from men not because they thought differently, or even because they were socialized differently, but because they were tokens and knew it.

Kanter also examined the *power structure* of the corporation—that is, not the official hierarchy but who really had discretionary authority to make things happen and who had someone else looking over their shoulders. She found that low-power bosses tended to be more controlling, less free with rewards, and less able to command the loyalty of subordinates than high-power bosses, for everyone knew that low-power bosses ultimately had to answer to someone else. High-power bosses could afford to give people more latitude and could command more generous rewards for a job well done. Both the men and the women in this company often complained that female managers were too cautious, too controlling, and too stingy with rewards, and many saw this as a female failing. Kanter suggested that these characteristics are typical of low-power bosses and that in fact the problem was that so many of the female managers had a lot of responsibility but little real power. Their behavior was characteristic not of their gender but of their position.

Finally, Kanter noted that the organization had a distinct *opportunity structure*. Some people got ahead and some did not. Everyone knew who the "water walkers" (those on the fast track) were and who was getting old in their positions and "dead in the water." They also knew that certain positions often led to promotion whereas other positions had capped opportunities for advancement (no one ever went from a position in the personnel office to senior vice president, for instance). People in positions with good advancement prospects tended to plan their promotions; they got to know superiors and mentors and worked on winning their regard—that is, they learned how to maneuver their way up the corporate ladder. People in low-opportunity, dead-end positions didn't bother. With few upward prospects, they tended to help others, to focus on competence in their current positions, and to voice fewer plans for the future. Again, this was a difference not in genders but in positions. Yet, once again, it was women who tended to be in the positions with limited advancement options. Kanter's proposal to the company involved changing the corporate structure: bringing in women in groups so that they are not left as tokens, giving managers the full power and authority that their titles suggest, and making sure that no position in the company is seen as a dead end.

Both of the books in which these researchers describe their work gained considerable attention when they were published. Kanter's *Men and Women of the Corporation* (1977) is still widely read, especially in the academic world, and Kanter went on to head Harvard Business School. At the time, however, Hennig and Jardim's *The Managerial Woman* (1977) was more influential in the business world. It isn't hard to guess why: Kanter's conclusions implied that the entire corporate structure needed to change; Hennig and Jardim only called for seminars to retrain individual women—a much less threatening proposition.

Change in matters related to gender has come slowly to most organizations. The hindrances often lie not just in a company's formal structure, what shows on an organizational chart, but in what Kanter called the **shadow structure**: the places where employees build alliances, establish confidences, and manage their reputations. A large organization, like the world system, has a "core" and a "periphery." The shadow structure can help move men to the center of dense networks of information exchange and keep women talking to one another on the peripheries of power (Ibarra 1992; McGuire 2002). This process may be blatant, as has been charged in a spate of lawsuits citing meetings held during male-only golf outings or in "strip bars" and "gentlemen's clubs" where women were not welcome or comfortable, but often it is more subtle. McGuire (2002) relates the case of a vice president of a large firm who admits that he gets promotions for his favored subordinates, largely men, by asking other vice presidents to put them on candidate lists, saying that only "losers" go to Human Resources. Commenting on this, McGuire notes, "Women, in effect, face a double-edged sword: they obtain less instrumental help than men do from their informal network members, but if they turn to formal outlets to meet their instrumental needs, they risk being further marginalized" (p. 318). She concludes that real change happens only when organizations change their structural arrangements, both formal and informal, as well as the assumptions about gender and competence that may underlie these structures.

Work and Family: The Double Burden and the Second Shift

Changing gender roles in the workplace are tied to changing roles in the home and family, with changes in one realm often forcing changes in the other. For example, women's growing economic independence has made it easier for them to leave abusive or unsatisfying marital relationships. This, along with other social and legal changes, is one factor in rising divorce rates. At the same time, high divorce rates mean that women are often left as sole custodians of children, often with limited child support and sometimes with limited income-earning potential of their own. The result is what has been termed the **feminization of poverty**, with the group most at risk of poverty being single mothers and their dependent children (Sidel 1996).

Because women's expanded roles in the workforce have generally not been accompanied by any relaxation of expectations for their family and domestic activities, many women today face the "double burden" of home and work responsibilities. Women still often have primary responsibility for the care of their children as well as the care of older adults in the family. As noted above, women also still shoulder the largest portion of the upkeep of the home, even when they are working full-time (Stapinski 1998; South and Spitze 1994). Arlie Hochschild (1989) has termed this the "second shift" and has noted the anger and frustration that many women feel when they find they are still doing the major portion of housework on top of their paid work. Again, the reasons for gender differences in the share of housework done are complex. Some have suggested that women's expectations for household cleanliness are higher than men's and that women often feel responsible for the state of their homes in a way that men do not. Perhaps "differential socialization" is at work here as well, although it also seems that many women still have less power and privilege in the home than do their male companions.

Although men put in slightly more time in paid employment than women, women spend twice as much time as men in housework and caring for family members. The result is what some have called "the leisure gap," in which men have more leisure time left over at the end of the working day than do

Average hours per day spent:

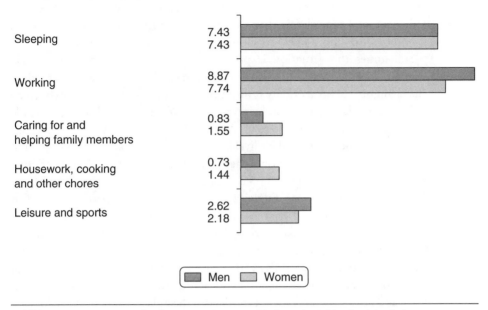

Sleeping 7.43 / 7.43

Working 8.87 / 7.74

Caring for and helping family members 0.83 / 1.55

Housework, cooking and other chores 0.73 / 1.44

Leisure and sports 2.62 / 2.18

Men Women

Exhibit 6.5   Division of Labor: Time Spent on Various Activities by Married Men and Women Ages 25 to 54, with Children, Who Work Full-Time

Source: Bureau of Labor Statistics (2004).

women (U.S. Bureau of Labor Statistics 2004; see Exhibit 6.5). Men may not feel that they have much leisure time—with long working hours and more home responsibilities than their fathers had, they may feel a time crunch—but the squeeze remains the greatest for women.

This is a cross-cultural and almost worldwide phenomenon. Around the world, women spend more total hours in work than men, and mothers work the most of all (Scarr et al. 1989). In Japan, a majority of women are in the paid workforce, but they face wage discrimination that is the greatest in the industrial world and barriers to upward mobility that are blatant by Western standards. Once married, they are still expected to be devoted wives and mothers that give all their effort to the home, even after a full day of work. Latin American women, now entering the paid labor force in ever greater numbers, likewise face what they call the *doble jornada,* or double day's journey. *Machismo,* that swashbuckling blend of male authority and male privilege that has long been an important part of Latin American cultures, appears to be breaking down as more Latin American men interact with their children frequently and help around the home (Gutmann 1996), but the main domestic responsibilities still fall to the women, even those who are primary wage earners. In Western Europe, where gender role changes have been most pronounced, the gaps between men and women in wages and in workload are smaller but still present. Like North American and Latin

American men, European men are more likely than ever to play and interact with their children but no more likely to participate fully in those children's daily care, and more likely than ever to help their wives or female partners at home but no more likely to shoulder all domestic tasks equally.

Sometimes this situation is pretty clearly a story of continuing male privilege. In other cases, it is also a story of male alienation and vulnerability. The changing global economy is displacing many men, especially low-income and working-class men, from their traditional roles even as it is overburdening women with new and double roles. Around the world, men who depended on strong backs and arms for their livelihood have been displaced by automation in industry and mechanization in agriculture, and by the constantly shifting nature of global production. Steelworkers in Gary, Indiana, Bethlehem, Pennsylvania, Manchester, England, and the former East Germany are idled by new more automated Japanese plants and by products that replace steel with molded plastic parts from Guangzhou, China. Dockworkers in Baltimore, London, and Gdansk, Poland, are idled by container ships that load and unload the world's wares with a single crane. Small farmers in Minnesota, Brazil, the Philippines, and South Africa are idled as corporate-controlled heavy machinery works the land in their place. Cattlemen in Montana watch their ranches turn to condos while cattlemen in sub-Saharan Africa watch their rangeland turn to desert. Sometimes they find alternative work: Security guards are in high demand in crime-plagued cities like Gary, Indiana; St. Petersburg, Russia; and Johannesburg, South Africa. Security seems to be manly work although it is typically low paying. Other men may try criminal activity themselves, or a variety of odd jobs—day labor, seasonal or informal construction work, or small repairs and sales. The alternative is often chronic unemployment: sitting around the tavern, the pub, the street corner, the village center, or the diner amid abandoned buildings drinking coffee, beer, vodka, or homemade brew and talking about hard times with other men. These men don't feel privileged; they feel humiliated and disgraced, or alienated and cheated. As a speaker at a church conference in Washington, D.C., noted, they're not deadbeats, just dead broke. With high rates of suicide, alcoholism, disease, and violent crime, sometimes they are also soon dead. In *Families on the Fault Line* (1994), Lillian B. Rubin quotes the unemployed Tony Bardolino:

> I was just so mad about what happened; it was like the world came crashing down on me. I did a little too much drinking, and then I'd just crawl into a hole, wouldn't even know whether Marianne or the kids were there or not. She kept saying it was like I wasn't there. I guess she was right, because I sure didn't want to be there, not if I couldn't support them. (P. 219)

Ironically, if there are women in the lives of these displaced men, they are often overburdened. They may have become the major wage earners for their families. Assembly plants in export processing zones in Mexico, the Caribbean, and across East Asia are often reluctant to hire unemployed men who may tolerate too little and demand too much. These plants prefer to hire young women. Strong backs aren't needed to assemble electronic components, just nimble fingers, keen eyesight, and endless patience. Young women bring these skills and the willingness to work without complaint. Other plants draw on traditionally female skills such as sewing seams in textile plants. Women in these plants often earn very little by Western standards, but they may become the major wage earners for their families, replacing the income lost when fathers and husbands are unable to find work. Patricia Fernandez-Kelly (1983) quotes Teresa in Juarez, Mexico:

> There were about seventy women like us sewing in a very tiny space. . . .
> When I was sixteen I used to cut thread at the shop. Afterwards one
> of the seamstresses taught me how to operate a small machine and I
> started doing serious work. Beatriz, my sister, used to sew the pockets
> on the pants. It's been three months since we left the shop. Right now
> we are living from the little that my father earns. We are two of nine
> brothers and sisters (there were twelve of us in total but three died
> when they were young). My father does what he can but he doesn't
> have a steady job. Sometimes he does construction work; sometimes
> he's hired to help paint a house or sells toys at the stadium. You know,
> odd jobs. He doesn't earn enough to support us. (P. 119)

Even when the women are engaged in full-time work, it is not easy for the men to step into what they have always thought of as female roles: taking over domestic tasks and child care. So these also remain primarily female activities. The men are left with no roles and seek alternative forms of gratification or just leave altogether. The women are faced with double and triple roles and may feel crushed by the added burdens. The men may be as likely to feel resentful of the women as grateful. Somehow the world has changed, and it is hard to know who to blame. The women may blame the men for not "doing something for themselves." The men may blame other racial groups: Displaced white men in both the United States and South Africa frequently blame black workers, displaced black men in both the United States and East African cities frequently blame Asian entrepreneurs. The men may turn to ultranationalist politicians and leaders—in the United States, in Russia, and recently even in Austria and Fiji. They may also blame women—or perhaps specifically feminists—for undermining the status they once knew.

Middle-class and upper-middle-class career men with children face stresses of their own. They often are expected to have the same single-minded devotion to their careers as men of the 1950s who were supported by their "domestic women," but now men are also expected to invest themselves more in the duties and demands of parenting. They face their own sets of unrealistic demands and guilt as they confront what Hochschild (1997) calls the "time bind." The very expectation that child rearing is primarily a female endeavor can limit men's options; even when an employer offers paternity leave, the very few men who take it may be subject to suspicion or ridicule (Hochschild 1997). When British Prime Minister Tony Blair and his wife had a baby, the joke was who should take the leave; as a well-paid attorney, she earns far more than he does as prime minister! The Finnish prime minister did take a few days of paternity leave, in part as a way of encouraging other male Finns to consider doing the same. Yet Sweden is being forced by more conservative members of the European Union to reduce its generous maternity and paternity leave policies to bring them in line with new budget-cutting priorities. Finding more equitable, more satisfying ways of combining work and family remains one of the challenges of this century for men and women, for their employers, and for their governments.

Gender and Class around the World

Women in many countries have made occupational gains, moving into professional career positions. This trend may be slow in some places, but it seems hard to reverse. In Afghanistan, women had to give up most professional positions, as well as all political power, under the rules of the hardline Islamic fundamentalist Taliban government. But even there, exceptions remained: Because women were forbidden to expose any portion of their bodies to men outside of their families, women usually received medical treatment from female physicians. Women in Iran currently face many political and social restrictions, but many have found ways to retain their professional niche. The traditional and the modern have had to accommodate one another. Growth in overall women's education continues to lag behind in many Middle Eastern societies, yet it is not uncommon to see well-educated Iranian women talking to business clients with their cell phones slid up under their veils as they drive through Tehran's traffic.

Women's role in government continues to expand, although slowly. Women occupy about 10% of the seats in the world's parliaments and legislative bodies, somewhat less in Latin America but almost 20% in East Asia (United Nations Development Program 2004). The largest countries in South Asia, India and Pakistan, have both had female leaders despite the perception of these societies as heavily imbued with traditional gender roles. It is true that both of these women, Indira Gandhi and Benazir Bhutto, were the daughters of well-regarded male leaders, but having a well-known name has also certainly helped the political ambitions of many men. Women have also gained greater prestige and recognition around the world for other accomplishments, although many of these are still related to women's traditional roles: royalty (Princess Diana), humanitarian work (Mother Teresa), and various forms of entertainment, where the emphasis is still often on women's sexuality. Yet many women are increasingly gaining new voice and confidence, as we will see in Chapter 12.

Women, especially rural women, are still the poorest of the world's poor, especially because they often bear heavy family responsibilities on their meager resources. In some urban areas women are now more employable than men but they see few benefits from this situation, as their new incomes often go to support their struggling families. Urban women have made substantial movement into middle-class employment but have found it hard to continue to move up to top positions. Men still dominate in positions of power and privilege, although many poor rural and urban-industrial working-class men have lost ground. They may be inclined to blame this loss on women's gains, but most have lost out to global technology and global marketing, which are still controlled largely by wealthy and powerful men.

KEY POINTS

- Global economic and social changes are having major impacts on the lives of both men and women, creating both new opportunities and new strains.
- Men retain much privilege and power, but working-class men have been especially vulnerable to job losses due to deindustrialization.
- The accepted roles of women are expanding around the world, with more women in professional careers, but on the whole women remain concentrated in some of the lowest-paying occupations.
- In all professions, women continue to earn less than men on average and have limited access to the top positions, a phenomenon known as the glass ceiling.
- Women's expanded roles in the paid labor force have not been matched by lessened responsibilities for home and family, and many women experience a "double burden" of expectations.
- Low-income men may find they are unable to provide for their families and feel that they have little to offer. High-income men may face a double burden of their own as they try to balance old career expectations with new expectations concerning fathering.

FOR REVIEW AND DISCUSSION

1. How has masculinity conferred both privilege and vulnerability over time? How has this affected the lives and life chances of both men and women?
2. How have gender roles and expectations changed over time? What aspects have been slow to change? How have these changes both benefited and burdened women?
3. What factors have been cited as contributing to the glass ceiling for women? What changes might promote the advancement of women in careers?
4. What factors have displaced men from traditional roles? What factors have overburdened women in their old and new roles? How are men and women coping with these changes around the world?

MAKING CONNECTIONS

In the Marketplace

Browse through a toy store in a mall or the toy department in a large department store. Is there a "pink aisle" filled with toys targeted at girls and a "khaki aisle" filled with toys targeted at boys, or does the store stress nongendered toys? What themes and expectations are stressed in toys for boys and toys for girls, and how do these differ? Play, among other things, is socialization and rehearsal for adult roles. What types of adult roles are children rehearsing when they play with these toys?

Next, visit a children's clothing store or the children's department in a department store and browse through the infant and toddler clothing, then go on to school clothes and teen fashions. Note the differences between the clothing for boys and the clothing for girls: colors, patterns, styles, fabrics, prevailing themes, and common accessories. What messages do the clothes convey about the meanings of masculinity and femininity? How might wearing this clothing affect a child's self-image, imagination, and behavior? How might the clothing affect the way the child is treated? How might the clothing affect the types of activities the child is likely to pursue? Think about styles and trends in adult clothing: Do they reinforce or challenge this early socialization?

Finally, the harder question: How does this socialization interact with the structures of careers, opportunity, and inequality in our society? Are certain images and behaviors more likely to lead to financial rewards and success?

In the Media

Examine the portrayals of men and women in advertising, whether on television or in print. If you look at magazine advertising, try to compare the ads in a variety of publications targeted at different audiences. What gender roles are emphasized? Do the people in the ads appear to be prosperous, poor, or in between? Are women most often shown in romantic, family, or career roles? What about the men? What attributes are stressed? For example, are the women smiling and playful while the men look stern and strong, or are there no consistent differences in how men and women are portrayed? How are the men and women shown relating to one another? Are those of one gender more often shown in positions of leadership, power, or dominance? Erving Goffman presented the classic study of advertising's portrayals of men and women in his book *Gender Advertisements* (1979). The fashions and styles in the book's pictures now look very out of date (except for a few that are now being revived), but Goffman's assessments of how people in ads are placed and positioned, and how they appear to relate, are as insightful as ever. You may want to compare Goffman's interpretations with your own. Media specialist Jean Kilbourne offers a current depiction of women in advertising in her series of filmed lectures titled *Killing Us Softly 3* (2000). You may also want to compare Kilbourne's critiques of media portrayals of men and women with your own. Do the images you found reflect a gender-unequal society? Do they reinforce gender inequalities?

7

Status Prestige

Superfluous wealth can buy superfluities only.

> —Henry David Thoreau, *Walden* (1854)

The man who dies rich thus dies disgraced.

> —Andrew Carnegie (1835–1919)

If you gain fame, power, or wealth, you won't have any trouble finding lovers, but they will be people who love fame, power, or wealth.

> —Philip Slater, *Wealth Addiction* (1980)

In a capitalist society, class is foremost about money. In an agrarian society, class is about family and land. Yet in both kinds of societies the class stratification system also includes a subjective element: how one lives and how one is seen by others. This is what Weber called *status*—a person's level of prestige within a community of recognition. Status is subjective, and it is community specific. If you have high status in one community or with one primary "audience," you may face a rude awakening when you find that somewhere else no one is impressed with your family name, no one has heard of your prestigious college, or no one cares about your particular accomplishments! The communities that confer status are not always local entities. A highly respected sociologist may command a prestigious place in a sociology conference anywhere in the world and yet be completely unknown elsewhere; such a person's community-specific prestige is of little help to him or her in gaining an upgrade on an airline seat at the gate in his or her hometown.

The Quest for Honor

Prestige can be conferred in many ways. Hunter-gatherer groups give recognition to the greatest hunters, storytellers, healers, or councillors. The big men of tribal societies often don't have much personal accumulated wealth or much coercive power. What they have is prestige. As long as people will follow them, they are leaders. As long as people honor them with gifts, they can share those gifts with others and gain greater prestige for their generosity. As long as they are honored with privileges, they will have comfortable lives. Prestige in such cases is based on personal charisma, bravado, and leadership: the brave-hearted warrior, the wise councillor, the spellbinding orator. Prestige built on charisma can be translated into other forms of privilege: corporate leadership with a high salary, political leadership with extensive power, or just a greater likelihood of getting one's own way.

Occupation can confer prestige. Most Americans have a fairly clear sense of the prestige, or lack thereof, that various occupations carry. With a few alterations, the rankings displayed in Exhibit 7.1 are quite consistent around the world. In general, the higher income a job provides, and the more education it requires, the higher it is ranked, although working conditions and an occupation's level of a sense of power or dignity also seem to be factors in these subjective classifications. Taken together, education, occupation, and income constitute **socioeconomic status**, or SES—an individual's standing in a ranking of positions. The concept of SES is similar to the idea of class, but it emphasizes a continuous ranking rather than a grouping of like-situated people. It is interesting to note that although related occupations in some fields may confer similar status, they often generate very different income levels. In some cases, the income gap within an occupation may be greater than the gap between occupations.

The scale of occupational prestige is a ranking of social status based on people's subjective interpretations. Certainly money matters, but status distinctions go beyond just income. The income of a unionized senior janitor in a school building may exceed the income of a new teacher in that building, and the plumber who comes to the school to fix the leaking bathroom pipes may earn more than both. Yet, of the three, the teacher has the most prestigious occupation based on education and type of work. Rituals of social honor are an important element of status: A judge occupies an extremely prestigious position that is reflected in the high bench, the long robe, the power of the gavel, the rising of everyone in the courtroom as the judge enters, and the deferential address of "your honor." In the United States, Supreme Court justice—the highest-level judgeship possible—is rated the most prestigious occupation. Yet an illegal drug dealer standing before the bench in any courtroom may have (momentarily, at least) a higher income than the judge. If the drug dealer wants to be prestigious, however, he may have to become a "drug lord," giving gifts, jobs, and protection to people within his neighborhood. As it does for the big man of a tribal community, this may accord him some prestige within his own community—although his high status there is not likely to be much recognized beyond that community.

Exhibit 7.1 shows some examples of occupations that have been ranked at different levels of prestige. These rankings were computed from surveys that asked a national sample their "own personal opinion of the general standing that [each] job has". Respondents gave each job a rating of "excellent", "good", "average", "somewhat below average", or "poor". (See Nakao and Treas 1994.) A complete listing of all 503 Census occupations is also available. Annual earnings, average education, % minority, and % female are from the 1990 Census.

High family standing also confers prestige. In agrarian times, family was the key element of an individual's status. Wealth was inherited along family lines. A person's family made him or her a member of the nobility or a peasant. Titles as well as land were inherited. Family mattered a great deal in early America as well, although gradually mercantile elites (trading families) came to join the ranks of the landed elites. Titled elites—earls, baronesses, and the like—were always scarce in the colonies because such people tended not to immigrate, so the field was more open for competitive entry into elite status. The southern United States, with its land-based aristocracy supported by plantation slavery, had the strongest elite identity. Titles were appropriated: *sir* and *lady* from the days of chivalry and *colonel* from the time when an aristocratic officer elite dominated the military. White suits for men and great hoopskirts for women made it clear that these people did no manual labor. Debutante balls, popular also in the North but a more enduring tradition in the South, served to help ensure that elite families associated together. For elites to exist, there must also be nonelites, and so other terms were common for the landless and those of less certain lineage: "the no-account lot" (Davis, Gardner, and Gardner 1941) and the awful but enduring "poor white trash."

The mercantile families of New England started with few trappings of prestige, but over time their wealth accumulated and—most important for prestige—became "old money." The original sources of this money mattered little—William Ryan (1971) derisively calls it "slave and rum money," for these were the early pillars of New England trade—and in time the wealth established the family names as prestigious. Note that in both North and South, wealthy families may have originally made their money literally on the backs of others, either trading or owning slaves, yet over time their wealth became associated with endowments to great universities—Harvard and Yale in the North, Virginia and William and Mary in the South—and philanthropy. Old-money families also paid careful attention to social interaction. In the North, Boston may have been, in the words of John Collins Bossidy, only "home of the bean and the cod" but nonetheless, "the Lowells talk to the Cabots and the Cabots talk only to God."

Yet if money can, in time, buy prestige, then new money can buy new circles of prestige. This was the great threat of the era of early industrialization in the latter half of the nineteenth century, when those of "uncertain" family

Prestige		Annual earnings	Average education	% minority	% female
86	Physicians	$116,185	19.7	8.6%	20.7%
85					
84					
83					
82					
81					
80					
79					
78					
77					
76					
75	Lawyers	$82,339	19.5	6.0%	24.5%
74	Professors	$42,887	17.9	8.6%	39.2%
73					
72					
71	Engineers	$45,360	15.8	6.7%	9.8%
70					
69	Clergy	$24,005	16.6	10.1%	10.4%
68					
67					
66	Registered nurses	$31,595	15.2	12.0%	94.3%
65	Accountants	$35,787	15.3	11.2%	52.7%
64	Teach, elementary	$28,078	16.4	15.2%	78.4%
63					
62					
61	Computer programmers	$35,828	15.2	9.6%	32.5%
60	Police, public service	$32,862	14.0	18.7%	12.0%
59	Financial managers	$47,766	15.1	8.8%	46.0%
58					
57	Health technicians	$21,970	13.8	18.9%	71.0%
56					
55					
54					
53	Firefighting	$33,372	13.3	15.1%	2.7%
52					
51	Managers	$50,349	14.4	8.3%	31.4%
50	Electricians	$30,335	12.7	12.5%	2.5%
49	Real estate sales	$41,190	14.4	7.0%	50.4%
47	Supervisors: production	$33,541	12.6	16.6%	17.7%
47	Bookkeepers	$19,514	13.2	11.2%	89.6%
46	Secretaries	$19,383	13.2	13.4%	98.7%
45	Plumbers	$29,029	12.0	15.7%	1.5%
44					
43	Bank tellers	$14,958	13.1	16.5%	89.8%
42	Guards	$21,971	12.6	31.2%	16.6%
42	Nursing aides	$16,570	12.3	37.4%	87.2%
40	Auto mechanics	$23,320	11.9	18.2%	1.8%
40	Farmers	$22,697	11.8	3.0%	14.4%
39	Carpenters	$24,344	11.9	14.9%	1.7%

Exhibit 7.1   Occupational Prestige

Source: Ratings based on Nakao and Treas (1994). Earnings figures based on the U.S. Census, 1990.

Exhibit 7.1 (Continued)

Prestige		Annual earnings	Average education	% minority	% female
39	Receptionists	$15,642	13.0	17.9%	95.7%
36	Hairdressers	$16,089	12.7	17.0%	89.6%
35	Assemblers	$21,016	11.4	30.4%	43.3%
34	General office clerks	$19,857	13.1	22.4%	82.3%
33					
32	Bus drivers	$23,680	12.1	30.5%	48.1%
31	Cooks [inc. short order]	$14,191	11.3	32.8%	47.6%
30	Truck drivers, heavy	$25,935	11.7	21.7%	6.0%
29	Sales, apparel	$18,633	12.7	17.8%	81.3%
29	Cashiers	$15,615	12.1	24.6%	79.1%
28	Waiters & waitresses	$13,149	12.4	13.9%	80.5%
27					
26					
25					
24	Laborers, exc. const.	$20,322	11.4	30.7%	22.0%
23					
22	Janitors	$18,577	11.1	36.7%	31.5%
21	Parking lot attendants	$17,809	11.9	42.9%	10.3%
20					
19	Vehicle washers	$17,321	11.2	36.2%	12.4%
18					
17	Misc. food preparation	$13,217	11.0	34.3%	49.8%
16					
15					

background could quickly accumulate great fortunes. Scots, Irish, Germans, Jews, and others with names like Rockefeller, Kennedy, Woolworth, and Carnegie could far outstrip "the rum and slave" money of the Astors, Lowells, Abbots, and Cabots. One attempt to preserve the importance of family, and hence of old money, was the creation of the *Social Register* in 1887. This publication was to be a guide to the right families; people who were not included in the register might have money, but they would be "awkward in the ballroom" or at least "make everyone else awkward." Over time, the *Social Register* has had to add more families, including new-money families, but its emphasis is still on heritage, family ties, kinship, and marriage.

Max Weber ([1922] 1979) wrote about competitive struggles for prestige, noting that out-groups try to wrest prestige from older in-groups. Excluded from the old order, they might found their own clubs, associations, foundations, colleges and universities, secret societies, and publications, and strive to increase the prestige of each of these. But prestige cannot be for all: A community in which everyone is equally prestigious is one in which no one is particularly prestigious. Thus elite groups always strive for some form of social closure: a way to monopolize the markers of prestige and deny access to others. The establishment of the *Social Register* was an attempt at social closure, as were attempts to deny Catholics and Jews entry into elite clubs and schools. Today, upwardly mobile African Americans, Hispanic Americans, and Asian Americans may find the same difficulty in gaining access into certain circles. Wealth and income may expand to make room for more, but prestige is essentially a fixed commodity: If too many climb to the top, it starts to look like the middle!

A high level of education may also confer prestige. In many ways, the titles attached to educational credentials have replaced the titles of nobility. A roll of parchment can change a person's title ever after from *Mr.* or *Ms.* to *Dr.* Résumés listing the right schools and honorary societies can replace genealogies listing the right ancestors. Preindustrial societies often took great pains to define and categorize their members' lineages—as seen in great ancestor lists in China, oral traditions in the Pacific, and genealogies in the Bible and in the *Social Register*—as family was the key to a person's position in society. We similarly note and categorize types and levels of degrees obtained and institutions attended because the prestige these can confer may be translated into desired positions.

Individuals may also gain prestige through philanthropy, or the giving of time and money to worthwhile causes. The big man in a tribal community increased his prestige by presenting gifts to others, often in lavish ceremonies. In modern society, prestige is similarly enhanced by charitable giving (especially in the form of named endowments), by service on community boards and commissions, and by many other acts that are seen as contributing to the community. Many Seattle cultural and social organizations are dependent on the "Amazon and Microsoft money" that comes from philanthropic young multimillionaires. There is a clear social benefit in this: Those who have gained great wealth and now want prestige as well must return some of their wealth to the community to get it. There is also a danger in this, in that it can give the wealthy unprecedented power to control the community. Philanthropy has been the hallmark of the devout and the truly thankful and generous, but it has also been the hallmark of corrupt politicians and syndicated criminals.

Although I have focused on elites in this discussion, prestige, often in the form of respectability, is very important to lower-income groups as well. In a classic study, Davis et al. (1941) found that American southerners were eager to show that they were "poor but honest folk" rather than part of the "no 'count lot." Elijah Anderson (1990) found that the residents of a poor, predominantly black neighborhood he studied were eager to distinguish between "respectable people" and "street people." Older, hardworking individuals, the "old heads," had long commanded respect and social honor there, although a disintegrated community structure threatened this position. Similarly, Mitch Duneier (1992) noted that a key theme among the working-class black men with whom he ate and talked was the issue of respectability. Within their small communities of neighbors and coworkers, they held some measure of social honor for their moral standards, their hard work, their experience, and their practical wisdom. They were also interested in distancing themselves and their identity from those they viewed as having none of these.

_____ Socialization: Acquiring Marks of Distinction

For millennia, prestige has been the property of families. Parents in elite families hope that their children will follow in their footsteps and do nothing to "dishonor" their family names. Poor parents may at least hope that their children will have marks of respectability that distinguish them from habits and companions that would similarly dishonor their families. For this reason, families often socialize their children in many ways, both explicit and subtle, to assume their expected class position. A fascinating example of this comes from Melvin L. Kohn's (1977) research on the attitudes and hopes for children held by families of different social classes. An interesting pattern emerges, as illustrated in Exhibit 7.2.

Do Kohn's findings surprise you? Why would poor parents care more than wealthy ones about their children's honesty as well as appearance? Some might see the roots of insider trading and corporate dishonesty in this picture! A more likely explanation, however, is that parents attempt to prepare their children to assume expected roles. The children of the upper classes are more likely than children of lower-class families to step into positions that allow them flexibility, autonomy, and opportunities for innovation, thus upper-class parents are concerned about their children's creativity and ambition. Working-class parents, in contrast, are likely to have jobs that reward punching a time clock and displaying responsibility and trustworthiness. They hope for the same for their children. Poor parents may fear for their children's well-being because of the dangers they face, from streetwise peers or from the police, and so they place great stress on honesty and appearance—the marks of "respectability."

Many Americans hope that the public schools will be the great equalizer in American society, yet schools may also socialize children to anticipate their class positions. So-called prep schools have long provided elite socialization for the wealthy. More recently established "college-prep" classes in public schools provide socialization into the expectations of the middle and upper-middle classes. Vocational classes may feature not only different subject matter, but also different expectations of students and ways for students to show competence and earn respect. The darkest side to these distinctions is that children from the poorest communities and least respected families may get the message from schools that there just "ain't no makin' it" (MacLeod 1995) and may come to believe that they can gain social respect only by winning the approval of the local "gang" or alienated, "turned-off" peer group. The subjective nature of this process also means that race and gender assumptions may intermingle with class distinctions. Stereotyping still leads counselors to steer boys to math and science and girls to literature and arts. Asian American

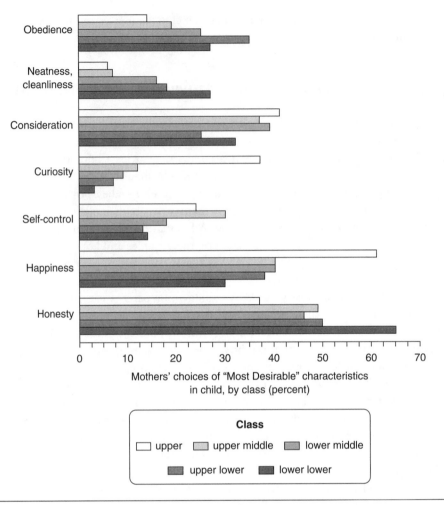

Exhibit 7.2   Parental Aspirations by Social Class
Source: Data from Kohn (1977).

youth may be steered by both parents and school personnel toward math, African American youth may be encouraged to work toward sports scholarships, and Latino students may be pushed toward "shop" or vocational classes. Given that children often ultimately meet parental and school expectations—both positive and negative—these can be powerful influences on future occupational prestige and class position.

Children are socialized not only to have work habits appropriate to their class but also to have corresponding consumption habits. Early socialization determines what Pierre Bourdieu (1984) calls the **habitus,** the habits of speech, work, lifestyle, and fields of interest and appreciation that determine where an individual feels comfortable, knowledgeable, and "at home." These can also become a person's marks of "distinction." Prestige is related not just to what one can afford, but also to what one can "appreciate." According to

Bourdieu, we cannot delineate "good taste" and "bad taste" apart from social position. Our cultural appreciation corresponds to the socialization of our class upbringing.

The poor and working classes are associated with tastes and interests often referred to as *lowbrow*. Their tastes in art may be scorned by other classes, as in jokes about glow-in-the-dark velvet Elvis paintings. Their activities, such as gambling or bowling, may have broader appeal but carry little prestige. Likewise, the music associated with the poor and working classes may be scorned (as rap often is) by the members of other classes or grudgingly accepted, or even popular (as blues and country music are), but appreciating such music is thought to require no particular "refinement" or sophistication of tastes. Of course, middle-class youth sometimes adopt lowbrow styles (such as in "grunge" clothing or rap music) as an expression of their rejection of middle-class blandness, but they usually reject or "outgrow" these styles just as quickly. The point is that style is not considered to be lowbrow because it is poor quality (rap has provided both outrageous examples of racism, sexism, and commercialism and some of the most biting social commentary of the past decade) but because it is associated with the rural or urban poor.

Middlebrow interests are those that typify the middle class. Middlebrow activities include the most popular sports, especially watching as spectators. Middlebrow music includes "pop" and "easy listening," even if the first has elements of blues rhythms and the second of classical melodies. Bourdieu categorizes photography as middlebrow art.

Highbrow tastes are those that require relatively extensive and exclusive socialization to appreciate. Moderately highbrow interests include impressionist art, golf (especially at a private club), and classical music. Truly highbrow are those that require still more "sophistication" to appreciate and participate in, such as abstract modern art, polo, and modern symphonic music or classical opera (preferably in a foreign language). Bartók is more highbrow than Mozart largely because fewer people enjoy and appreciate his music, thus it is more elite and a greater mark of distinction.

Highbrow tastes mark an individual as a member of an upper class, and as *belonging* in that class. Such tastes can be a justification for privilege: "Others would not truly appreciate what we have anyway." They can also be a way of defining a prestige community. The accumulated ability to appreciate "high art" and sophisticated cultural expressions constitutes what Bourdieu (1986) has dubbed "cultural capital." Like the more technically oriented "human capital," it can be an element in a person's ability to convince teachers, college recruiters, employers, and club members that he or she is refined, well educated, and well socialized. Is cultural capital worth more in a social-prestige-conscious country like France than it is in the more money-conscious United States? Michèle Lamont (1992) pursued this question by comparing the residents of Paris and Indianapolis. She found that cultural capital matters just as much in Indiana as in France, but that cultural capital for Americans is likely to include a wide range of professed interests rather than just highbrow tastes. The Americans in her sample, perhaps afraid of seeming elitist, showed the extent of their socialization or "enculturation" by becoming what Lamont terms "cultural omnivores." They knew the classical composers but also claimed affinity for country-western music or rhythm and blues. They played golf and also watched basketball. The mark of their distinction was not just elite tastes but broad tastes that showed they could appreciate great diversity. In fact, some cultural forms can become "high-lowbrow," originating among the poor but being adopted by elites. Jazz music, for example, has done this on both sides of the Atlantic.

Association: Who You Know and What You Know about Who You Know

Prestige has long been about families, but it is also very much about communities. Appreciative communities confer prestige. Being part of the right inner circle is also a key way to gain, maintain, and pass on prestige. Parents often take great care in selecting their children's schools and, later, in influencing their children's choices of institutions of higher education, not just because of what their children will learn in particular schools but also because of the people they will meet. The importance of association in schools grows as children get older: By high school it affects who they will date, and their college choices may influence who their lifelong friends, associates, and maybe even spouses are. Colleges often maintain alumni associations and clubs to build on and continue this pattern of association, with the most elite schools having the most elite— and often the most close-knit—clubs and alumni associations.

Many other community groups and organizations also bring together people of similar social class. Membership in such organizations may help maintain someone's standing in the community and is often a source of both personal and business contacts. The capitalist class may gather locally at the Summit Club, the Community Foundation, the Historic Preservation Society, or the Philanthropic Club, as well as on numerous boards of directors for both businesses and community service agencies. Members of this class also meet nationally at exclusive retreats, hunting and fishing clubs, and other select gatherings. G. William Domhoff (1974) describes one of the longest-established and most prestigious of these:

> Picture yourself comfortably seated in a beautiful open-air dining hall in the midst of twenty-seven hundred acres of giant California redwoods. . . . You are part of an assemblage that has been meeting in this redwood grove sixty-five miles north of San Francisco for over a hundred years. . . . You are one of fifteen hundred men gathered together from all over the country for the annual encampment of the rich and the famous at the Bohemian Grove. (P. 1)

Bohemians of the 1970s and 1980s included President Ronald Reagan, then Vice President George H. W. Bush, Attorney General William French Smith, Secretary of State George P. Schultz, former President Richard Nixon, former President Gerald Ford, Supreme Court Justice Potter Stewart, and the presidents and chairmen of many of the largest U.S. corporations, banks and investment firms, and media enterprises.

Members of the upper-middle class meet one another at country clubs, business organizations, and civic clubs. An old small-town adage claims, "The Rotaries own the town, the Kiwanis run the town, but the Moose have the most fun," reflecting the class composition of each of these groups, from small capitalist to managerial middle and upper-middle class to a more often working-class group. Only the last of these is likely to meet in a bowling alley.

Like civic groups, churches are often class segregated. Individual churches often have core clusters of members who share a particular social class, and everything from the decor to the music may reflect the position and tastes of this group. This is less likely for groups that have fewer local choices of places of worship: Synagogues and African American churches, for example, sometimes bring together a relatively wide range of classes, and Islamic centers in the United States may include the greatest class range of all. Despite the inclusive creeds of most religions, places of worship are also usually ethnically and racially divided as well, given their separate histories. A given denomination may contain a wide range of types of churches, but they also tend to order themselves by average income (see Exhibit 7.3). This tendency for institutions that preach the blessedness and special welcome of the poor to remain class segregated as well as race segregated gives some idea of the power of class-based association and socialization.

There are important differences across social classes not only in the types of organizations to which individuals belong but also in the quantity of organizations in which they have membership. The farther up a person is on the hierarchy of social class, the more likely it is that he or she belongs to many different formal organizations and social circles. Compared with those lower down, those near the top receive many more invitations to serve and meet and organize, and the benefits they gain from these associations may be greater. Near the bottom of the class hierarchy, the demands of basic living are greater and the opportunities for participation in organizations fewer. Low-income people are at least as likely as upper-income people to describe themselves as very religious, but they are less likely to belong to a church. They are just as likely as upper-income people to be concerned about their communities but are less likely to belong to civic groups. They may describe friends as being very important but may know only their own family members and a few neighbors and coworkers. This pattern of limited interaction among people with low income makes sense given their limited access to transportation, resources, and time, but it also means that, compared with higher-income groups, they are likely to have fewer avenues of new information and opportunity, fewer ways to build "cultural capital," and less voice in community decision making.

Religious Group	Median Annual Household Income	Percent College Graduates
Jewish	$36,000	46.7
Unitarian	$34,800	49.5
Agnostic	$33,300	36.3
Episcopalian	$33,000	39.2
Eastern Orthodox	$31,500	31.6
Congregationalist	$30,400	33.7
Presbyterian	$29,000	33.8
Buddhist	$28,500	33.4
Hindu	$27,800	47.0
Catholic*	$27,700	20.0
Lutheran	$25,900	18.0
Mormon	$25,700	19.2
Methodist	$25,100	21.1
Muslim	$24,700	30.4
Jehovah's Witnesses	$20,900	4.7
Baptist	$20,600	10.4
Pentecostal	$19,400	6.9
Holiness	$13,700	5.0

Exhibit 7.3   Socioeconomic Profiles of American Religious Groups: 1993

*Varies greatly by ethnicity.

Source: Lachman and Kosmin, 1993. © 1993 by Seymour P. Lachman and Barry A. Kosmin. Used by permission of Harmony Books, a division of Random House. Inc.

Lifestyles of the Rich and the Destitute

For those who follow the ideas of Karl Marx, social class is defined by the social relations of production. In many ways, social status or prestige is defined by the social relations of consumption. Certain addresses, labels, models, and styles confer prestige. The businessperson selecting a suit or a car, the teenager selecting the right label from the right store, and the political candidate attempting to project the right look are all intentionally seeking to cultivate particular images. This works to the extent that others, whether or not they are cultivating their own images, generally recognize certain markers as clues to a person's social standing.

Residence

The single most powerful expression of social status may be one's address. For most of us, our residence is our single largest investment, and it contains and conveys the most about us. Residence not only confers prestige (or humiliation) but it is also a big factor in association: who we meet, greet, and try to emulate.

Take a drive, literally or mentally, around your community. Where would you find the capitalist class? If yours is a midsize community or larger, it quite likely includes a boulevard of huge historic homes in classical lines and gothic stone, or in English, French, or Spanish manorial styles. These homes are set back behind stone walls, wrought iron fences, and vast lawns. The oldest may sit near the road, but those of the automobile age are likely set into hills and woods, often overlooking the area's most impressive body of water. If you live in a large city, you may be familiar with several of these boulevards and clusters of homes reflecting different times and styles. Beacon Hill in Boston is an example, as are Beverly Hills and parts of Wilshire Boulevard in Los Angeles. Such residential areas are not limited to the coasts, however; they can be found in midsize midwestern communities as well: Summit Avenue in St. Paul, Minnesota, or Meridian Street in Indianapolis. In a small community, you might find only one or two such homes, and these may have been converted: the castle of the lumber baron now serving as city hall, or the home of the first industrialist now serving as a bed and breakfast.

You will likely have to travel further out to find the homes of the newer money of the upper-middle class. In the biggest cities you will find a very wealthy and cosmopolitan few overlooking Central Park and the East River from apartments in New York and looking down on Lake Michigan from Chicago's Gold Coast. But most you will find in groomed and sometimes gated suburban subdivisions with names that evoke the woods and hills of elite English manors: Kensington Heights, Stratford Woods, Nottingham Manor. The streets are curves and culs-de-sac, and any lakes or streams may be human-made. Often you can find new construction. Two decades of growth in this class has led to a building boom of $300,000 to $800,000 homes (depending on location). The styles often imitate or at least evoke the boulevards—Tudor, brick French provincial or Spanish arches—but the lines are adjusted to accommodate two- to four-car garages. If you live in a small community, there may be no subdivision but only a single "trophy house" in similar style cut into the orchards or corn or wheat fields.

The middle class fills neighboring suburbs and a few convenient and "desirable" urban neighborhoods. Large apartment complexes house those who are looking to move or still trying to accumulate the down payment for a house. The gradation into working-class status can be seen in older bungalows and compact two-bedroom homes on small lots with tidy grass but little landscaping. Rural working-class housing includes older farm homes and the occasional double-wide trailer.

The working poor in your community probably do not own property. Even if home is a trailer in a trailer park, it is likely to be rented. The rental properties of the working poor cluster throughout transitional parts of the city, or next to the railroad tracks and warehouses in a smaller community.

Many working poor still live near the sites of old industrial complexes, but increasingly, many live outside the city in clusters of prefabricated and aging housing in a pattern that might better be termed "peri-urban" than suburban.

Finally, the underclass occupies rental housing in the least desirable neighborhoods, often with substantial security problems. Homes in these neighborhoods include apartments in old houses that have been subdivided, rent-subsidized apartments, and apartments in government housing projects. High-rise "projects" such as St. Louis's Pruitt-Igo and Chicago's Cabrini Green and Robert Taylor Homes were first built to accommodate an influx of rural-to-urban migrants and replace the unsafe "kitchenettes" that had sprung up to fill the demand. They are now being abandoned, in some cases physically leveled, in favor of more dispersed housing that many hope will not be as conducive to crime and drug activity. In your community even dispersed public housing, especially older apartments and duplexes, may be quite evident to you in its use of drab colors: dark brown brick and factory green paint. The least fortunate of the poor will be found in homeless centers, in motels that rent rooms by the week, or moving between the homes of friends and relatives. Only a handful of the homeless will actually be on the streets, although these few are likely to catch your attention in major cities.

Fashion

Nowhere do lifestyle and image converge as clearly as in the clothing we put on our bodies. We are told to "dress for success" and select the proper "power suit." We are told that "clothes make the man." Meanwhile, young girls learn about womanhood while playing with "fashion dolls" such as Barbie and their fashion accessories. Later, they are confronted by a billion-dollar industry marketed by dozens of popular "fashion magazines." Fashions themselves change, but fashion itself is always in fashion.

In an intriguing book titled *The Language of Clothes* ([1981] 2000), Alison Lurie notes that humans have always used clothing to make statements about social position. In ancient Egypt, clothes very much made the man or woman, for wearing clothing was the domain only of those of high status; commoners and slaves worked largely nude. From long Greek tunics to Roman togas to the elegant long robes of renaissance Florence, using a lot of fabric in one's clothing was a sign of power and honor. Certain colors and styles were reserved for emperors, kings, and nobility. One way to keep high-income outsiders—merchants, Jewish moneylenders, foreign traders, and others—out of the inner circle of honor was to deny them the right to wear certain types of clothing, even if they could afford them. Social theorist Georg Simmel ([1904] 1957) analyzed shifts in fashion as attempts by high-status groups to maintain a superior social distance over up-and-coming groups. Fashion for Simmel was a product of class distinction and the attempt to monopolize social honor.

Today, styles are less likely to be dictated by law, but we still robe judges, strip prisoners, and indicate military rank with differing uniforms and insignia. People can also show their social positions in other ways. In his classic book *The Theory of the Leisure Class* ([1899] 1953), Thorstein Veblen noted the ways people use consumption as a means to establish social position. He called this **conspicuous consumption** because it is done less for its own purpose than to be seen. Lurie finds many ways to apply Veblen's ideas to clothing. For example, "conspicuous waste" was once seen in puffy sleeves and long trains at a time when fine cloth was very expensive. "Conspicuous multiplication" is seen in the desire to wear a different outfit each day; this reaches extremes in entertainers who never wear the same clothing twice. Clothing can also speak of "conspicuous leisure" in the sense Veblen noted. The three-piece white suit of the southern gentleman was noteworthy in that a man would find it difficult to do any manual labor while wearing it: Colonel Saunders would have a hard time frying chicken, let alone splitting wood, in such a suit. Likewise, women wearing hoopskirts couldn't get near an open fire or a messy kitchen counter. Wearing these clothes was evidence that the wearer supervised others who did all the manual labor. Even in the modern workplace, a senior executive may be able to show up in his golf shirt, noting conspicuous leisure, but middle managers or even clerical workers or sales staff would never dress that way.

The prestige signature of fabrics and colors reserved for elites has given way to literal signatures worn on the outside of designer and name-brand clothing. These have evolved from small signatures stitched into the pockets of designer pants to shirts that display the bold emblems of their high-status manufacturers as their only decoration. High-status labels are especially important to those whose prestige communities, and the status hierarchy within those communities, are in constant flux, such as young people, particularly teenagers and preteens. Labels are also particularly useful for distinguishing high-status leisure clothes from those that carry less prestige. A pearl-gray three-piece suit may make its statement with color, tailoring, and fine fabric, as in the past, but how does one differentiate between high-status and lower-status blue jeans, T-shirts, khakis, and polo shirts? The key is the signature, or the logo.

Clothing also makes lifestyle statements, from the gold jewelry of movie stars and gang members to the "subdued and refined elegance" of that pearl-gray suit and its companion evening gown. The latter communicate not only high income but also highly cultivated tastes, an assertion of cultural capital.

Transportation and Leisure

Prestige and lifestyle also dominate where we go and how we get there. In modern American society we often get our first impressions of others not from their clothes but from their cars. An advertisement in the 1950s contended,

"We now judge a man by the car he keeps." In many ways, automobiles simply enlarge the prestige claims that were once made solely by clothing. These can be rich and brazen (the yellow Cadillac) or wealthy yet subdued (the gray Mercedes-Benz). Cars can speak about lifestyle, values, and attitudes, from sports cars to luxury vans to small gas misers. Their labeling is also in signatures, in the form of hood and trunk ornament logos. Vehicles also demonstrate the attempt to bring prestige to leisure. The pickup truck was once a standard of working-class practicality. The very popular sport utility vehicle (SUV) starts with a truck chassis but adds a wide range of luxury amenities and prestige labeling to accompany the high sticker price. The prestige label is not only in the manufacturer (e.g., Land Rover) but also in the supposed designer (e.g., Eddie Bauer Edition). The SUV with a roof rack and trailer hitch has become a symbol of a lifestyle of conspicuous leisure.

Leisure itself often follows a prestige hierarchy. Few but members of the old upper class are captivated by polo; Prince Charles was very pleased that Prince William took up that activity. Golf has captivated many, from the capitalist class to the middle class, yet it retains its upper-middle-class image of physicians, attorneys, and businesspeople escaping the office for the links. In the United States, major professional sports dominate the leisure interests of the middle and working classes, and sometimes the aspirations of poor children. These differences reflect differences in disposable income but also differences in taste and association. Polo is expensive to play, but one can often watch a local polo match for much less than the ticket price for a seat at a professional basketball game. Yet it may be hard for a working-class polo enthusiast to return to work on Monday and convey this enthusiasm to buddies: "Ah, you should have seen the final shot in the first chucker!"

Travel destinations also reflect disposable income and lifestyle expectations as well as values and tastes. The capitalist class flies and seeks private retreats. The upper-middle class may also fly to resorts and go on ocean cruises. Car travel with stays in motels and camping becomes more popular in the middle. Travel diminishes drastically for classes below the middle: For the working poor, vacation time may be almost nonexistent, and those in the underclass may have neither the transportation nor the housing security to be away from home for long. The forced simplicity of camping has limited appeal to those who are primarily concerned about the condition of their permanent housing.

Tastes in Transition

Meet the Bobos

They would never drive a Cadillac, but they might spend just as much on a luxury SUV. They don't eat T-bone steaks, but they think nothing of spending $5 for an organic wheat-berry muffin. Never mix them a martini, but they will gladly take a double-tall nonfat latte. They would absolutely never wear mink, but neither would they go out into a drizzle in New York City without a triple-layer, third-generation Gore-Tex shell parka rated to withstand the gales that assault the slopes of Mt. Everest. They would never show up for work in a Brooks Brothers suit; rather, they often look like refugees from the Norwegian Olympic team. Who are these new trendsetters? David Brooks (2000) calls them "Bobos"—short for *bourgeois bohemians*. He sees this group as combining the countercultural tastes of the 1960s with the entrepreneurial spirit and "I've earned the best" ethic of the 1980s and 1990s. Brooks argues that the status codes of the Bobos now dominate American social life. The Bobos have created new upper-class lifestyles to replace the void left by the downsizing of those 1950s corporate executives and "organization men." The new group is more diverse and enamored of the high-tech options of the information age.

Are Brooks's Bobos really so different from past elites? They are still clearly concerned about status. It is unclear what practical purpose a shiny black Lincoln Navigator SUV serves except as a fine way to get from Aspen to Vail in a snowstorm, yet these behemoths and their cousins fill the beltways that link new high-priced suburbs to new edge-city office buildings. Earlier elites also sometimes found it fashionable to imitate more rustic lifestyles. Nineteenth-century Englishmen who could easily afford lives of silk clothes and satin carriages became enamored of tweed riding clothes, horses and hounds, and all the trappings of rigorous rural life, all the while maintaining townhouses in London. The townhouses now are Greenwich Village or Boston's Back Bay on one coast or on the waterfront of Seattle or in the former groves of Silicon Valley on the other, but the blending of lifestyles to establish a distinctive aura is still present. The consumption is still conspicuous even as it stresses the informed tastes of "cultural omnivores" and a measure of social responsibility.

What effects these new patterns of consumption may have remains an open question. Bobos may prefer to spend their vacations in adventure travel and ecotourism rather than relaxing at high-rise beachfront hotels. They may forgo trips to Neiman Marcus in favor of logging on to the online marketing site Novica to decorate their homes with items bought "direct" from indigenous craftspersons in distant locations. These trends could provide new opportunities as the new wealthy use their consumer power to promote cultural and natural preservation, or they could just extend the reach of high-power consumerism to new areas. Maybe the Bobos might even be persuaded in time to trade in their gas-guzzling SUVs for high-priced but trendy high-mileage hybrid vehicles, and the planet will breathe a bit easier.

The Millionaires Next Door

The Bobo phenomenon is just one example of the shifts in patterns, and quantity, of consumption that have taken place as a result of the rapid expansion of the upper-middle class and new capitalist class in the 1980s and 1990s (Stanley 1996). The newly wealthy have created a boom in new construction. Families that once had cottages on the San Juan Islands and in the Cascades of Washington State have found that they can no longer afford the taxes as property values have skyrocketed with the coming of the "Microsoft millionaires." Even in the Midwest, families that once had modest lakeside cottages with tire swings along the Great Lakes find that these are too expensive to maintain amid soaring property values. The cottages are sold as "teardowns" and replaced by $500,000 homes for early retirees and telecommuters and as large second homes for well-off residents of Detroit and Chicago (and, in at least one case, Switzerland). Meanwhile, urban-based stockbrokers, attorneys, and other professionals buy up neighboring orchards and vineyards in a manner reminiscent of London-based elites and their country estates. New estates and old combine residence, leisure, and business in settings that are secluded yet made conspicuous by large gates and signs with custom logos as well as now by designer labels on wine bottles and customized home pages on Web sites.

In the 1880s, 1890s, and into the twentieth century, all across the United States the old money of the pre-1860 era gave way to the new money of industrial barons and their modern castles. In the 1980s, 1990s, and into the twenty-first century, the "old" money of the heirs of these barons has been giving way to the new money of postindustrial information-age barons and their postmodern castles. Says one retail consultant, "Communities have changed, Long Island's old guard has sold many of those old homes. The land has been subdivided. Instead of one $10 million home, you have a lot of $3 million homes owned by people in their 30's and 40's" (quoted in Trebay 2000). These new residents don't need frequent trips to New York's Madison Avenue, for nearby is Long Island's Americana shopping center (they never say *mall* here), devoted to Tiffany, Armani, Gucci, Chanel, and other trendy luxury brands. No Bobos here; it's elegant extravagance in place of "refined rusticity." "You can't believe the money people are throwing around these days. It's liquid millions," claims one retailer (quoted in Trebay 2000). Once again, a new surge of money has brought a new wave of conspicuous consumption.

The Overspent American

Juliet Schor (1998) argues that conspicuous consumption has shifted its center from the upper classes to the middle and upper-middle classes. Americans like display but have often been suspicious of ostentation, excessive displays of wealth. Only in times of surging economic optimism—briefly in the 1880s, again in the 1920s, and once again in the 1980s—have the new rich found it acceptable to flaunt their wealth. Such displays are acceptable in times of optimism perhaps in that they carry the implication "Next year you could be where I am." In other times, the truly well-off may prefer the subtler approach of the Bobos or the millionaires next door. The rich may prefer a quieter life, or, even if they wish to show off, taboos against being "uppity" require them to cultivate the subtle art of showing off without looking as though they are.

Why have other groups become so consumption conscious? Schor believes this shift is rooted in **reference groups,** the people to whom we compare ourselves. If your reference group is your neighbors, there may be some "keeping up with the Joneses," but quite likely the Joneses are not much richer or poorer than you are. Increasingly, however, we don't know our neighbors, and so we draw our inferences about what makes a good lifestyle from the popular media—television, magazines, and maybe the Internet— where the reference groups are often high income. Unless a television show is specifically about low-income people in hard times (such as *The Waltons*), the setting is most often upper-middle-class or above. Soap operas often feature wealthy families, but even when this is not a stated theme, they typically take place in beautifully furnished settings. This is true in Latin America as well as in the United States. The furnishings, accessories, and lifestyles most of us are most familiar with are not those of our peers, or even of real people, but those of actors and models who have been dressed and placed to project particular images. Schor argues that trying to match these images of the "good life" has driven many Americans deeply into credit card debt and a lifetime of financial problems. Even though we have more than ever before, we feel we still don't measure up to an imagined other. Overconsumption and consumer debt grew rapidly in the United States in the 1980s, and this seems to be becoming a worldwide phenomenon. The reference group for much of the developing world in regard to consumption is the U.S. and European middle class—or, more often, the actors and models who portray an idealized version of this class.

Communities of Recognition

It is easy to disparage prestige seeking as foolish snobbery, yet most of us are concerned with the regard of the communities we live in. The popular television program *Cheers* touted the appeal of a place "where everybody knows your name." I have neither high-prestige clothes nor cars, yet it is pleasant to be out in my small community and be greeted frequently by students, former students, colleagues, and community people with a "Hey, Dr. Sernau, how's it going?" Certainly this is better than life in community of strangers or being greeted with hostility, suspicion, or a heavy hand on the shoulder: "Sernau, you weasel, where's the 20 you owe me?" Many people seek marks of distinction and most seek at least evidence of respect. These measures of **social exchange,** rather than just market exchange, can matter a great deal in determining the quality of someone's current life experiences and future life chances.

KEY POINTS

- Social prestige, what Max Weber called *status,* includes the subjective aspects of stratification, including how much honor, respect, and deference is given to someone.
- By means of socialization, children tend to adopt the values and outlooks of their parents and their immediate communities, and these perspectives in turn often prepare them to continue in their class positions.
- People of similar class backgrounds tend to associate with one another by means of residence, schooling, and involvement in religious and civic organizations. The higher the class standing of a person, the denser his or her network of associations is likely to be and the more prestigious the organizations in which he or she is involved.
- Social status is communicated through habits and tastes as well as through patterns of consumption, including choices of residence, fashion, and leisure.
- New economic growth has created new elites who in turn have created new, distinctive lifestyles and new patterns of conspicuous consumption.

FOR REVIEW AND DISCUSSION

1. How have people attempted to gain, establish, and defend their prestige and social positions over time?
2. How are children from differing social classes socialized differently? How does this shape their *habitus* and their future life chances?
3. How do people show their social class and status through consumption? Think of examples from fashion, housing, leisure, and related areas of consumption. How is this affected by a person's past socialization and present associations?

MAKING CONNECTIONS

On Campus and in Your Community

Pick one area of consumption and examine it as Lurie ([1981] 2000) does with fashion. Areas of consumption include fashion, leisure, sports, music, housing, automobiles, and restaurants; do not choose an area of production (such as occupation). Talk with people of differing ages and backgrounds about what they consider to be status symbols or marks of distinction in the area of consumption you have chosen. Do different types of status communities emerge? How do people demonstrate their class positions in this area, and how is taste correlated with class? Do you find examples of Veblen's concept of conspicuous consumption?

In the Marketplace

Visit an "upscale" or exclusive shopping center or department store (such as Macy's) and compare it with a discount department store. Observe what is constant and what is different between the two (aside from the prices). Are there differences in colors, fabrics, styles, places of manufacture, or care instructions on the clothing for sale? How are the styles of advertising and customer service different in the two kinds of stores? Are there ways in which the more expensive store communicates its "superiority"? Are there ways in which the less expensive store tries to imitate upscale stores or to establish its own identity? Venture beyond clothing into sporting goods, furniture, lawn furniture, and home accessories. What aspects of different lifestyles and differing expectations are captured in these products? Do you find specialty stores or departments that appeal to particular status communities, such as trendsetters, old money, or "Bobos"?

8

Power and Politics

We can either have democracy in this country or we can have great wealth concentrated in the hands of a few, but we can't have both.

—U.S. Supreme Court Justice Louis Brandeis (1856–1941)

The man of great wealth owes a peculiar obligation to the state because he derives special advantages from the mere existence of government.

—Theodore Roosevelt (1858–1919)

The poor have sometimes objected to being governed badly. The rich have always objected to being governed at all.

—G. K. Chesterton, English essayist (1874–1936)

In one sense, all inequality is about power. This power is, as Max Weber noted, the ability to do what you want no matter who wants to stop you. According to Weber, "class" is power in the economic realm, "status" is power in the social realm, and what he called "party" (*Stand*) is power in the political realm. Often the exertion of power is subtle and hidden: the power to purchase, to impress, to command attention. When power enters the political arena, however, the conflicts around it become more open and vivid, the opposition more clear and direct, and the exertion of power more obvious. This chapter is about the balance of power: between workers and owners, between the elites and the masses, between men and women. It is also about the exertion of power and the perception of power. The outcomes of current power struggles are based on today's social inequalities and will in large measure determine the shape of tomorrow's inequalities.

People Power and Powerful People

In a democracy, the electoral process is intended to be the arena in which political power struggles take place. Yet around the world, people who live in democratic societies complain about behind-the-scenes power brokering, the power of special interests and privileged cliques, and the influence of money, both legal and illegal, in the political process. Even as they enjoy greater freedom of information and expression than they knew in the days of the Soviet Union, Russians and Eastern Europeans are disillusioned by elected officials' inability to bring meaningful social and economic benefits to workers. Western Europeans are finding that their governments are not immune to the charges of corruption and graft that plague developing countries. The Germans seemed genuinely shocked when their longtime ruling party was found to have been seriously involved in graft; the French were less so under similar circumstance, for they have long been sure that their country is glorious but the government is full of scoundrels. Likewise, Americans proudly speak of freedom and democracy in general but a large proportion never actually vote, often giving as the reason their pessimism that the electoral process will bring significant reforms or benefits. Who truly holds political power, and why do the masses so often seem unable to use elected government to better their condition?

Class Consciousness

According to Karl Marx, progress in the popular struggle requires that the workers first develop class consciousness; that is, they need to see their common plight and develop a sense of solidarity. They then need effective means of mobilization, to be able to act on their own behalf. Although Marx was suspicious of the ability of elected governments to effect real change—he believed governments are created to do the bidding of capitalist interests—in *The Communist Manifesto* he did call on the workers to support progressive reforms wherever they were to be found. In Europe, workers created parties with platforms that were intended to represent their interests. In France, the workers' party was the Socialist Party—never very socialist but always pro-union, with strong union backing. In Great Britain, it was the Labour Party, which rose to dominance in the 1960s. In many other European countries, social democratic parties played the role of workers' advocate. By the mid-1890s, a decade after Marx's death, his coauthor Friedrich Engels had abandoned the call to revolution and instead looked for the social democratic political movement to bring workers' rights steadily and incrementally (Engels 1895, in McLellan 1988).

Workers' parties were also begun in the United States. In 1904, 1908, and 1912, labor leader Eugene Debs ran for president on a socialist ticket with strong union support. Yet a true workers' party never took root in the

United States. Workers in the middle of the 1800s looked to the newly formed Republican Party for reform. Abraham Lincoln had strong pro-union as well as pro-Union sentiments. He decried the working conditions he found on a trip to New Orleans and argued that one reason to abolish slavery was to protect the rights of free labor, which otherwise saw wages depressed by competition from slave labor. At the beginning of the twentieth century, Theodore Roosevelt brought a populist message to the Republican Party (and later to his own "Bull Moose" Party), promising workers a "square deal." Workers in the 1920s, however, found little sympathy from the pro-business governments of Harding, Coolidge, and Hoover. When Franklin Roosevelt, a wealthy distant cousin of Theodore, rebuilt a new coalition for the Democratic Party, many workers were attracted by his promise of a "new deal." The strong ties forged between labor and the Democratic Party in the 1930s have in some measure continued to the present. Since the AFL-CIO began making political endorsements, that organization has always supported Democratic presidential candidates. Yet the Teamsters found it more advantageous to forge ties with Richard Nixon, a Republican, when he ran for president in 1968 and 1972. Although union endorsements went to Jimmy Carter in 1980 and to Walter Mondale in 1984, in those elections many in the working class, especially white men, became "Reagan Democrats" and supported the Republican ticket. As Mondale accumulated endorsements and lost votes in 1984, many began to wonder whether such endorsements actually help or hinder candidates. Although Bill Clinton won a few of the Reagan Democrats back in 1992, American voters continue to be reluctant to vote along class lines.

This reluctance is part of what has become known as "American exceptionalism" (Lipset 1996). Americans have long cherished the ideal of a common bond among all citizens that transcends class lines, and they tend to carry this ideal into the voting booth. They believe that government is a vehicle for promoting the common good rather for promoting class interests, and so they seek "reformers" from both the left and the right. A fact not lost on those who develop campaign platforms is that income, even more than age, is a key factor in the likelihood that a given American will actually register and vote. Although the largest numbers of eligible voters are in the lower economic brackets, voters in the upper brackets are much more likely to get to the polls and vote; this gives them disproportionate weight in elections (see Exhibit 8.1).

Black Power?

Americans have also long been divided by factors other than social class. Whereas Europe has clearly been divided by class, the United States has been divided more clearly by color, and race and ethnicity continue to be important factors in American society. Black voters first turned to the party of Lincoln,

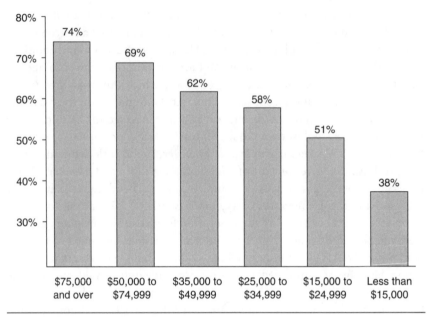

Exhibit 8.1   Voter Turnout by Income

Source: Based on data from U.S. Census Bureau, 2002.

"the great emancipator." Southern Blacks gained voting rights after the American Civil War with the passage of the Fifteenth Amendment, which guarantees each citizen the right to vote regardless of "race, color, or previous condition of servitude." Their right to vote was enforced by federal troops stationed throughout the former Confederacy. Under the watch of "radical Republicans," many African Americans began to vote and to hold public office. As former Confederate states reentered the Union, their delegations to the U.S. Congress contained that body's first black members. Given that there were no black congressional representatives in the North, this situation may have been engineered out of spite and a desire to humiliate the defeated South, but these newly elected African Americans served with diligence. Southern whites, however, complained of "bayonet rule" and northern "carpetbaggers" trying to run the South for profit as well as revenge.

As President Ulysses S. Grant, a former Union general, came to the end of his term amid scandals and controversy, the presidential campaign pitted Republican Rutherford B. Hayes against Democrat Samuel Tilden. To this day, the rightful winner of that election is disputed: The popular vote was very close, probably favoring Tilden (charges of fraud make even this uncertain), and the electoral vote depended on which southern delegations were considered "legitimate." The country was on the verge of a second civil war.

The controversy ended with an implicit compromise: Hays would be president, but he would remove federal troops and much federal authority from the South. Almost overnight, the political situation in the southern states changed. Southern blacks were kept from voting by poll taxes, contorted literacy tests, and outright intimidation, including by the growing Ku Klux Klan. Black state and national representatives lost their offices, and African Americans were largely kept to the margins of the political process for the next 30 years. During the economic growth of the 1920s, the Republican Party became associated with a pro-business agenda that offered little to black voters.

Then, in 1932, African Americans were brought into Franklin Roosevelt's New Deal coalition as Roosevelt carefully worked to cultivate the black vote without alienating the traditional Democratic base. The Great Depression, which brought hardship to so many across traditional divides, provided the opportunity to build a multiracial and multiethnic coalition of working-class voters. Urban blacks in northern cities, many of whom would not even have been allowed into the Democratic Party in their southern home states, now became loyal Roosevelt Democrats. Following World War II, Truman integrated the armed forces and gave initial support to the emerging civil rights movement. The Republican Eisenhower administration continued cautious support for this movement, but it was Lyndon Johnson, a southern Democrat, who most vigorously fought for civil rights legislation, culminating in the Civil Rights Act of 1964 and the Voting Rights Act of 1965. White backlash against these measures helped fuel the presidential candidacy of George Wallace in 1968. In 1963, as the Democratic governor of Alabama, Wallace had placed himself in the doorway of an auditorium at the University of Alabama to block the university's integration, bowing finally only to federal troops sent by President Kennedy—National Guardsmen whose rifles and bayonets recalled the federal intervention of the 1860s. As the candidate of the American Independent Party, Wallace based his presidential campaign on "law and order" and "traditional values," phrases that seemed to be code for the conditions of the old racially divided order. He won almost 14% of the vote. The winner in 1968 was Republican Richard Nixon, who realized the power of Wallace's support and devised what became known as the "southern strategy." This strategy included appealing to southern voters with Wallace-like phrases about law and order that also appeared to carry a racial message. Southern states that had been Democratic since the Civil War moved toward the Republican Party, but black voters, alarmed at this new appeal of Republicans to southern conservatives, became ever more solidly Democratic. In 1980, Republican Ronald Reagan won the presidency with a sweep of southern states and won over many working-class whites, the so-called Reagan Democrats, but most black voters remained supportive of Jimmy Carter, who, like Johnson, was a white southern Democrat who had made racial bridge building a hallmark of his campaigns.

Reagan's vice president and successor, George H. W. Bush, tried to broaden the ranks of Republican support, but his most successful, and most controversial, campaign advertisement echoed the "southern strategy" of the 1960s. The ad focused on the release from a Massachusetts prison of Willie Horton, a violent criminal who then went on to commit other crimes. The ad blamed the release of such a danger to society on the "liberal" policies of Bush's opponent, Massachusetts governor Michael Dukakis. In fact, the program under which Horton had been released was instituted by Dukakis's Republican predecessor. Despite this, and despite the fact that the ad made no mention of race, the ad clearly presented white victims targeted by a black convict, released under a governor who was "soft on crime," and promised "law and order" under a Bush presidency. Although the ad did seem to win support for Bush from whites fearful of violent crime, it also further alienated black voters. In 1992, they turned in large numbers to another white governor with a strong record of racial bridge building, Bill Clinton of Arkansas. As president, Clinton not only appointed more black judges, cabinet members, and heads of federal agencies than any previous president, he also seemed to have an inner circle of black advisers, people like Vernon Jordan, which would have been unheard of in the past. Many felt that African Americans were not just being given token positions, but had finally been given access to the power of the White House. When white support for Clinton started to fade following scandals, many black leaders, including religious leaders, were among the first to call for forgiveness for the president and movement away from personal scandal back toward his agenda for racial reconciliation and broad-based economic development.

The presidential election campaign of 2000 was noteworthy in that both Democrats and Republicans actively courted black voters. In 1996, Republican nominee Bob Dole had called on all bigots in the party to find the clearly marked exits and leave. Yet fiery rhetoric from Pat Buchanan and Newt Gingrich made many wonder whether the "southern strategy" of appealing to white voters with thinly veiled references to black criminals and welfare queens was really dead. In 2000, Gingrich and Buchanan were gone, replaced by General Colin Powell, the very popular black head of the Joint Chiefs of Staff during the Gulf War, along with black musicians, rapping multiracial students, and many references to racial inclusion from nominee George W. Bush, who seemed eager to extend his father's political base with no hint of the legacy of Willie Horton. Nonetheless, the racial divide remained strong: In 2000 only 3% of Republican delegates to the Republican National Convention were black, compared with 20% of delegates to the Democratic National Convention.

The 2000 presidential election also revived concerns among some African Americans that their political participation was being discouraged and even actively thwarted. In the state of Florida, where the presidential vote was extraordinarily close and contested, a vastly disproportionate number of voters who were turned away from the polls because of irregularities and faulty registration lists were African American. Other barriers remain as well. Almost every U.S. state denies convicted felons the right to vote, in most cases for the rest of their lives. In many communities young black men have very high arrest and incarceration rates, mostly for drug-related crimes. Not only can these men then not vote while incarcerated, but those convicted of felonies are barred from voting for life. During the 2004 election campaign, voter registration drives aimed at registering new black voters in "battleground" states such as Pennsylvania and Florida encountered many black men who were ineligible to vote at all.

Brown Power?

Compared with their appeal to black voters, Republicans have had more success in reaching across other ethnic divides as well as religious divides. As the Republican Party of the late 1800s and early 1900s became increasingly anti-immigrant and staunchly Anglo-Protestant, most of the new immigrants—Irish, Polish, and Italian Catholics, Russian and German Jews, and Greek Orthodox—found their political home within the Democratic Party. By the 1980s some were defecting, however, given rising income levels and differences over social-moral issues such as abortion, but the major urban areas of the Northeast and Midwest that attracted these immigrants tend to remain Democratic strongholds. At the same time, the Republicans began to attract conservative Protestants in the South. These former "Dixiecrats" had been Democrats since the time of the American Civil War and wanted no part of the party of Lincoln even though they increasingly shared common conservative political sentiments. By the 1980s, more concerned about social-moral issues and controversies such as affirmative action than about old affiliations, many white southerners gathered under the Republican banner. Religious divides continue to be strong in American politics and can easily confound class logic: The most conservative religious groups may draw support from working-class membership even though these groups support social conservatism rather than an agenda of liberal economic reform.

The ethnic groups that have gained the most new attention from political parties in recent years are those grouped together as Hispanics. One hears little anymore of "black power," but political analysts have started to talk about "brown power," noting that Hispanics are the fastest-growing "minority" in the United States. Latinos have so far not been a major force in U.S. politics. Although he speaks Spanish, Dukakis got only a small boost in his campaign from his attempt to court Latino communities. Mexican Americans, by far the largest Latino group, have long been part of the Democratic coalition. Heavily concentrated in the working class and working poor, they were also brought into Roosevelt's New Deal and have largely remained with the Democratic Party, but they have not had the numbers to carry the large states of Texas and California. Cuban Americans concentrated in the Miami metropolitan area have been a vocal presence in U.S. politics since the 1960s but have also been politically divided, some joining other Latinos as Democrats and some, including many small business owners and ardent anticommunist, anti-Castro activists, leaning toward the Republicans. By 2000, the numbers of Latino voters in the United States were large enough for both presidential candidates to take notice. Al Gore became very concerned about winning Cuban American support in Florida and New Jersey. George W. Bush, who has Latino in-laws, made great efforts to win over the full range of Hispanic Americans. He was more successful in winning Latino voters in 2004. Overall, his gains with this group were small, but the support of a few more Latino voters who agreed with his religious and military initiatives was crucial to his winning some closely contended states. Today, most Hispanics seem pleased that decades of political invisibility have given way to recognition and hope that this can be turned into meaningful political power.

The Gender Gap

American voters have also developed a large gender divide, seen most strongly in recent elections. In 1992, George H. W. Bush secured a narrow majority of the male vote but lost large shares of the female vote to Bill Clinton. In 1996, Bob Dole appealed to at least half of the male voters but lost the election decisively due in part to his poor showing among female voters. The so-called **gender gap** in voting has become as big an issue as class and racial divides. Women voters are more likely than men to support government programs intended to benefit education and families, as well as to support gender-based affirmative action, and so are more likely to be drawn to Democratic candidates. Working-class men, especially those who have lost jobs or seen their wages decline, are more likely to feel betrayed by the liberal progressive agenda and so are more likely to support Republican or third-party candidates.

The Clinton administration appealed directly to female voters as it did to black voters, with many more appointments of women to key posts than ever before. Clinton promised a government that "looked more like America" and indeed appointed the first women to hold the positions of secretary of state, attorney general, and secretary of health and human services. Further, Hillary Rodham Clinton greatly expanded the role of the first lady (an increasingly archaic-sounding term) beyond that of the fashionable Nancy Reagan or the "grandmotherly" Barbara Bush to one of political influence, heading the administration's health care reform campaign as well as efforts on behalf of children. She also promptly became a highly controversial figure, perhaps in large measure because of many Americans' general suspicion of influential women (Nancy Reagan's influence over her husband was also the subject of suspicion in the 1980s, even though Mrs. Reagan was not a very political woman). In a still well-remembered remark made during her days as first lady, Hillary Clinton stated that she didn't want to "stay home and bake cookies." After Bill Clinton left office, she went on to become a U.S. senator

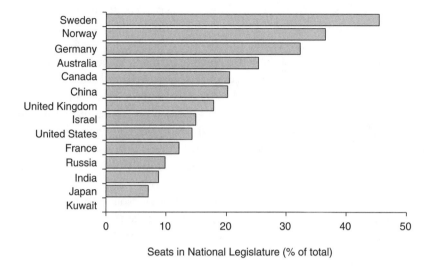

Exhibit 8.2 Women in Government

Source: United Nations, Human Development Report (2004).

from New York, a position in which she still serves. She remains intensely controversial in spite of having pursued fairly centrist policies and bipartisan programs in the Senate: She still appears on lists of the "most admired" women in America as well as on lists of the "most disliked." In the 2004 election campaign, media attention focused on Democratic candidate John Kerry's wife, the wealthy and outspoken Teresa Heinz Kerry. Before Kerry's run for president, she was best known for highly regarded philanthropy as head of the Heinz Foundation, but in the context of the campaign her wealth, influence, and background (born in Mozambique, she speaks five languages) struck some as suspect. Many Americans found her to be a less reassuring presence than the smiling and pleasant Laura Bush.

Women have been slowly gaining prominent positions in U.S. government, but the United States lags behind many other countries in this area, with women vastly underrepresented in legislative positions (see Exhibit 8.2). Although the United States has had two female secretaries of state in recent years (Madeline Albright under Bill Clinton and Condoleezza Rice under George W. Bush) and two women are currently justices on the U.S. Supreme Court (Sandra Day O'Connor since 1981 and Ruth Bader Ginsburg since 1993), women in high offices are still the exception. Most of the women in government in the United States, like women in business, remain confined to secondary positions in which they report to male supervisors or chief executives.

Religion, Region, and Values

Since the early days of immigration, religion in the United States has been correlated with ethnicity and class. New England and midwestern Protestants once formed the core of the Republican Party, reflecting the interests of a northern business class. Southern Protestants tended to be strong Democrats. Roman Catholic immigrants, largely northern and working-class, eventually found a home in the Democratic Party, sometimes recruited first at the local level by Democratic Irish and Italian American mayors. In the 1920s, when some parts of the Republican Party became associated with anti-immigrant sentiments and the preservation of "Protestant America," Jewish Americans and Roman Catholics, particularly those of recent immigrant background, became strongly affiliated with the Democratic Party, bolstering Roosevelt's New Deal coalition of the 1930s.

Much of this has changed. In a great regional reversal only recently completed, the Northeast has become the stronghold of the Democratic Party and the South the stronghold of the Republican Party. A strong urban-rural divide has also emerged that is exactly the reverse of the affiliations of the late 1800s. Urban dwellers, including many voters of color, perhaps more tolerant of diverse lifestyles and certainly concerned about urban decay, have tended to support Democratic candidates. Rural voters, on the other hand, once part of the New Deal coalition, have become strong Republican supporters. In the 2004 election, Kerry drew enough votes in Philadelphia and Pittsburgh to offset a big deficit in rural Pennsylvania, but he did not carry enough votes in his strongholds of Cleveland and Columbus to offset losses in rural Ohio.

Religious influences continue to shift as well. In 2004, Protestants tended toward Bush and the Republicans, but this masks a divide: Mainline Protestants tended toward Kerry, whereas conservative evangelical Protestants tended toward Bush. Evangelical African Americans continue to vote Democratic, but by smaller margins than in the past. Among Hispanic voters, Catholics continue to vote more heavily Democratic, but a growing evangelical and Pentecostal minority lean Republican. In the 2004 election, the Catholic vote was split as Catholics tried to balance their Democratic heritage against strong antiabortion statements from some bishops and clergy that made some skeptical of Kerry, a Roman Catholic himself. Only the Jewish vote and that of Americans who place their religious affiliation in the ambiguous "other" category were strongly Democratic in both 2000 and 2004.

Some observers have suggested that a "religion gap" exists, in that frequency of church attendance has become one of the strongest predictors of voting patterns in recent years. This religion gap may reflect the importance given to "moral values" in the 2004 election. It is always hard to know what really motivates voters, but exit polls found that slightly more people selected "moral values" as their most important criterion in choosing a candidate than selected "concern for the economy." Of those who stressed moral values, 80% voted for Bush. Of those who stressed the economy, 80% voted for Kerry. This huge divide was not anticipated, especially given that both candidates had offered plans for the economy and both spoke of faith and values.

In relation to income, voting broke along predictable lines in 2004: Higher-income voters favored Bush and other Republican candidates, and lower-income voters favored Kerry and other Democrats. An "age gap" was apparent, with younger voters, many newly registered, favoring Kerry. The

	Bush	*Kerry*
Vote by Gender		
Men	55	44
Women	48	51
Vote by Race		
White	58	41
African-American	11	88
Hispanic	44	53
Asian	44	56
Other	40	54
Vote by Age		
18–29	45	54
30–44	53	46
45–59	51	48
60 and older	54	46
Vote by Income		
Under $15,000	36	63
$15,000–30,000	42	57
$30,000–50,000	49	50
$50,000–75,000	56	43
$75,000–100,000	55	45
$100,000–150,000	57	42
$150,000–200,000	58	42
$200,000 or more	63	35
Are You a Union Member?		
Yes	38	61
No	54	45
Are You a Working Woman?		
Yes	48	51
No	53	46
Vote by Education		
No high school	49	50
High school graduate	52	47
Some college	54	46
College graduate	52	46
Postgrad study	44	55
Vote by Class (self indentification)		
Upper Class	56	39
Upper-Middle Class	43	54
Middle Class	48	49
Working Class	51	46

Exhibit 8.3 Voting by Various Characteristics

Source: CNN Election 2004.

Note: About 1% of votes went to Ralph Nader or others.

Exhibit 8.3 (Continued)

	Bush	Kerry
Vote by Religion		
Protestant	59	40
Catholic	52	47
Jewish	25	74
Other	23	74
None	31	67
Vote by Church Attendance		
Weekly	61	39
Occasionally	47	53
Never	36	62

"education gap" was subtle, with a clear tilt to Kerry apparent only among those with advanced degrees. Income and employment were less powerful predictors than might be expected, as it appears that many voters made their decisions based on issues of perceived values and personal character rather than voting their "class interests." Who would receive the biggest tax cuts, along with plans for education and Social Security, somehow caught the public imagination less than news about "gay marriage" and a proposed "defense of marriage" constitutional amendment.

The New Suburban Majority

Together, the multiple divides of race, ethnicity, religion, region, and gender can prevent any sense of worker solidarity in the electoral process, leaving candidates to scramble to assemble winning coalitions of factions and interests. The 1990s saw the United States cross an important threshold: A majority of Americans, and particularly a majority of voters, are now suburban; there are more voters in the suburbs than there are in rural areas and central cities combined. This suburban contingent is largely but not exclusively white, predominantly middle-class and above, and particularly concerned about security and continued prosperity. Although few in this group are true members of the capitalist class, most have some accumulated wealth in mutual funds or other assets tied to the stock market. Talk of diversity notwithstanding, they have become the group most targeted by political campaigns of all types.

Discussion of values should not overshadow the continued effect of income in voting, as seen in the 2004 presidential election:

Among those making less than $50,000, Bush actually lost ground, as his
performance fell from 21 percent in 2000 to 20 percent in 2004. Among
those making over $50,000, Bush's performance jumped 3 points, from
28 percent to 31 percent. And most of this improved performance
was concentrated among the wealthiest of voters, those making over
$100,000. . . . In 1980, Ronald Reagan famously asked, "Are you
better off now than you were four years ago?" Among many of these
upper-middle-class and wealthy voters in 2004, the answer is clearly
yes. Bush's tax cuts were concentrated among those at the upper end
of the income scale. Furthermore, since the last election, the share of
the national income of voters who make over $54,000 per year has
increased, from 72.8 to 73.2 percent. Not only did these voters disproportionately benefit from Bush's policies, but Kerry ran on a
promise to raise taxes for many of them. (Klinker 2004)

Fear of taxes among prosperous suburbanites, who vote in large numbers,
may have mattered as much as in the 2004 election as social issue concerns.

At the same time the suburbs have been growing more politically powerful,
urban areas have undergone a major loss of power. From 1850 to 1950,
U.S. mayors, many of whom had strong ethnic identities—Irish, Italian, and
German in particular—wielded considerable power and often used that power
to assist the progress of "their people," including their own ethnic groups.
Since the 1970s, there has been a steady stream of black mayors in major U.S.
cities, along with a few Latino mayors in the Southwest. Yet these mayors
have been able to do little to improve the lot of poor urban blacks and Latinos.
They have little money to spend on ambitious programs and little real power
to command patronage jobs they can give to "their own." In the past few
decades many U.S. cities have experienced bankruptcy or near bankruptcy.
With industry, retailers, and other employers leaving the cities, mayors must
often assume the role of traveling salesman, trekking around the country and
the world trying to attract new businesses and new investors to their cities.

As the above discussion shows, the question of who holds power is more
complicated than who votes, or even who holds office. Who does hold
power? That is, who really rules?

Who Rules? The Power Elite Debate

How concentrated is power in the United States, and how much is the political system open to differing perspectives? Two conflicting answers to these
questions have crystallized in the past several decades. In many ways they
echo the "conflict versus functionalist" debate on inequality.

In 1956, C. Wright Mills contended that the United States as a whole is
dominated by the **power elite,** which consists of the corporate, political, and

military leaders who dominate and control what he termed the "military-industrial complex." Mills's book on this topic, titled *The Power Elite,* was extremely influential, and since that time the idea of a power elite has been central to the arguments of critics of the American political system. The suspicion that political power is dominated by the members of a closely guarded elite has become known as the *elite perspective.* It echoes the conflict perspective's stress on the power of dominant groups. Mills's formulation also owes a lot to the ideas of Max Weber: Elites may have multiple bases of power, but they have a common desire to bring social closure and to monopolize political power.

Mills's perspective has its critics, some of whom come from the political left. Paul Sweezy (1968), for example, doesn't dispute that the military-industrial complex exists, but he does dispute the idea that the military is an important power base. The military is, after all, more or less under civilian control in the United States and most advanced industrial nations. Sweezy contends that the government is actually under corporate control, at least in making major economic decisions. Key corporate leaders can finance powerful lobbies, make huge campaign contributions, exert great influence over government agencies, and even hold key cabinet and administrative posts. Sweezy agrees that there is a power elite, but he asserts that it is rooted in the ruling class. Ultimately, it comes down to the cynic's golden rule: Those who hold the gold make the rules. This is also clearly a conflict perspective, but one more rooted in the ideas of Karl Marx, who contended that the ruling ideas are inevitably the ideas of the ruling class. In advanced industrial societies, this is the capitalist class.

This view is certainly quite different from that presented in high school civics books, which typically argue a **pluralist position**. Pluralists see U.S. society as divided among varying interest groups, all of which are jockeying for power and position. The decisions that are ultimately made depend on who can build winning coalitions of voters and supporters. This perspective fits much better with a functionalist view of society, in which multiple parts are seen as working in concert to create an ultimately effective overall system.

Not all pluralists are so optimistic, however. In his book *The Lonely Crowd,* published in 1953, David Riesman proposed a different view. Certainly elites may exist, he acknowledged, but they do not always have a common agenda, and they may even exist as "veto groups," checking one another's power. Checks on power are part of the American ideal, but Riesman worried that the United States had become so fragmented that no one could effectively lead. This is a picture of political gridlock of the kind that has at times paralyzed Congress. According to Marxist class theorists, such political inaction may allow capitalists to wield the real power in the economic sphere. Note that pluralists do not need to envision a perfect democracy—political analysts from Max Weber to Joseph Schumpeter (1949) have noted the limits of democracy. Pluralists do not deny the existence of elites; they merely contend that elites may have differing agendas. To rule, elites must have not only access to power but also a way to coordinate their aims.

In *Who Rules America?* (1967) and his subsequent books, G. William Domhoff contends that the controlling elites do have a common agenda, even if it is only that of preserving their hold on power and privilege. He maintains that elites coordinate their activities as they interact at elite clubs, select alumni associations, and other exclusive social gatherings, and as they work together through interconnected associations, foundations, and boards of directors. So common is the phenomenon of "interlocking directorates," in which prominent executives and financiers sit on the boards of multiple corporations, that Domhoff's charts of network affiliations look like dense spider webs binding together a powerful, wealthy elite (see Exhibit 8.4).

Organization	Size[*]	Organizational Interlocks	Centrality
1. IBM	18	34	1.00
2. Conference Board[†]	31	53	0.87
3. General Foods	16	24	0.81
4. Chemical Bank	24	36	0.79
5. Committee for Economic Development[†]	200	119	0.78
6. New York Life	25	36	0.77
7. Yale University[†]	18	23	0.66
8. Morgan Guaranty Trust	24	40	0.65
9. Consolidated Edison	14	22	0.63
10. Rockefeller Foundation[†]	19	25	0.62
11. Chase Manhattan	24	33	0.62
12. AT&T	18	35	0.60
13. U.S. Steel	17	30	0.59
14. Sloan Foundation[†]	16	25	0.59
15. Caterpillar Tractor	11	19	0.59
16. General Motors	23	31	0.54
17. Citibank	27	37	0.53
18. Pan American	23	25	0.52
19. Council on Foreign Relations[†]	1,400	154	0.52
20. Metropolitan Life	29	30	0.51
21. Metropolitan Museum[†]	44	37	0.47
22. Equitable Life	37	42	0.47
23. Mobil Oil	13	14	0.46
24. MIT[†]	74	54	0.45
25. American Assembly[†]	19	26	0.44

Exhibit 8.4   Interlocking Directorates, Foundations, Think Tanks, and Media

[*]The size of an organization is the number of directors or trustees on the controlling board of the organization, except in the cases of the Committee for Economic Development and the Council on Foreign Relations, where all members are included. Organizational interlocks are the number of connections an organization has to other organizaions in the corporate community through sharing one or more directors with other organizations. The centrality of an organization is a mathematical expression of both the number of organizational interlocks and the degree to which those interlocks are with other organizations that are highly central to the overall network.

[†]Nonprofit organization

Source: Data from Salzman and Dornhoff, 1983, pp. 205–216.

The debate over the existence of a power elite also extends to the local level. Examining the decision-making process in New Haven, Connecticut, Robert Dahl (1961) found a wide range of differing positions and agendas in contention with one another. Looking at the same city, Domhoff (1978) suggested that the real decisions had already been worked out in the back rooms and the common agenda largely set before the complex and contentious process that Dahl noted ever began.

So which is it, elites or plurality? Who holds the keys to power? It may be safe to say that forces pull in both directions. The two perspectives described above may be opposing poles in this process. John Walton (1970), who has examined elites and elite behavior in a wide variety of times and places, concludes that whether one finds support for the elite position or the pluralist position may depend largely on the measures one uses—reputations, affiliations, decision making, and so forth—and on the community involved. He suggests that different local communities have different patterns of power based on their individual economic and social histories.

Pure elite cohesion can be seen in some places: Central American countries in which both the economy and the political process are controlled by a handful of intermarried families is one extreme. Mexican students contend that when Yale-educated Ernesto Zedillo was chosen as the presidential candidate of Mexico's official party, Yale graduate students from elite Mexican families immediately began dividing up the cabinet posts. Yet elites can lose: Several elite-dominated Latin American governments (including Zedillo's own party) have been toppled. It would be hard to show that in 1992 Bill Clinton held more elite connections than the enormously well-connected George Bush. Elite theorists do note, however, that when challengers succeed in coming to power, they often have to make peace with the elites. When John Kennedy entered the White House in 1960, he immediately appointed to his cabinet a number of conservative business leaders—including several who were Republicans—as a way of "reassuring Wall Street" and the financial community (Burch 1980). Clinton proposed (and, in some cases, withdrew) controversial appointees for some socially oriented cabinet and administration positions, but the individuals he first appointed to the top financial posts all had strong corporate ties and business community endorsement. Although it may be helpful to recognize the power of elites, too much stress on control by hidden elites can lead to voter apathy and citizen inaction. The challenge is to find ways to limit and balance elite control.

Changing the Rules: Campaign Finance Reform

Attempts to establish limits on elite control have often taken the form of calls for campaign finance reform. The first wave of such calls came in the wake of the Watergate-related scandals and reporting on Richard Nixon's infamous lists of "generous" friends and "hostile" enemies. The reform law enacted in 1974 limits individual contributions to $1,000 per candidate and $25,000 in any given election. Contributions from **political action committees (PACs)** cannot exceed $5,000 per candidate. Candidates who abide by spending restrictions are eligible to receive federal matching funds. Political campaigns have found several ways to circumvent this law, however. The most popular of these has been the use of what has become known as **soft money**—that is, contributions that have been made not to particular candidates but instead to parties or "coalitions" that may run advertisements on issues. These "issue ads" are not regulated, and any amount may be spent on them, even if they run during a campaign and clearly encourage a particular vote: For example, "The current leadership is failing to address this issue and we need change now," or, conversely, "We need to keep strong leadership on this issue—we cannot trust the untried and dangerous ideas of challengers."

During the 1990s, some Democrats in Congress (who were being outspent by their opponents) proposed tighter campaign finance laws. In response, Republicans, who at the time were less likely than Democrats to be multiterm incumbents, proposed term limits legislation—that is, legislation that would limit the number of terms an individual could serve in a given office. Term limits, they argued, would prevent longtime incumbents from using the power of their office to raise money and continually be reelected. Of course, without any accompanying finance reform there was the danger of removing one set of "fat cats" only to replace them with a new set of even "fatter cats."

Personal wealth has in fact become another key way in which candidates avoid spending limits. There is no limit on how much of a candidate's own money he or she can spend on a campaign. Billionaire Ross Perot ran for president by spending large amounts of his own money. Similarly, in the 1996 and 2000 presidential elections, Steve Forbes (the son of multimillionaire publisher Malcolm Forbes, whose *Forbes* 400 lists the country's wealthiest people) ran in the Republican primary without spending limits, as he used his own money. When Ross Perot was hesitant to make a third presidential run on the Reform Party ticket, billionaire Donald Trump considered the possibility of running using his own vast resources.

Continued attempts to reform campaign finance laws have been caught in congressional gridlock between the two main parties, and the ability of billionaires in the Reform Party to effect real reform seems questionable, except perhaps in regard to a change from allowing the use of donated corporate wealth to allowing the use of personally accumulated corporate wealth. During the 2000 presidential campaign, Senator John McCain gained considerable attention, and electoral support, by taking up the reform message within the Republican Party. The Democratic and Republican presidential nominees, Al Gore and George W. Bush, then began to vie for this support, each claiming to be the "real reformer." Pluralists note that differing coalitions may oppose one another in the fund-raising process: conservative think tanks running advertisements challenged by liberal environmental organizations, business coalitions battling ads sponsored by labor union donations, and so forth. Nonetheless, the political process continues to be dominated by fund-raising and issues of money. George W. Bush was the clear early front-runner for the Republican nomination, not because he initially had extensive voter support but because he began the race with far more money than his leading opponents could muster, an incredible $60 million at the beginning of the campaign.

Beyond Bush's personal campaign "war chest" was $137 million in soft money that the Republican Party had raised. Party officials claimed this reflected the contributions of thousands of small donors, but two-thirds of it in fact came from just 739 contributors who each gave more than $100,000 (Van Natta and Broder 2000). Of these, 600 were members of Club 100—contributors who had given at least $100,000 each and helped fund the earlier campaigns of George H. W. Bush. They were now, however, small-time contributors and began to complain of being neglected. The true "royal treatment" went to the Regents, a group of 139 who each gave more than $250,000. The Regents golfed with George W. Bush and his advisers at exclusive country clubs, attended extraordinary dinners at places such as Tiffany and Company, which was closed to all others for the occasion, and had special access throughout the Republican National Convention. "You pay a little more, you get a little more," said the Republican finance chairman (quoted in Van Natta and Broder 2000).

Although Republicans garner more money than Democrats in this process, it was Democrat Bill Clinton who first mastered the use of the presidency as a vehicle for raising soft money. Clinton invited large donors to join him for private White House coffees, and the largest donors could find themselves overnight guests in the executive mansion's Lincoln Bedroom.

Political candidates all contend that they in no way sell political favors and that campaign money does not alter their policies. Clearly, however, generous contributions buy a great deal of access to power and attention to the contributors' issues. Sometimes they buy even more, such as prestigious positions. President George H. W. Bush appointed as ambassador to Spain a non-Spanish-speaking Florida real estate developer who had given millions to the Republican Party. The founder of the Regents is another Florida financier, a shopping-center magnate, who was appointed ambassador to Australia. Clinton appointed as ambassador to France a New York investment banker who had given $600,000 to the Democrats.

Giving and patronage reach their highest levels in presidential politics, but they are also found in other national political arenas and in state legislative politics. Major donors often gain privileged access to candidates. Further, lobbyists can use invitations to "policy forums" in exotic locations to gain access to legislators. Large donations come not just from individuals but also from PACs, many of which have enormous resources to spend in their attempts to influence both legislation and campaigns. During the 2000 election year, organized labor gave $60 million (most, but not all, of it to Gore and other Democrats). This amount was matched by almost $60 million in gifts from the American Medical Association and other health-related organizations. Lawyers and lobbyists gave $69 million. Energy companies and agribusiness each gave more than $40 million (almost all of it to Bush and other Republicans). The top donors were groups representing insurance, finance, and real estate with more than $150 million in donations (Center for Responsive Politics 2000).

Being associated with certain causes can be particularly lucrative for candidates and officeholders. Congressman Thomas Bliley of Virginia, an influential Republican on the House Commerce Committee, became known as the "Phillip Morris Rep" for his ardent support of tobacco companies and opposition to both clean indoor air laws and class action lawsuits. Tobacco companies not only donated more than $120,000 to his campaign coffers (see www.tobacco.org) but also sent him and his wife on an all-expenses-paid trip to England; the couple flew to London on the Concorde (at a price of more than $12,000 per ticket), stayed at a luxury hotel ($1000 per night), and attended the tennis championships at Wimbledon ($3000). Of all these, only accepting the Wimbledon tickets constituted a violation of House rules. Bliley was never penalized for the rule violation, and he retired from the House of Representatives in 2004 to become a tobacco lobbyist (which some believe he was from the beginning). In previous decades, Democrats in the House who were strategically placed on key committees were able to wield great power through their control of the national purse strings. Now, power often lies in the ability to raise funds to reward friends and punish enemies, a talent used to great effect by Texas Republican Tom DeLay.

Public support for campaign finance reform is overwhelming, but there is less agreement on what form it should take. Some oppose the public funding of campaigns, calling it "campaign welfare." During the 2000 presidential election campaign, the Reform Party, which was established in large measure to advocate for campaign and lobbying financial reform, found itself in an internal struggle over who could claim the party's $12.5 million in public campaign support. Consumer activist and one-time Green Party candidate Ralph Nader proposed strictly limiting the amount that campaigns can spend and replacing campaign spending with public service news coverage, public debates, and other unpaid ways of spreading the candidates' messages. This proposal has been challenged on the grounds that limiting spending amounts to limiting "free speech," although it may seem odd to classify television commercials at $30,000 per minute as "free" speech. A change to campaigns with little advertising but many public forums for presenting the candidates' views and the issues remains an illusive but intriguing possibility.

In the Senate, Arizona Republican John McCain and Wisconsin Democrat Russ Feingold jointly authored legislation intended to end the greatest campaign finance abuses. Passed in 2002, the McCain-Feingold Bill banned previously unlimited soft-money contributions from wealthy individuals and organizations to political parties while raising the limits on individual contributions to $10,000; sought to eliminate phony "issue ads" that were really candidate endorsements; restricted foreign donations; required greater disclosure of spending, especially by unions; and barred candidates from converting campaign money to personal use (Hoover Institution 2004). Ways around the law's restrictions remain, however. During the 2004 presidential primaries, George W. Bush (running unopposed for the Republican nomination), John Kerry, and Howard Dean all gave up public financing in order to avoid the limits this would impose on their own fund-raising. Candidates who were dependent on public funding were hopelessly outspent. Interest groups that avoided endorsing any candidate were able to spend large amounts on ads and campaigns that attacked the positions of political opponents. MoveOn, an Internet-based progressive organization, vigorously attacked Bush administration policies and records throughout the campaign. At the same time, a group calling itself Swift Boat Veterans for Truth attacked John Kerry's Vietnam War record with a flurry of ads paid for by wealthy donors.

Voter cynicism continues to grow regarding the purpose and truthfulness of campaign advertising, yet few effective alternatives have been proposed, and those that are proposed rarely receive the support of both parties. The growth of the Internet has created new, more open ways of informing the public and organizing political movements, but it has also created a new avenue for false rumors and well-funded paid advertising.

Global Power: Who Really Rules?

The great economic trend of the late twentieth and early twenty-first centuries has been toward globalization. The great political trend has been toward democratization. Since 1970, the number of democratic governments in the world has doubled. Rare now is the national state that doesn't at least claim its government is democratic. As Anthony Giddens (2000) observes:

> Democracy is perhaps the most powerful energising idea of the twentieth century. There are few states in the world today that don't call themselves democratic. The former Soviet Union and its East European dependencies labeled themselves "people's democracies," as communist China continues to do. Virtually the only countries that are explicitly non-democratic are the last remaining semi-feudal monarchies, such as Saudi Arabia—and even these are hardly untouched by democratic currents. (P. 86)

Giddens contends that these two great forces—globalization and democracy—are closely linked. Globalization is a democratizing force. Information now flows so quickly and so freely between so many places that no authoritarian government can control it. The revolutions of Eastern Europe in 1989 spread by fax, e-mail, and television, with ideas and images too numerous to be stopped by any censor, traveling far too fast for any column of tanks to challenge. Globalization undermines much traditional authority, including that of authoritarian governments.

There is an irony in the spread of democracy, however, that Giddens calls the "paradox of democracy." Even as people in newly "free" countries fill the streets to cheer new democratic changes, people in the older democracies are less likely than ever to even turn out to vote. Voter participation in the United States, the world's oldest modern democracy, is among the world's lowest and is declining. In part, people are disgusted by issues such as campaign financing. But increasingly they are also likely to argue that who is in government, and maybe government itself, just doesn't matter very much anymore. This sense can come quickly within a democracy. When South Africa held its first free and open elections in 1994, more than 90% of the electorate stood in long lines to cast their first and decisive vote, bringing Nelson Mandela to the presidency. Just six years later, another South African presidential election was marked by widespread apathy and low voter turnout. Many South Africans noted that changing their government had not brought the country the sweeping economic and social changes it needed. Like many black mayors in the United States, Nelson Mandela had high prestige and received high marks for his efforts and personal standards, but his real power to effect change was limited.

Globalization has moved power from national governments to transnational bodies. Foremost of these is the United Nations, which has seen its role in conflict intervention as well as its role in addressing humanitarian needs increase substantially. Other international bodies have also gained power. More powerful than any nation's ministry of finance is the International Monetary Fund as it makes lending and currency policy decisions. European governments are warily but gradually conceding power to the European Union.

The real powerhouses of the twenty-first century, however, are not transnational governmental organizations but multinational corporations. In part, this simply reflects their enormous reach and financial clout. The assets of the largest corporations often exceed the gross national products of many countries, including large countries. In dollar terms, many of the largest entities in the world are not nations but corporations (see Exhibit 8.5). Automobile and oil companies top the list, but technology, tobacco, and finance are also prominent. With Wal-Mart surpassing General Motors in 2004, retail has also become a dominant force.

In their book *Global Dreams* (1994), Richard J. Barnet and John Cavanagh provide scores of examples of how a few corporations are able to use their enormous financial clout and their control of earth-spanning technologies to control a global commercial culture that can now penetrate the "remotest" regions. (Today, the word *remote* is accurately used only in reference to the devices consumers use to operate other devices at a distance, especially to scan hundreds of television channels—most of which are owned

	Country/ Corporation	GDP/sales ($mil)		Country/ Corporation	GDP/sales ($mil)
1	United States	8,708,870.0	31	Indonesia	140,964.0
2	Japan	4,395,083.0	32	South Africa	131,127.0
3	Germany	2,081,202.0	33	Saudi Arabia	128,892.0
4	France	1,410,262.0	34	Finland	126,130.0
5	United Kingdom	1,373,612.0	35	Greece	123,934.0
6	Italy	1,149,958.0	36	Thailand	123,887.0
7	China	1,149,814.0	37	*Mitsui*	118,555.2
8	Brazil	760,345.0	38	*Mitsubishi*	117,765.6
9	Canada	612,049.0	39	*Toyota Motor*	115,670.9
10	Spain	562,245.0	40	*General Electric*	111,630.0
11	Mexico	474,951.0	41	*Itochu*	109,068.9
12	India	459,765.0	42	Portugal	107,716.0
13	Korea, Rep.	406,940.0	43	*Royal Dutch/Shell*	105,366.0
14	Australia	389,691.0	44	Venezuela	103,918.0
15	Netherlands	384,766.0	45	Iran, Islamic rep.	101,073.0
16	Russian Federation	375,345.0	46	Israel	99,068.0
17	Argentina	281,942.0	47	*Sumitomo*	95,701.6
18	Switzerland	260,299.0	48	*Nippon Tel & Tel*	93,591.7
19	Belgium	245,706.0	49	Egypt, Arab Republic	92,413.0
20	Sweden	226,388.0	50	*Marubeni*	91,807.4
21	Austria	208,949.0	51	Colombia	88,596.0
22	Turkey	188,374.0	52	*AXA*	87,645.7
23	*General Motors*	176,558.0	53	*IBM*	87,548.0
24	Denmark	174,363.0	54	Singapore	84,945.0
25	*Wal-Mart*	166,809.0	55	Ireland	84,861.0
26	*Exxon Mobil*	163,881.0	56	*BP Amoco*	83,556.0
27	*Ford Motor*	162,558.0	57	*Citigroup*	82,005.0
28	*Daimler Chrysler*	159,985.7	58	*Volkswagen*	80,072.7
29	Poland	154,146.0	59	*Nippon Life Insurance*	78,515.1
30	Norway	145,449.0	60	Philippines	75,350.0

Exhibit 8.5   World Economic Power

Source: Data from Anderson and Cavanagh, 2000.

Exhibit 8.5 (Continued)

		Country/ Corporation	GDP/sales ($mil)		Country/ Corporation	GDP/sales ($mil)
	61	**Siemens**	75,337.0	81	New Zealand	53,622.0
	62	Malaysia	74,634.0	82	**EOn**	52,227.7
	63	**Allianz**	74,178.2	83	**Toshiba**	51,634.9
	64	**Hitachi**	71,858.5	84	**Bank of America**	51,392.0
	65	Chile	71,092.0	85	**Fiat**	51,331.7
	66	**Matsushita Electric Ind.**	65,555.6	86	**Nestle**	49,694.1
	67	**Nissho Iwai**	65,393.2	87	**SBC Communications**	49,489.0
	68	**ING Group**	62,492.4	88	**Credit Suisse**	49,362.0
	69	**AT&T**	62,391.0	89	Hungary	48,355.0
	70	**Philip Morris**	61,751.0	90	**Hewlett-Packard**	48,253.0
	71	**Sony**	60,052.7	91	**Fujitsu**	47,195.9
	72	Pakistan	59,880.0	92	Algeria	47,015.0
	73	**Deutsche Bank**	58,585.1	93	**Metro**	46,663.6
	74	**Boeing**	57,993.0	94	**Sumitomo Life Insur.**	46.445.1
	75	Peru	57,318.0	95	Bangladesh	45,779.0
	76	Czech Republic	56,379.0	96	**Tokyo Electric Power**	45,727.7
	77	**Dai-Ichi Mutual Life Ins.**	55,104.7	97	**Kroger**	45,351.6
	78	**Honda Motor**	54,773.5	98	**Total Fina Elf**	44,990.3
	79	**Assicurazioni Generali**	53,723.2	99	**NEC**	44,828.0
	80	**Nissan Motor**	53,679.9	100	**State Farm Insurance**	44,637.2

Sources: Sales: *Fortune*, July 31, 2000. GDP. World Bank, *World Development Report 2000*.

by a handful of media conglomerates—that are filled with corporate-controlled products; it hardly applies to any location on the planet.)

Monopoly Power

In *When Corporations Rule the World* (1995), David C. Korten notes the long history of suspicion concerning power, particularly the monopoly power of large corporations. In 1776, Adam Smith referred to corporations several times in *The Wealth of Nations*, but always as negative forces, massive entities whose monopoly power could undermine the benefits of free competition that Smith hoped the free market would provide. Indeed, the corporations of his time typically used royal charters and special privileges to protect their vast regional monopolies. Thomas Jefferson denounced the abuse of royal monopolies in another document of the same year, the Declaration of Independence. This alliance of state and corporate power was avoided in the United States until the years following the American Civil War. Not long before his death, Abraham Lincoln worried about the new trend:

> Corporations have been enthroned. . . . An era of corruption in high places will follow and the money power will endeavor to prolong its reign by working on the prejudices of the people . . . until wealth is aggregated in a few hands . . . and the Republic is destroyed. (Letter to Colonel Elkins, November 21, 1864, quoted in Korten 1995:58)

In a remarkable twist of law, the Fourteenth Amendment, written to protect the legal rights of African Americans and former slaves, was turned to protect the rights of corporations, now given the standing of "legal persons." Throughout the 1880s and 1890s, corporations controlled directly and indirectly by John D. Rockefeller, Andrew Carnegie, Cornelius Vanderbilt, J. P. Morgan, James Mellon, and a handful of others came to utterly dominate both the U.S. economy and the U.S. government. In his 1934 book *The Robber Barons,* Matthew Josephson wrote of how during this time "the halls of legislation were transformed into a mart where the price of votes was haggled over, and laws, made to order, were bought and sold" (quoted in Korten 1995:58).

The power of corporations to dominate the national agenda waned a bit during the Progressive Era at the beginning of the twentieth century, as antitrust laws attempted to challenge corporate monopolies, but returned again in the 1920s and 1950s during times of expanded U.S. economic dominance. Yet 100 years later, in the 1980s, corporate power again fully dominated the U.S. government agenda. This was no longer just national corporate power; it was multinational corporate power. These corporate giants not only dominated the national agendas of the United States, Great Britain, Germany, and Japan during this time, they utterly eclipsed the national power of smaller countries around the world. Nation after nation in Southeast Asia, East and West Africa, and across Latin America shifted to policies that promoted participation in international "free" trade yet also protected the place of the large multinationals in that trade through land giveaways and tax advantages.

The global economy has created both new opportunities and new concentrations of power. Opportunity is the ability to exploit new means of achieving economic gain: new places, new resources, new positions within the social structure. Power is the ability to control the distribution of resources. People seek both simultaneously, but, as Weber realized, there is always the problem of social closure: The existing power structure seeks to close the opportunity structure.

The trend of societal change has been toward greater productive technology, from the potter's wheel to the computer, and toward larger economies of scale, from the band to the globe. Both of these increase opportunities, with new ways to produce and new places to distribute one's production. Both also can increase the concentration of power, as one can control powerful technologies (whether productive, such as bigger machines, or just coercive, such as bigger bombers), and one can come to control global markets. When the concentration of power increases, as in agrarian empires, inequality increases. When new opportunities are created, as with new industrial technologies, inequality can diminish. Yet opportunities can also succumb to power concentrations. Computer technology, for example, can create vast new opportunities, such as Internet entrepreneurship, but it can also create new monopolies in the hands of few who control the systems, as seen in the battles over Microsoft's dominance.

Large corporations hold enormous redistributive power. It would seem that as more societies are brought into the global economy and socialist economies become "market" economies, we would also see the final triumph of markets over redistribution. In fact, what we have seen so far is redistribution on a grander scale. People are not free to travel the world and trade face-to-face, so global merchants and financiers act as the new redistributors. They deal continually in money and markets, but they control those markets, standing, like the "big men" redistributors of an earlier time, at the nexus of a network of exchange. The new big men can also claim a share for their services, but now this is called profit. Both Smith and Marx saw market capitalism as something quite new, but in this description, corporate capitalism is starting to sound like a new verse of a very old song. The new capitalist bankers are clearly redistributors, standing between investors and debtors. A "free market" may be free from the interference of government redistributors, but it is not free from the control of corporate redistributors.

With the exception of the simplest open-air markets, or their "farmers' market" imitators, trade is rarely a face-to-face transaction between producer and consumer. It is usually mediated by a party with particular access to information and resources in the social network. Historically, the wealthiest tycoons have been the most privileged and aggressive of these go-betweens. The highest incomes in the world do not go to the most talented entertainers or others, but to the heads of large, well-connected corporations or, in nations where the government is a key intermediary, to the influential, well-connected heads of state. Even those who are well paid for their talents—such as entertainers, sports stars, and medical specialists—draw huge incomes only when they are placed at a key nexus of media contracts, government and privately supported structures, and multimillion-dollar money flows. Likewise, the wealthiest nations have often not been those with vast resources, but those that hold privileged or powerful positions in large networks of trade. From village markets to global markets, the path to wealth is a central position in human transactions, and the reality of poverty is exclusion from both power and opportunity.

Is it possible to have markets that are not dominated by extraordinary concentrations of power? Can the breadth of opportunity be increased? Can elective power be used effectively to balance concentrations of market power? And can the new electronic technologies be used to increase empowering connections between people—fax-to-fax, at least, if not face-to-face? These are the global challenges for the twenty-first century, which we consider in Part III.

KEY POINTS

- Both class consciousness in voting and voting along class lines have long been stronger in Europe than in the United States.
- The U.S. electorate is divided by social class, but also by gender, race, ethnicity, and religion.
- The political power of blacks and Latinos is increasing, but the single largest voting bloc in the United States is white, suburban, and middle-class.
- Power elite theorists contend that a small handful of people with special access to power through their wealth or position actually control the nation as well as local agendas. Pluralists contend that decisions must be worked out among many competing groups.
- Campaign finance reform has emerged as a major political issue in response to the huge amounts spent by individuals and PACs in campaigns and lobbying.
- Globalization and neoliberal policies that include free trade and small government have weakened the power of many governments while at the same time giving enormous power to large multinational corporations.

FOR REVIEW AND DISCUSSION

1. Why has voting in the United States not been more sharply divided along class lines? How are class, race, and gender intertwined in U.S. voting patterns?
2. What are the arguments for and against the idea that United States is dominated by a power elite that controls decisions? In what ways is power concentrated? How is wealth used to influence elections?
3. What are the major shifts that have occurred in democracy and the distribution of power in the world with globalization? What factors have affected the amount and extent of influence exerted by large corporations, both nationally and internationally?

MAKING CONNECTIONS

The growth of the Internet has created some incredible concentrations of wealth in the hands of many new millionaires and quite a few new billionaires. It has also opened up new and easy ways for people to be informed about flows of money. This is especially true in the area of money and politics: Internet users can now easily access information that they formerly could have obtained only by scrutinizing reports, briefs, and dusty volumes.

You can find out about the policy positions, view the voting records, and read the public statements of more than 13,000 candidates for public office and elected officials, both local and national, at the Project Vote Smart Web site: www.vote-smart .org. The American Political Science Association has named this site as one of the best on the Internet; it is consistent and nonpartisan. The site allows you to list your positions on key issues and then compare them to those of candidates and elected officials. It also allows you to search the legislation and voting records of office-holders on a wide variety of issues.

A number of good Web sites currently track campaign finances and the flow of political money from PACs and lobbyist donations. Be cautious about accepting as fact any information offered by sites supported by particular interest groups, as such sites often present a biased picture. The Center for Responsive Politics operates a good nonpartisan site at www.opensecrets.org. This site allows you to access information on the money collected and spent by major candidates as well as who gave what. You can focus on your own congressional representatives as well as their campaign opponents and see who is supporting them. Links also allow you to look at the tobacco lobby, the gun lobby, and dozens of other major interest groups: how much they have given and to whom. Even more detailed information is available at the Federal Election Commission's information site: www.tray.com. Select any official, candidate, or organization and explore the links between voting records and campaign contributions. Also available on this site are recent issues of the *Ouch* newsletter, which reports on links between money and votes.

Begin with the Project Vote Smart site and explore the voting records of your national and state representatives on issues of inequality: minimum wage legislation, tax changes, assistance to the poor, health care coverage, and similar issues. What positions have your elected officials taken and how have they voted on these issues? (During an election year, you may want to compare candidates.)

Go to the Center for Responsive Politics and FEC sites and look at the donations given to your local and national representatives and candidates by major corporations, labor unions, and other interest groups. Which groups gave the most? Do you find connections between your representatives' voting records and the pattern of contributions?

PART III

Challenges of Inequality

9

Moving Up

Education and Mobility

One is struck by the sheer beauty of this country, of its goodness and unrealized goodness, of the limitless potential that it holds to render life rewarding and the spirit clean. Surely there is enough for everyone within this country. It is a tragedy that these good things are not more widely shared. All our children ought to be allowed a stake in the enormous richness of America.

—Jonathan Kozol, *Savage Inequalities* (1991)

Getting Ahead

Miami has two groups that casual observers might simply note as "black" yet are very different in their backgrounds and expectations. One group is made up of poor, Miami-born African Americans. This group's members have been largely left out of Miami's growth and prosperity. They have felt themselves pushed to the margins of a white-controlled city, and they continue to feel marginalized in an increasingly Latino city, with the added bitterness that comes with believing that newcomers have gotten a better welcome and a surer hand up than they who helped build this southern metropolis. Some are still hopeful about change, but many have concluded that there just "ain't no makin' it in Miami." The other group is composed of recent Haitian immigrants and refugees. For them, Miami is the city of hope. They fled the endless poverty of the poorest nation in this hemisphere and the killings and repression that have often accompanied recent struggles there. In background, they are some of the poorest of the poor, but many are sure that in Miami they will find the American dream. They are certain that they are poised to move up. A few are critical of their native-born black neighbors, saying they have given up the fight to advance, are too ready to complain, too ready to expect help from somewhere else. The Haitians' Miami-born black neighbors in turn say that the newcomers are just naive—they don't know the realities of being poor and black in Florida, they will soon learn the harsh realities. Who is right in this exchange? Has one group been too quick to give up, or has the other just been slow to wise up? These are key questions for communities such as Miami, for whole societies, and now increasingly for the global division of labor: Who gets ahead and how?

Social Mobility and Social Reproduction

Social mobility refers to the movement of people both upward and downward in a stratified system. **Social reproduction** refers to the tendency of people to remain in their social classes of origin and to replicate the experiences of their parents. A closed stratification system allows for no social mobility and demands social reproduction. Hereditary slavery is an example of a closed system, unless significant numbers are able to win or buy their freedom. A caste system, such as the one that prevailed in India for centuries, can also be closed, in that one remains in the caste of one's birth for life. Yet even in India, and certainly in recent decades, caste has been a greater determiner of social standing and prestige (as in the case of European royal family lines) than it is has been an absolute determiner of economic standing. Medieval Europe was a largely closed system in which people were born into their lifelong "estates," but by the late Middle Ages, growing towns and cities were confronted with the dilemma of the rich merchant or moneylender who may be of low birth or an outcast group, but nonetheless grows richer than many petty nobles, some of whom may be in his debt. An **open system** allows for free movement up and down the society's levels of stratification. In fact, in a purely open system, *class* could become an obsolete term, because there would be no stable communities of common social standing. Just as few societies have ever been completely closed, almost none have been completely open. "Equal opportunity" may be the ideal, but nowhere is being born into wealth and power—or, conversely, into poverty and oppression—an irrelevant issue in a person's future success.

Even as Americans have been fascinated by royal titles and elite families, they have long held to the ideal of open society in which advancement is based purely on merit. In the late 1800s, popular Horatio Alger stories with titles like *Luck and Pluck* and *Sink or Swim* told the stories of poor but honest and hardworking boys (rarely girls) who started as newspaper boys and become newspaper tycoons, or began by mopping floors and soon rose to the top of the firm. African Americans of the time might have had a hard time relating to these stories, but they told their children the story of Joseph in the Old Testament: He was sold into slavery, but through cleverness, perseverance, and faith, he rose to become second only to Pharaoh as ruler of Egypt.

In 1973, David L. Featherman and Robert M. Hauser (1978) conducted a large national study to examine the openness of U.S. society. They compared the occupational categories of sons with those of their fathers and also compared their findings with those from a similar study conducted in 1962. Mothers and daughters were left out of this research, in part because not all were in the labor force, and also because previous studies had produced unclear findings concerning whether daughters benefit most from high-status mothers or high-status fathers. The study left out many other complexities as well, but it provided a glimpse into the openness of the American system. Featherman and Hauser found a great deal of social reproduction (see Exhibit 9.1). The largest numbers of sons worked in occupational categories similar to those of their fathers. There was some movement both up and down, but in the 1960s and 1970s more movement was in an upward direction: sons moving into higher-status occupations. Strikingly, in the early 1960s this was not true for black men, who often found themselves in lower manual-labor positions, regardless of their fathers' occupations. That is, not only were most black fathers in lower occupational positions, but even those in better positions had a hard time passing this advance on to their sons. This situation had improved markedly by the early 1970s, but black men on the whole were still less able than their white counterparts to pass higher occupational status onto their sons.

The fact that these studies found more upward than downward occupational mobility also tells us something about the U.S. economy in the decades that immediately preceded the studies: It was creating more room at the top. A stratification system can offer several types of mobility. **Circulation mobility** occurs when some move up and others drop in the system. A closed and nongrowing economy can offer only this type of mobility: For someone to move up, someone else has to move down and make room at the top. If Horatio Alger's shoeshine boy rises to the top office, then someone else, maybe the lazy son of the president, must lose position and take his place shining shoes.

Structural mobility occurs when economic growth and expansion increase the room at the top, so more people can move up. Sales of Horatio Alger's books were fueled by many Americans' anticipation that the industrial expansion of the United States at the end of the 1800s would mean new millions, new fortunes, and new positions of power and prestige, maybe even for the readers of the stories. The greatest single structural expansion of the U.S. economy occurred following World War II, when the United States dominated the global economy and led the world in industry and technology. For roughly the 30 years between 1945 and 1975, U.S. workers came to assume that continued economic expansion would ensure that they would be better off than their parents and that their children would be better off still. Not everyone shared in this growth, but the American occupational structure was expanding. Many during this period shared the experience of my own family, as I related in Chapter 3: a farming grandfather who shifted

Father's occupation	Mobility of White Men, Son's Occupation (in percent)						
	Upper White-Collar	Lower White-Collar	Upper Manual	Lower Manual	Farm	Total	Row%
1962							
Upper white-collar	54	17	13	15	1	100	17
Lower white-collar	46	20	14	18	2	100	8
Upper manual	28	13	28	30	1	100	19
Lower manual	20	12	22	44	2	100	27
Farm	16	7	19	36	22	100	29
Total	**28**	**12**	**20**	**32**	**8**	**100**	**100**
1973							
Upper white-collar	52	16	14	17	1	100	18
Lower white-collar	42	20	15	22	1	100	9
Upper manual	30	13	27	29	1	100	21
Lower manual	23	12	24	40	1	100	29
Farm	18	8	23	36	15	100	23
Total	**30**	**13**	**22**	**31**	**4**	**100**	**100**

Father's Occupation	Mobility of Black Men, Son's Occupation (in percent)						
	Upper White-Collar	Lower White Collar	Upper Manual	Lower Manual	Farm	Total	Row%
1962							
Upper white-collar	13	10	14	63	0	100	4
Lower white-collar	8	14	14	64	0	100	3
Upper manual	8	11	11	67	3	100	9
Lower manual	7	9	11	71	2	100	37
Farm	1	5	7	66	20	100	48
Total (N = 1,122)	**5**	**8**	**9**	**68**	**11**	**100**	**100**
1973							
Upper white-collar	44	12	8	36	0	100	4
Lower white-collar	20	21	13	46	1	100	4
Upper manual	16	14	16	54	0	100	10
Lower manual	12	12	14	61	1	100	46
Farm	5	7	17	63	8	100	36
Total (N = 3,493)	**12**	**11**	**15**	**59**	**4**	**100**	**100**

Exhibit 9.1   Occupations of Fathers and Sons

Source: Featherman and Hauser (1978). Reprinted by permission.

Note: Some percentages do not add to 100% because of rounding error.

to industrial work, a father who went from industry to white-collar service, followed by son in professional employment. Successive generations moved up, not because we were all smarter than our parents (although of course we thought we were) but because the economy needed more people in higher-skilled positions. This trend has weakened, although not disappeared, since 1975. What has changed is that in an era of stagnant wages and concern over global competitiveness, many who are now entering skilled white-collar employment may not be assured of greater earnings than previous generations who worked in industry. Upward occupational mobility is no longer an assurance of ever greater incomes. Further, the corporate downsizing that has accompanied restructuring and efforts at greater international competitiveness has meant that the rate of downward mobility has increased. More white-collar managers and professionals are "falling from grace," as Katherine S. Newman (1988) puts it. Part of the tremendous stress they experience comes from the fact that this now common experience is not part of the prevailing American ideal. To people who are accustomed to a pattern of more rather than less room at the top, downsizing can feel like personal failure.

Reproductive mobility can occur when lower-income groups have more children than upper-income groups. In the United States, it is not true that the poor have more children than others, although this has been the stereotype. It is true, however, that many working-class families have had more children than families in higher-income groups, especially two-income, upper-middle-class families. If a two-career couple decides not to have children, their retirement from the workforce creates spaces, if you will, that can be filled by other people's children. We occasionally hear inspiring tales of people who were born poor and moved up. We rarely hear tales of people born rich who have moved down. Those in the latter group may be less inclined to tell their stories, but it is also true that there are likely fewer of them because of past structural and reproductive mobility. Some children from lower-income families may look forward to the eventual opportunities afforded by reproductive mobility, but many others face the more immediate risk of poverty as deindustrialization and job loss leads to downward structural mobility.

Finally, **immigration mobility** occurs when a social system is open to outsiders who mostly enter near the bottom. This has also become part of the assumed American experience. Most immigrants who have arrived in the past have not been the elite members of their home societies. Given that they had to pay for their passage to the United States, they often were not the poorest of their societies either, but they typically occupied positions near the bottom of the U.S. stratification system when they arrived. Immigrants to the United States often began their lives here in poor neighborhoods: Irish shantytowns, Jewish ghettoes, Italian slums, multiethnic working-class communities located "across the tracks" or "back of the stockyards," and so on. Their entrance often served to advance other groups that were already here.

Irish American laborers soon became foremen supervising Eastern European laborers, and so forth. Each new arrival near the bottom could serve to advance others. Today, immigration mobility continues as poor immigrants, such as the Haitians in Miami and the growing Latino immigrant population across the country, take the least desirable jobs, allowing others to move into more desirable occupations as the economy and labor market expand. But it is no longer true that almost all immigrants enter the labor market near the bottom. East Asian and Caribbean entrepreneurs, South Asian physicians and software programmers, and many others now enter at upper levels of the system. This "brain-drain immigration" can benefit the overall economy by providing skilled professionals (trained at someone else's expense) and new investment that will create jobs. It also means, however, that these immigrants may enter "over" some native-born groups in the stratification system. Low-income African Americans, such as those noted in Miami, often experience frustration when the arrival of new groups does not push up their position. When new immigrants seem to monopolize the labor market above them, poor African Americans may perceive this as part of a societal pattern of holding them down. In the burgeoning immigrant cities at the beginning of the twentieth century, such as New York and Boston, the newest group was often the least favored and the most discriminated against: a harsh welcome, indeed, but the group got relief when a newer group arrived. In the burgeoning immigrant cities at the beginning of the twenty-first century, such as Los Angeles and Miami, new immigrant groups are sometimes preferred as employees over long-standing groups such as low-income African American men (Nee, Sanders, and Sernau 1994). This can add to the resentment and frustration of members of the latter groups, who may conclude that there "just ain't no makin' it."

Status Attainment

This situation is a good reminder that the first question—How much openness and opportunity is there?—must always be followed by a second: And for whom? Who gets ahead and why? In the 1960s, researchers Peter Blau and Otis Dudley Duncan made a systematic attempt to answer this question; they studied what they called **status attainment,** or how an individual reaches a particular socioeconomic status position. The book in which they reported their findings, *The American Occupational Structure* (1967), was as well-known for the methodology it described as for its conclusions. The problem with untangling the ingredients that help people move up in society is that the causal factors themselves are also interrelated. For example, is someone helped by having a well-educated parent or a wealthy parent? Which matters more? How are these factors interrelated?

Like Featherman and Hauser, Blau and Duncan also worked with data on sons and fathers. They used illustrations they called *path diagrams* to

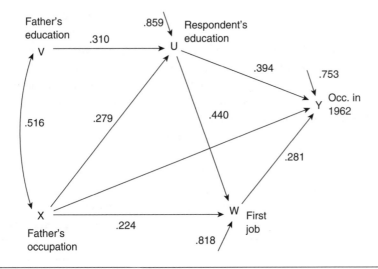

Exhibit 9.2   Example of a Path Diagram

Source: Blau and Duncan 1697. Copyright 1967 by John Wiley & Sons, Inc. Reprinted with permission.

portray the relationships among the different factors associated with socioeconomic status (see Exhibit 9.2). Scholars still often use path diagrams to illustrate complex processes. The numbers that appear next to the arrows in these diagrams are correlation coefficients: how much one aspect moves in accord with the other. Perfect correlation would be indicated by a one, and no correlation between the two items would be indicated by a zero. The diagram from Blau and Duncan's work shown in Exhibit 9.2 indicates that a son's attainment is affected by several factors. Father's education helps son's education directly (perhaps through role modeling) and indirectly by means of father's occupation (maybe dad pays for son's college). Father's occupation again helps the son after college (maybe dad knows a good business contact) and so forth. The relationships are all nontrivial, but many are not very strong. Note that the strongest factors come from outside forces that are not included in the model (maybe the son wastes time in college or makes a major mistake on his first job).

In two books, *Inequality* (1972) and *Who Gets Ahead?* (1979), Christopher Jencks tried to unpack some of Blau and Duncan's findings. Instead of using occupation as his final dependent variable, the thing to be explained, Jencks used income. This was a significant shift, for he found only a .44 correlation between occupation and income: Many people with the same job title or category have very different earnings. Jencks directly addressed the problem of the arrows from unknown causes. Family background, school success, even a crude measure of IQ all mattered, but even put together with job category, they all explained only one-fourth of the variation in income. Jencks wasn't afraid to label the remaining three-fourths of variation: luck. Much of the income difference seemed to be the result of sheer luck: One person started the right business at the right time, another happened to be in the wrong place at the wrong time. He tried to separate ability from family effects by testing brothers. Test scores explained about 40% of future school success but only about 10% of income difference. Take heart if you don't seem to be earning what you think you should be worth—intelligence and ability as measured on these tests was only a very weak predictor of future income level. Schools, in this, seemed more "fair" than the labor market: They rewarded talent more consistently. Jencks came to an unpopular conclusion from his studies. He suggested that equal opportunity is not enough to bring greater equality. The problem is not that some people are much smarter than others; rather, some are simply much luckier. Jencks concluded that to have a more equal society, we would have to equalize rewards as well as access; in his words, we would have to penalize failure less and reward success less, and give more second chances. His studies drew considerable attention in the 1970s, but his conclusions had little effect on policy in the 1980s; they seemed thoroughly at odds with the mood of the times, as Americans wanted to believe that hard work is always rewarded and riches are always earned.

Education: Opening Doors, Opening Minds

Educational Access and Success

One of the troubling aspects of Jencks's findings was his suggestion that there is no educational "fix" for inequality. Americans have often looked to education as the solution to the problems of poverty and economic disadvantage. When my students discuss social problems, they most often propose "more education" as the solution. As an educator with a classroom full of people who may be sacrificing a great deal to get a higher education, I certainly don't want to minimize their view of the importance of education. Education is undeniably valuable—it is valuable in its own right, even if it never gains the learner a bit of extra income. It is also a potent force for higher earnings—but only if there is broad access to high-quality education and if this education is linked to good job opportunities.

By themselves, the returns for education—that is, the income advantages received for each year of education—are impressive. Exhibit 9.3 shows the average earning gains in the United States for various educational degrees. The first big gain comes with high school completion. Although it is true that a high school diploma in itself may count for less in the job market today than it did in the past, it is still often the key credential that allows an individual to obtain other certificates and training. The main reason for the big income difference that comes with a high school diploma is not that high school graduates are guaranteed good incomes, but that they qualify for other programs. Also, those who do not graduate from high school often have long and frequent bouts of unemployment and erratic employment.

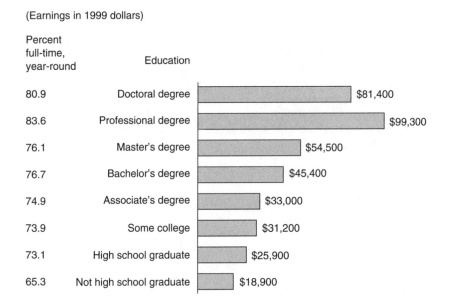

(Earnings in 1999 dollars)

Percent full-time, year-round	Education	
80.9	Doctoral degree	$81,400
83.6	Professional degree	$99,300
76.1	Master's degree	$54,500
76.7	Bachelor's degree	$45,400
74.9	Associate's degree	$33,000
73.9	Some college	$31,200
73.1	High school graduate	$25,900
65.3	Not high school graduate	$18,900

Exhibit 9.3   Earnings by Educational Degree

Source: U.S. Census Bureau, Current Population Surveys, March 1998, 1999, and 2000.

Degrees from vocational schools and two-year programs bring still greater returns, but the next big jump comes with college graduation. Again, as many soon-to-be graduates worry, a college degree is no guarantee of high-wage employment. It is often, however, the key credential for getting one's foot on the right ladder, one that will eventually lead to higher earnings. The single biggest payoff comes with a degree from a professional school: law, medicine, graduate business, or the like. Such a degree is often needed for entrance into one of the highest-paying professions. The return on a Ph.D. is also solid, but not as great, given that most people with doctoral degrees go into research and teaching positions, which pay less than upper-level executive positions and self-employment in professional practice. The differences in earnings associated with educational levels translate into huge income differences over a lifetime, as Exhibit 9.4 illustrates.

No matter how high the tuition, higher education pays big returns—if you finish. Not only college entrance rates but college completion rates vary considerably across groups. As Exhibit 9.5 shows, gender appears to be less of a factor in college completion than race or income. In the 1940s, women were close to parity with men in college completion, with 100 degrees going to women for each 100 going to men. During that time, of course, many American men were at war. In the 1950s, many men got degrees with help from the GI Bill, which funded education for returning war veterans. Women married younger and were less likely to complete college degrees. With every decade since that time, however, women's completion rates have

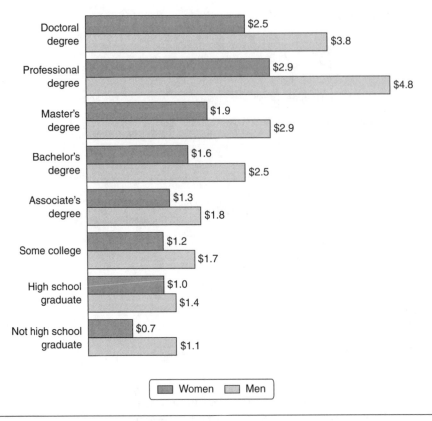

Exhibit 9.4   Lifetime Earnings (In millions of 1999 dollars)

Source: U.S. Census Bureau, Current Population Surveys, March 1998, 1999, and 2000.

grown, until now more college degrees go to women than to men. Women are no longer disadvantaged in college completion, but they are disadvantaged in returns on education. Every year in college, and each degree, brings them less return in the labor market than their male counterparts enjoy.

Members of some racial/ethnic groups have not fared as well as women in terms of college completion: Black college students are less likely than nonblacks to graduate, and Latinos are less likely than non-Latinos to graduate. Progress in this area has been slow and erratic, and some observers are concerned that as universities move away from affirmative action programs the gap between races may increase. It grew in the 1980s in part because of changes in federal funding for education. Remember, we are focusing here on completion rather than entrance. For low-income groups, staying in college may be more difficult than getting in. Note that the students who are most disadvantaged for college completion are those who are low income, regardless of race. Higher education for blacks, Latinos, and low-income men does help close the income gap (more powerfully than for women relative to men), but the graduation gap persists.

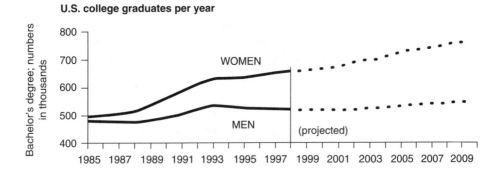

U.S. college graduates per year

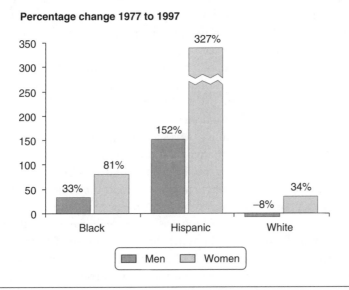

Percentage change 1977 to 1997

Exhibit 9.5   College Completion for Women, Blacks, Hispanics, and Low-Income Students

Source: U.S. Department of Education, National Center for Education Statistics, 2002.

The reasons for low graduation rates among blacks, Latinos, and those from low-income backgrounds of any race or ethnicity are many. Students from these backgrounds may feel like tokens, isolated from their communities, growing apart from their families, and not well accepted by campus communities dominated by middle- to upper-class whites. Decades of low graduation rates for these groups mean that these students are likely to have few role models and mentors (this is changing slowly for women). They are likely to be trying to balance heavy work and family responsibilities with the demands of their class loads. Black, Latino, and low-income students are much more likely than white and higher-income students to be employed full- or part-time and to have dependent family members. They are also less likely to have family resources they can draw on and more likely to be dependent on financial aid. Students who cannot attend college full-time or who are struggling to do well in their classes also find their financial aid placed at risk. Many persevere and emerge triumphant graduates in spite of the odds against them, but many do not; they get worn down, they are needed elsewhere, they decide that perhaps they do not belong in college after all, and they drop out along the way.

Unequal Education and "Savage Inequalities"

A major part of the struggle that low-income college students face, particularly many students of color, stems from the fact that they received their earlier education in urban (or, in a few cases, rural) public schools that were overcrowded and underfunded and so provided them with few of the extra opportunities their better-off classmates enjoyed before college. Differences in education often begin at a very early age.

In a book that stirred national attention, Jonathan Kozol (1991) describes the "savage inequalities" of American education. He did not need to compare public schools with elite private "prep" schools to find these inequalities. In public schools in the wealthier areas of Long Island, New York, such as Manhasset, he found shining new school buildings and new additions to school buildings, libraries better than those at some colleges, gyms with indoor tracks, and swimming pools. Every child had access to the latest computer technology. Everywhere at these schools the tools of success were available, along with an unspoken but clear message to students: "You matter, you are worth this expense, you are going somewhere." At the middle-class schools of Fordham in the north Bronx, Kozol found more concern about crowding and safety, but the schools remained oases of green and positive places to be. In the poorest areas of the South Bronx, however, he found schools shrouded in razor wire, the buildings collapsing, with plaster and water falling on students. School libraries in this area were almost devoid of books, let alone computers. The teachers, many of whom were hurried recruits completely untrained in the fields in which they were teaching, struggled to maintain order. In schools such as these, not only are the tools of success missing, but so are any positive messages to students. These schools seem to say, "You don't matter, you're not worth much expense, this is a warehouse to hold you until someplace else is willing to take you."

What Kozol found is not just some perversity of New York. In a study following up on Kozol's work, Jeffrey Hayden (1996) found the same divergence among schools in Ohio. A suburban school system with a booming tax base built the Perry Educational Village, a complex that looks a bit like a cross between a liberal arts college, a new shopping mall, and a business complex. The pride and centerpiece of the community, it speaks eloquently about the importance of education. Yet urban Cleveland schools are in such distress the state had to take them over, and still they have no textbooks for some subjects and occupy decaying, sometimes hazardous buildings. As

Kozol points out, no businessperson would consent to work a week under such conditions, yet we expect both teachers and students to work and thrive in these neglected schools. The schools in poor rural counties fare no better than those in inner cities, with classes meeting in old coal bins and students needing to put on jackets to make the trip to the next building to use a bathroom or to travel downtown to find lunch at the local gas station. The libraries in some of these schools feature encyclopedias so old that they suggest "someday we will get to the moon" and computers (if they have any at all) so obsolete they can't run any current software or be used for Internet access.

All across the United States, similar contrasts can be found within many metropolitan regions. In well-off suburbs, school visitors find bright new buildings with the latest facilities surrounded by grassy athletic fields and big circular driveways where buses and minivans drop off children with backpacks full of new books and supplies. Often the students are greeted by teachers and administrators at the door as they arrive in the morning. A few miles away, in central-city settings, children enter their schools by walking through narrow gates in razor-wire-topped fences and into blackened brick structures with wiring, plumbing, and heating systems that no longer meet building codes because of decades of neglect. As the children approach the single unlocked doorway through which they can enter, they are often greeted by literary references carved into the stone and concrete entryways urging them to aspire to great things. A few do so aspire, but the entire atmosphere of their learning environment suggests that an excerpt from Dante might be more appropriate: "Abandon all hope, ye who enter here."

According to Kozol, the fundamental reason these contrasts exist is not mismanagement but simple inequities in funding. Public schools in the United States are funded primarily by local property taxes, which draw on local wealth. Poor communities have low property values, and even if they tax themselves at very high rates, they cannot generate much income for local schools. Schools on Long Island and elsewhere in New York suburbs spend between $11,000 and $15,000 per student annually; in New York City (where costs are higher), this figure drops to $5,000, and in some places in the rural South it is just over $2,000. Kozol notes that school funding is one of the few places where, in precise dollar amounts, we tell children essentially what we think they're worth. In poor districts, the message is clearly "not much." Given that urban and rural schools are much more likely than their suburban counterparts to be housed in old buildings that are expensive to maintain and difficult to upgrade (many cannot be wired to handle modern computers, for example), and that they spend more on special-needs children, the gap is even greater than the dollar amounts indicate. Mismanagement may also play a role, certainly, but it is interesting that suburban districts are almost never under state scrutiny for mismanagement and poor urban districts frequently are.

Those who believe that the problems of inequity in schools are rooted in bureaucracy, mismanagement, and lack of responsiveness have offered "school choice" as a solution. They argue that if parents are able to choose their children's schools, poor-quality schools will lose students to better ones, and so will need to reform or close. Those who can afford to send their children to private schools already have options, and proponents of school choice assert that such options can be extended to all parents through vouchers that they can redeem at the schools of their choice. Critics of school-choice proposals note that the vouchers these plans provide rarely cover the total tuition costs of private schools, so poor parents may still not be able to afford to send their children to these schools while wealthier parents may receive just enough through vouchers to influence them to leave the public schools, taking their money and their support with them. It is unclear whether voucher programs would ever enable poor parents with limited transportation resources, limited information to assess schools, and limited options within their neighborhoods to avail themselves of high-quality school choices. Proponents of school choice suggest that new schools with new specialties would spring up, yet it is also unclear whether any such new schools, especially private ones, would be eager to specialize in educating poor, high-need children with few resources (Coleman 1992; Astin 1992; Sernau 1993).

Currently, many states are experimenting with charter schools—that is, schools that are freed from complex state regulation as long as they meet the terms of their charters. Such schools, which are typically run by parents or teachers, and sometimes by private interests, have offered interesting possibilities in many locations, but critics argue that they drain resources from regular public schools. Some observers have also wondered, If charter schools do better without state regulations, would not all schools benefit from greater flexibility and revision in state regulation?

Kozol contends that the only way to address the inequalities that exist in schools is with equitable funding—that is, funding that does not vary from district to district. His preference would be for the United States to fund the public schools out of the national wealth, as is the practice in most countries. The United States, however, has a long tradition of local control and suspicion of national efforts. Even given this tradition of local control, the constitutions of most U.S. states place the responsibility for public education on the state and not on individual communities. This has been the basis of lawsuits across the country demanding more equitable funding for schools.

The challenge is that schools with high-need students in high-need neighborhoods may need more resources and programs than students in other neighborhoods, even beyond spending parity. A growing number of urban schools around the country have begun to experiment with full-service school programs. Schools participating in such programs don't close at 4:00 P.M.; instead, they are open well into the evening and on weekends, offering recreational activities, health and pregnancy (and pregnancy-prevention) services, counseling, vocational training, literacy and enrichment programs, computer training, and other services to students and sometimes to other neighborhood members as well. These programs are financed by special grants or by partnerships between community businesses and social service agencies.

Debates over educational reform continue to rage, in particular the debate about who is responsible for the problems of the public schools. It is possible, however, to envision a multilevel approach to school reform that would involve local districts finding new and creative ways to involve parents, businesses, and their local communities; states working to equalize funding among districts beyond local property tax bases; and the federal government investing in the crumbling infrastructure of older urban and rural schools just as it does in highways and other infrastructure.

No Child Left Behind

In 1965, as part of the **War on Poverty,** President Johnson signed the Elementary and Secondary Education Act (ESEA), which created the Title I program, one of the longest-standing federal interventions on behalf of poor children. Title I committed the federal government to fund programs aimed at providing special assistance to children from "disadvantaged" backgrounds. The ESEA was reauthorized in 1994 under President Clinton, with a new commitment to Title I programs for poor children. The reauthorization in 2001, under President George W. Bush, was different, however. The new version has been dubbed the No Child Left Behind Act, from a phrase that President Bush had borrowed for his election campaign from the Children's Defense Fund—ironically, an organization that has often opposed his policies. This version of the act stresses school accountability and standards (as evidenced by test scores), shifts strategies to the state level, and provides for the elimination of "failing" schools (schools that do not meet standards) from funding programs. The emphasis is on the achievement of poor children, but the ideology behind the act, reflecting that of the president and the Republican congressional leadership, is based in the belief that the key problems of public schools lie in inefficient bureaucracies and public systems rather than in inequitable funding that leaves poor school systems unable to address their inadequate facilities, large class sizes, and special student needs.

The No Child Left Behind Act includes multiple provisions, almost all of them controversial (Rebora 2004). It requires the following:

- Annual testing in reading and math
- State "report cards" on student achievement
- Demonstration of teacher certification and proficiency to teach in core areas
- Competitive literacy grants known as "Reading First"
- A shift in Title I targeting toward districts with high concentrations of at-risk students
- Performance standards for individual schools, with funding rewards and penalties, and parents given the option to leave "failing" schools at district expense

Supporters of the legislation claim that "NCLB is an ambitious law and forces states to move faster and further to improve the achievement of every student. Perhaps the combination of NCLB's tight timelines and high expectations and existing state education agendas will prove successful where past reform efforts have fallen short" (Education Commission of the States 2004).

Opponents, including the major teachers' unions, contend that the law is flawed in both its funding and its mandates. They argue that schools are not given the resources they need to meet the act's high standards for rapid improvement (National Education Association 2004). Others worry that the act will undermine public education, especially in poor urban areas. Schools in low-income neighborhoods are not meeting the standards for improvement, meaning that funds in already overextended local budgets must go to pay for the movement of children to other schools. In time, many urban public schools could close, leaving their communities with few other options except private schools or privately run charter schools. Some see in this an indirect attack on public education and a way to force privatization. In Chicago, one of the nation's largest school districts has already put forward a plan that would close 60 "failing" public schools and open 100 smaller schools, some of which would be run by private companies and would not have teacher union contracts (Dillon 2004).

Because the No Child Left Behind Act is complex and its provisions are being implemented in stages, it remains to be seen whether the law will ultimately channel resources toward or away from schools in low-income areas. The effects of the greatly increased amount of standardized testing the act calls for are also uncertain: Will testing requirements propel all toward higher standards, or will they lead to less innovation as schools merely "teach to the test"? It is also unclear whether the law will make sure that "no child is left behind" or instead further discourage and disadvantage those who have traditionally not done well on standardized tests: low-income students, African American and Latino students, and, at least in some cases, girls.

Ladders with Broken Rungs

Raising the Bar: Human Capital and Gatekeeping

Can education be a means of lifting the economic prospects of everyone in a society? The answer to this depends in part on the uses of education. **Human capital theory** (Becker 1964) sees education and experience as investments in people that increase their productivity. Highly educated people are more productive than less educated people, and so they are paid more. The implications of this seemingly simple assertion can, in fact, raise some quite contentious issues: Are men paid more than women because, on average, they bring higher amounts of human capital to their positions? Are people really paid "what they're worth" in terms of productivity? A human capital approach leads to optimism about education's role in society: As people gain more education they become more skilled and capable and so more productive, and so they can command higher incomes. This implies also an "expanding pie": If we all produce more, everyone can earn more, so all can gain.

A darker view of the role of education is that it serves primarily as a "gatekeeper," providing credentials for high-status group membership (Collins 1977). According to this perspective, there are only so many positions at the top, and so some screening criteria must be used to keep too many people from getting to the top. In earlier times the most important criterion might have been family lineage: Only those of privileged or noble birth could hope to attain prestigious, powerful, and privileged positions. In much of the world we have grown uncomfortable with the idea of such class-based attainment, and so educational credentials have become a substitute. In ancient China, a state civil service system allowed people of great talent but low birth to move up, like the biblical Joseph, into powerful administrative positions (even as the emperorship itself remained hereditary). It is unlikely that these people brought tremendous technical human capital with them, for they showed their worthiness by memorizing Confucian verses and long excerpts of ancient poetry. Poetry memorization, then, became the "gate" through which only certain people—those with great formal memories, those with respect for Confucian tradition, those who had invested a great deal in the formal system—to attain positions of power.

Today, the educational requirements attached to employment are clearly increasing. Most jobs now require a high school diploma, many require technical training, and many employers in fields such as sales now require hirees to have a college degree. Why have the educational entry requirements continually been raised? One answer is that jobs are more technical now than ever before, and workers require more human capital to do them well. Another answer is that education is essentially a gatekeeper, and as more people get more degrees, the standards must be raised to keep the gate sufficiently narrow. This implies that getting more education may not help those on the bottom rungs of society, for as more people get more credentials, the standards continue to shift, and upward mobility remains out of reach.

Many fields have several ladders of promotion rather than a single ladder toward advancement, and educational credentials determine which ladder one is climbing. The clearest example is the military: Someone entering the army as a private can aspire to becoming a master sergeant but will never become an officer without specialized training, which is typically limited to college graduates. In business, an administrative assistant may move up the ranks but will likely never become a manager without some formal educational degree. In medicine, a nurse-practitioner, no matter how competent and experienced, can never become a physician without a medical school degree. Nor can a paralegal become a lawyer without a law degree. Often, one career ladder exists for nondegreed hourly wage workers and another for degreed salary earners. Sometimes multiple ladders are available; for example, universities employ hourly staff, professional staff, and faculty. Employees may be promoted within each of these categories, but promotions between categories are extremely rare, for a college degree divides the second from the first, and an advanced degree distinguishes the latter group. Multiple-ladder systems make education requirements absolute, and this can have wide-ranging effects. For example, the U.S. Army has a great many black sergeants but far fewer black officers. Many administrative assistants, paralegals, and nurse-practitioners are women with great expertise, human capital perhaps, but they often cannot advance above a certain level without leaving their positions to pursue full-time formal education in order to obtain the degrees that will allow them to change classifications—in essence, to change ladders.

The jobs that are most open to individuals with few or no educational credentials sometimes offer no **internal job ladders** at all. No matter how competent a worker is, there is no room for advancement. One can work hard and do a great job at the cash register, but all the managers have degrees; one can work hard and do a great job in the warehouse, but all supervisory personnel have degrees; and so forth. Horatio Alger's boys never seemed to face this problem. People in positions where they have no room for advancement quickly "get old" in their positions—that is, they can neither advance in their current workplaces nor build their résumés with promotions to impress potential new employers. Some fields have addressed this problem by establishing apprenticeships and journeyman training, by allowing workers the option to advance to semisupervisory roles (such as foreman), and so forth, but such opportunities for advancement are becoming increasingly rare, especially in rapidly growing occupations.

On Track, Off Track, Dead in Your Tracks

Within an organization not everyone competes in the same internal job market, and within schools not everyone gets the same education. Distinct patterns of career movement within an organization are known as *tracks*. In the company she examined in the study described in Chapter 6, Rosabeth Moss Kanter (1977) found that the employees themselves often knew clearly which of them were on the "fast track" (headed for the top) and which were "dead in their tracks" (headed nowhere). Being well mentored and trusted within an organization may place someone on an "inside track," whereas those without valuable close contacts—a group that may particularly include women and employees of color—are likely to find themselves on a much slower "outside track." Most companies deny they make such distinctions, but a few commentators, such as Felice Schwartz, have actually suggested they should. In a 1989 article in the *Harvard Business Review*, Schwartz proposed the controversial idea of a "mommy track" for women balancing career and family. Women who can devote long hours and their complete attention to their jobs could be placed on the fast track toward promotions, while expectations would be lower for women on the mommy track, especially in terms of time. Women on the mommy track would also advance less rapidly than their fast-track colleagues. Some observers immediately criticized Schwartz's suggestion as a return of the old double standard and a new excuse for organizations to avoid advancing women. Schwartz contended that the mommy track is just a realistic response to the current social reality that women often have primary family responsibilities, and she noted that a woman's placement on such a track need not be permanent. Kanter has suggested that "parent track" might be a better term, given that some fathers may also not want to spend 60 to 80 hours a week at the office.

In education, the practice of **tracking** students has long been controversial. Many schools devise tracks to respond to students' differing abilities: an advanced placement track or "gifted and talented" program for some, an "accelerated track" for others, and a "basic" track for still others. In high schools, such tracks may be phrased in terms of students' career expectations: a vocational track and a college preparatory track. Distinctions may begin much earlier, however, as when students are divided into reading groups early in their primary education. Although the early distinctions may be disguised by cute names, even young students often have a sense that the "bluebirds" are ahead of the "robins" and know who the "buzzards" are. Tracking can confer cumulative advantages or disadvantages, so that slow starts are never compensated for, and although the track a student is placed in may not be regarded as permanent, it can limit his or her options from early on, as when failure to take certain math courses prevents a student from taking advanced science courses later.

The practice of tracking can have gender implications. For example, boys, who tend to start reading a bit later than girls on average, are often overrepresented in the lowest-level reading groups in elementary schools. Some are pushed to catch up and do so, but others may take from this experience an early message that school is not for them. Boys of color and boys from low-income backgrounds are especially prone to accept that idea; they are also more likely than girls and other boys to be labeled as hyperactive and learning disabled from an early age. Some girls may fall behind the high-achieving boys in math and science by junior or senior high school, and typically few of them are encouraged to press ahead with their learning in these subjects; they may just be counseled into less demanding tracks, only to realize later that this has all but closed certain career options to them. This kind of tracking has important income consequences: Women dominate the fields of early-childhood education and home economics, which offer the lowest returns on a college degree, and are greatly underrepresented in engineering specialties, which offer the highest returns on a degree.

Countries and territories	Net primary school attendance (%) (1992–2002)		Primary school enrolment ratio				% of primary school entrants reaching grade 5	
			1997–2000 (gross)		1997–2000 (net)		Admin. data	Survey data
	male	female	male	female	male	female	1995–1999	1995–2001
Sub-Saharan Africa	58	54	89	78	63	58	65	82
Middle East and North Africa	82	74	95	86	83	75	93	—
South Asia	76	69	107	87	80	65	66	91
East Asia and Pacific	—	—	106	106	93	92	94	—
Latin America and Caribbean	91	91	126	123	96	94	77	87
CEE/CIS and Baltic States	79	76	99	95	88	84	—	96
Industrialized countries	—	—	102	102	96	97	—	—
Developing countries	74	70	105	96	84	77	79	89
Least developed countries	58	53	87	76	67	61	66	79
World	74	70	104	96	85	79	80	89

Exhibit 9.6   World Educational Enrollments

Source: Data from UNICEF, 2004.

Tracking often also has a strong racial and ethnic component, with black and Latino students (especially Latino students who begin with limited English proficiency) often placed in lower or "vocational" tracks. Students from working-class backgrounds are much more likely to find themselves in vocational tracks, in Europe as well as in the United States and Canada, than are students from middle- and upper-middle-class backgrounds. Classic sociological studies have shown how the expectations embedded in the educational experiences of students from different classes often shape student outcomes (see, e.g., Chambliss 1973). More recently, Kozol's examination of a wide range of public schools revealed great distinctions within as well as between them. In "advanced" classes, energetic teachers worked intensively with small groups of students who were often white and often from better-off backgrounds. In "basic" classes, ill-motivated, ill-prepared, or just frustrated teachers worked with large numbers of students, often students of color from low-income families and neighborhoods.

Given tracking's possible negative effects on students, many researchers, such as educational sociologist Maureen Hallinan (1994), have argued that schools must be very cautious in how they implement tracking so as to avoid bias. Other researchers, such as Jeannie Oakes (1994), have contended that the problems associated with tracking are so great that they negate the value of the entire approach. Instead of trying to place students in tracks, these critics contend, schools need to respond to the individual potential of each child, the "late bloomer" as well as the "fast starter."

Tracking of a sort is even more pronounced in many European systems, as well as in many developing countries that have adopted European models. In many systems, students take tests at quite early ages, when they are as young as 10 to 12 years old, that determine the educational tracks into which they are placed—that is, whether they will be prepared for a university education or a "vocation." These tests can be extremely important, placing very high pressure on students—as is the case in Japan and many Asian countries—for they can determine the course of a student's entire career. In Germany, only students with high scores enter *Gymnasium,* which prepares them for a university education. In Japan, the pressure placed on students to get into the finest universities is intense, but, ironically, once they have been accepted, Japanese students may be more relaxed than their American counterparts, for they are perceived as having "arrived." Their future success is virtually ensured by the clamor of companies that seek to hire graduates from these schools. (Some have noted a similar process at Harvard University, where gaining entrance is extremely difficult, but once there, students may feel less pressure, for they are assumed to have arrived at elite status.) Unlike Americans, Germans and Japanese have few second chances for educational success if they fail as students early on. The counterpoint to this is that German and Japanese students who are not prepared to attend major universities are not left to fend for themselves; rather, they are frequently channeled into vocational tracks that include quite sophisticated mathematical and technical training and/or apprenticeships that lead to skilled industrial employment. Both the Germans and the Japanese are eager for their industrial workers to have the training and high-order thinking skills needed to be productive on the job (Wilson 1996). This strategy served both countries well as they built up industrial systems with reputations for high quality and worker involvement. There is some evidence in the recent struggles of firms in both Germany and Japan, however, that their highly tracked systems may not be flexible enough to respond to the wild fluctuations of a changing global economy.

Compared with European and Asian countries, the United States and Canada send a considerably larger share of their populations on to higher education. The community and regional college system in the United States, along with various technical and business colleges, provides second chances to older, returning students as well as those who may not have been on the college-prep track in their early schooling. The downside to this is that Americans' school-to-work transition is often weak and individualized, with students needing to find their own way through a confusing and changing labor market. Further, students who do not pursue higher education may also acquire few of the technical, analytic, and computational skills demanded by any type of skilled industrial or service employment and may be left to drift from one low-skill, low-wage position to another, gaining few skills and very limited marketable "human capital" along the way. In recent years, the United States, from the Clinton administration on down through the state governments, has devoted some attention to improving the school-to-work transition, especially for those high school graduates who do not go on to college. An ideal system might combine the "skills for everyone and a place for everyone" approach of the German and Japanese systems with some of the flexibility and "second chances" offered by the U.S. system.

Unfortunately, many developing countries seem to be saddled with school systems that combine the worst elements of both. Mexico tracks students early, as do Germany and Japan, but leaves many students who do not excel to struggle on their own, as does the United States. Many school systems in African countries are based on French and British models that were established to legitimate and refine elites but offer little practical training. Students learn Latin and rhetoric, study Shakespeare or Molière, but gain little technical expertise. In terms of the categories we have examined, many African schools may be described as still functioning as gatekeepers, defining and refining elites, rather than increasing the national stock of human capital (Dore 1976). With high unemployment rates, the school-to-work transition is also a big problem in developing countries, and students (like those in poor inner-city and rural areas of advanced industrial countries) may not see clear mobility advantages to staying in school. Developing countries are also desperately struggling to extend educational opportunities to all children, including those in rural areas. Poor African countries have rich oral traditions and traditions of family-based education, but they have the world's lowest formal education enrollment rates. A few, such as Kenya, have only recently begun to offer free primary education. Even when education is free, the costs to students' families are often high; many cannot afford to pay for books, for example, and many cannot afford for children to be in school and thus unavailable to carry out family tasks. A poor girl in West Africa may need a sponsor to stay in school, but if her sponsor is a wealthy older man, the tuition may come with sexual expectations. In parts of the Middle East, the lack of formal education opportunities has been supplemented by Islamic schools such as the madrassa. These schools may offer only memorization of the Koran, however, and are sometimes funded by extremist groups. Pakistan has begun to make a large-scale national effort to move children of both genders into public schools, in part to counter the influence of the madrassa.

In many countries, the educational enrollment of girls continues to lag well behind that of boys. In this area, many Asian countries have made progress more quickly than Middle Eastern and African countries. UNICEF estimates that worldwide, some 121 million children are still out of school, 65 million of them girls. Only 74 percent of all boys and 70 percent of all girls attend primary school. Attendance rates are lowest in Sub-Saharan Africa, where only 58 percent of boys and 54 percent of girls are in school (see Exhibit 9.6). The education of girls is of vital importance, for development analysts are finding that the single best predictor of falling birthrates, declining infant and child mortality, and general well-being of much of a nation's population is the level of education attained by women (United Nations Development Program 2000).

Clearing the Bar: Career Trajectories

The above discussion of tracks and career paths illustrates why studies of mobility can be difficult and complex: In addressing mobility, we can't just look at an individual at a given point in time; rather, we need to see the entire dynamic picture of how that person got to that point, why, and where he or she may be going from there. Sociology in general has been moving from static descriptions to dynamic analyses (Tuma and Hannan 1984), and such analyses are especially important in the study of inequality. Mathematicians in the time of Galileo studied the trajectories of cannonballs to try to understand the nature of motion, and today physicists study the trajectories of subatomic particles to try to understand the nature of matter. A person's path through occupation and income levels can be viewed as a **career trajectory**, with the vertical axis indicating the attainment of prestige, power, or income and the horizontal axis showing the amount of time taken or the person's age (see Exhibit 9.7). Some people's career trajectories are very flat: They end their working lives earning about as much or doing about as much as when they began. Others have very steep trajectories as they climb to the pinnacles of their professions. The trajectories of still others, such as professional athletes, might be shaped like an inverted U (a parabola to mathematicians), with a steep rise, a peak, and then a decline (professional athletes who engage in lucrative businesses after they retire from sports avoid this decline).

Different professions are associated with widely differing earnings trajectories. Certain fields of engineering offer some of the highest starting salaries to college graduates, yet the earnings trajectories for engineering are fairly flat (except for those who start their own businesses or move into management): Senior engineers earn modestly more than their junior colleagues. Law offers the promise of a steep trajectory: Although growing competition means that new attorneys today cannot be assured of high incomes soon after they begin practicing, moving up from a small firm to a large firm or becoming a senior partner in a firm can make a trajectory rise sharply. A new

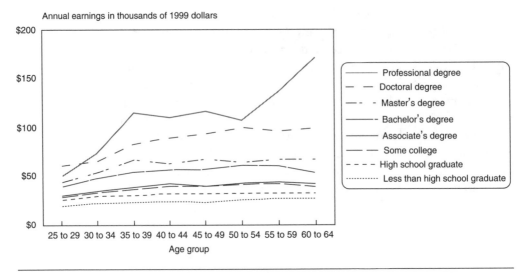

Annual earnings in thousands of 1999 dollars

Exhibit 9.7   Earnings Trajectories by Education

Source: U.S. Census Bureau, Current Population Surveys, March 1998, 1999, and 2000.

lawyer's earnings may start below those of a new engineer, but over time the lawyer's earnings will surpass the engineer's.

One reason women often have lower earnings than men is that their careers have flatter trajectories. The starting salaries offered to women straight out of college are quite comparable to those offered to men. Men, however, often have steep career trajectories, whereas women's tend to flatten out earlier and at lower levels than men's. The phenomenon of the glass ceiling seems to be at work in this, given that women enter potentially high-income professions but move into the upper echelons of those professions in relatively small numbers. The structure of organizations, as discussed in Chapter 6, may be at least in part to blame for this, as may be the challenges that women face in balancing family and career. Analysts have noted that women professionals often cut back on their commitment to their careers in their 30s (although this is becoming less common), working part-time or leaving the paid workforce for a while to raise children and tend to families. Most eventually return to their careers full-time, but only after having been "absent" during the period that for many men sees the steepest rise in career trajectory. Even if these women resume their upward climb, they never reach the same heights as their male counterparts.

The role of discrimination in this pattern cannot be discounted, however. Asian American men also seem to hit a glass ceiling and have flatter career trajectories than white men. Many begin in highly paid technical professions such as engineering and computing. These professions offer good starting salaries but relatively flat trajectories for those who do not make the leap into senior management, and it seems that many Asian American men, like many white women, do not make this leap. Both groups are often valued for their expertise and hard work but are then passed over for executive promotions. Self-employment has been one way out of this situation for Asian Americans, and it is now starting to become more common for white women as well. Black and Latino men and women are less likely than Asian Americans and whites even to have professional career trajectories in the first place, but when they do, they too tend to find their careers stalling in the middle as others get the top-level promotions.

In modeling trajectories, social scientists are still largely in the "Galileo" phase, just trying to describe them accurately. The "Newtonian" leap of identifying the forces or laws that create trajectories is still more difficult. It is likely, however, that when such forces are identified, they will include a familiar list of factors we've already examined: institutional and personal discrimination, stigma and stereotypes, social networks, outside responsibilities, and initial advantage and resources.

Blocked Opportunity: "Ain't No Makin' It"

You may be eager to insert one other element in the list of forces influencing career trajectories: What about personal effort and aspirations? What about working hard and aiming for the top? Certainly hard work and sheer determination don't hurt, but they also don't guarantee success. Your grandfather, especially if he was an immigrant or the son of an immigrant who rose from humble beginnings, may have insisted otherwise, like the Haitians in Miami, with whom we began this chapter. Aspirations matter, but high aspirations don't just happen, they are instilled—by successful parents who "made it" themselves, by teachers and counselors who see "something of themselves" in promising students, by employers and mentors who take note of certain individuals. And aspirations are sustained by small successes that promise better things to come.

Jay MacLeod describes the findings of his fascinating study of the residents of low-income housing in a book titled *Ain't No Makin' It* (1995). In his research, he looked at two informal groups of young men. The members of one group, the Hallway Hangers, who were white and primarily Italian American, began with low aspirations. They had seen family members beaten by a changing economy that held no place for them, they had seen their status actually diminish, and they had largely given up hope for advancement. They retreated into sexism (in the form of putting down the women in their lives) and a vague racism (blaming neighboring blacks for their situation). The Brothers, MacLeod's second group, was made up of African Americans who, interestingly, had higher aspirations than the Hallway Hangers. Many of the members of this group held menial jobs, but they had hopes of moving up. One of the most distressing aspects of MacLeod's book comes in his description of what he found when he followed up with these two groups after his initial research was completed: Neither the defeated Hangers nor the more optimistic Brothers had gotten very far.

The latter had done better than the former at staying out of trouble with the law and probably treated their women and families better, but they too had been beaten down by chronic unemployment, jobs with no advancement, and a societal structure that was unwilling to make room for them as they tried to move up. MacLeod concludes:

> But the leveled aspirations of the Hallway Hangers can be directly traced to the impermeability of the class structure. Moreover, the ample ambition of the Brothers has been drained away by the tilted playing field under their feet. Over and over we discover beneath behavior cited as evidence of cultural pathology a social rationality that makes sense given the economic constraints these young men face. Born into the lowest reaches of the class structure, the Brothers and Hallway Hangers variously help and hinder the inertia of social reproduction. Their individual choices matter and make a difference, but the stage is largely set. Even the Hallway Hangers, far from authors of their own problems, are victims of a limited opportunity structure that strangles their initiative and channels them into lifestyles of marginality, and then allows the privileged to turn around and condemn them for doing so. (Pp. 265–66)

Finding ways to move from misdirected blame toward confronting real barriers and changing the structure of society is a key challenge of our times, both for public policy and for social action. Around the world, opportunities for individuals to receive formal education are increasing. At the same time, however, the penalties for not having the right academic credentials are also increasing. And although more can begin to move down the path of formal education, for some that path is strewn with barriers and obstacles to completion. The question of who will be able to compete in the global marketplace is paralleled by the question of whether education itself should be a marketplace, with options for sale to consumers with means, or a right, perhaps a responsibility, that comes with national and global citizenship.

KEY POINTS

- An American cultural ideal emphasizes upward social mobility for the hardworking, but many families experience social reproduction; that is, children continue in the social class of their parents.
- Status attainment models examine how family background, education, and initial opportunities affect a person's later occupation and socioeconomic status.
- Added years of higher education bring substantially higher earnings on average, but students from low-income and nonwhite backgrounds have significantly lower college completion rates, and women receive much lower returns on their years of education than do men.

- In addition to the divide between public schools and private schools, great inequalities in funding, infrastructure, and equipment exist between well-off suburban public schools and underfunded schools in poor urban and rural districts.
- Some see education as an important source of human capital that makes workers more productive, benefiting both themselves and society. Others believe that current educational systems act more as gatekeepers, allowing some groups to gain the credentials needed for entrance to high-status positions while blocking the aspirations of others.
- Educational and career tracking, whether implemented formally or informally, can determine both the quality of the education a student receives and the likelihood of the student's future success. The stigma of poverty coupled with structural barriers to advancement can severely undermine the hopes and aspirations of people from low-income backgrounds.

FOR REVIEW AND DISCUSSION

1. What are some of the factors that can lead to social mobility? What are some of the forces that produce social reproduction? What makes a system open or closed?
2. How can education help or hinder social mobility? What are the controversies that surround school choice and school tracking? What proposals have been offered concerning school reform?
3. What determines a worker's job trajectory? How is this trajectory affected by race, class, and gender?

MAKING CONNECTIONS

National Center for Education Statistics

Statistics on education nationwide as well in individual states and even by school district are available from the U.S. Department of Education's National Center for Education Statistics at www.nces.ed.gov. Examine the statistics related to major concerns in educational quality and trends. How do the schools in your state or community compare with schools nationwide?

In Your Community

Are there experiments in education going on in your community, such as charter schools, magnet schools, full-service schools, and alternative schools? Contact your local school district office or your campus school of education and see if you can make arrangements to visit one of these schools, to talk with teachers or administrators, attend a program, or, ideally, help out with a program or class. (Always be sure you have proper administrative clearance before entering public or private school grounds, as campus security has become a sensitive issue.) Do you find the innovations promising? Is the school or program addressing an important community need? Can this program be replicated elsewhere? What is the relationship between this school and others in its district?

Tutoring

Opportunities abound in many communities for volunteers to tutor both children and adults in a variety of subjects. Taking part in such opportunities can be a great way for you to learn firsthand about educational challenges and the possibilities and difficulties of upward mobility. Find out what programs are available in your community from your campus community relations office, the local school district office, or the local public library. You should also be aware of national programs such as America Reads and America Counts, which may provide training and stipends to tutors; your campus should have information available on such programs. Spend time tutoring a child who needs extra help or tutoring an adult who is working toward a GED or improving academic skills. Or volunteer to work in a family literacy program that is seeking to improve reading skills for both parents and children. Some programs require volunteers to make ongoing commitments, whereas others, such as some "drop-in" or after-school programs, welcome volunteers to visit occasionally. What academic deficiencies is the program you are involved with trying to address? What did you experience regarding the diversity of needs, backgrounds, and learning styles of the program's students? What are the possibilities as well as the limitations of this program for promoting academic and career success?

10 Abandoned Spaces, Forgotten Places

Poverty and Place

> *The contrast of affluence and wretchedness, continually meeting and offending the eye, is like the dead and living bodies, chained together.*
>
> —Thomas Paine (1737–1809)

The landscape of urban America, and of an urbanizing world, brings rich and poor together—and keeps them worlds apart. According to Mary Schmitz,

A lot of people in Cabrini-Green don't leave the grounds much, but Stan Davis has long been more a traveler than most. So on a sunny spring day, fueled by curiosity, he makes a breathtaking journey few Cabrini residents have taken, across what must be the biggest income gap in Chicago. Five-foot-nine and wiry, a silver medallion of the Madonna glinting around his neck, Davis strides in his black jeans across Larrabee Street, over to Crosby Street and into the new country. It takes 30 seconds. He keeps moving, past the construction-site fence with the big "Not Hiring" sign. Toward the new $860,000 homes of River Village and the Montgomery Ward catalog warehouse reborn as the lofts of "domain." Out toward the new riverside promenade. The yacht office. The pricey Asian-French fusion restaurant Japonais. Then he turns back toward the corner package liquor store, where Cabrini men are talking, drinking and spooning microwaved chili from Styrofoam cups. (*Chicago Tribune*, July 15, 2004)

To understand poverty, we need to understand the context. There is no "face of poverty"; rather, poverty has many faces in many, often quite different, places. This is why I distrust the idea of a "culture of poverty," even though common experiences may well foster common lifestyles and strategies.

245

The world of poverty is as multifaceted and multicultural as the world of the rich. We just work harder at keeping it hidden.

Urban Poverty: Abandoned Spaces

Cities, U.S. cities in particular, have long been magnets for the desperate hopes of the poor and have often concentrated social problems even as they offered concentrations of opportunity. New York was the murder capital of the nineteenth-century world as it filled with immigrant ghettoes—Irish, Jewish, and Italian—before these spaces were ceded to newcomers including African Americans, Puerto Ricans, and new immigrants. Chicago of the early twentieth century was gangland: not just Al Capone and organized crime but also street gangs whose members were Polish, Slovak, and a dozen other various nationalities and ethnicities. Over the twentieth century, U.S. urban problems were particularly shaped by migration and deindustrialization. In the 1940s and 1950s, large numbers of rural African Americans and increasing numbers of Latinos migrated to U.S. cities in search of industrial employment, joining the white immigrant groups who had come with the same hopes at the beginning of the century.

Like their predecessors, the new urban poor were concentrated in low-income ghettoes. African Americans came to dominate Harlem in Manhattan and then to share the area of the Bronx directly across the Harlem River with poor Latino groups. Black migrants from Mississippi and neighboring areas came north to Memphis and St. Louis as well as Chicago. In Chicago, they were concentrated in an area known as Bronzeville, where many lived in dilapidated and overpriced "kitchenettes" carved from older homes. Their ability to move on was limited by fierce white resistance, but they were able to establish a thriving albeit often low-income community, as in Harlem. In time, growing pressure on housing forced the city to look for alternatives. When Mayor Daley drew sharp attacks from his white constituents for his proposals to establish public housing throughout Chicago, he instead chose to concentrate on building high-rise dwellings such as Cabrini Green and the Robert Taylor Homes. These promised low-income members of the community the possibility of affordable housing near to employment and free from unscrupulous absentee landlords. St. Louis built the huge Pruitt-Igo housing complex with the same intent. Low-income black migrants were concentrated in portions of north St. Louis and across the river in East St. Louis near employment in chemical and industrial plants. Migrants also worked their way westward to establish themselves in the Watts area of South Los Angeles.

Deindustrialization and the Changing Metropolis

All of the locations described above became what William Julius Wilson (1996) terms "institutional ghettoes." A combination of low income, public policy, and community hostility kept nonwhite urban groups concentrated in certain areas, often the least desirable portions of large cities. The same period saw the movement of many white urbanites toward the suburbs. The older white urban ethnic enclaves declined as their members began to intermingle in the growing suburbs and vacate the cities, leaving them to the new migrants. Influenced by the civil rights movement, many urban African Americans rose up to challenge the deprivations and isolation of the urban ghettoes in the late 1960s. Most of the attention to their protests went to places where unrest exploded into riots, as in Detroit and Watts.

By that time, however, the character of these ghettoes was beginning to change. Residential discrimination, although far from over, had at least begun to subside enough that better-off black residents could begin to leave the most troubled communities. Sometimes they took their businesses and organizations with them. At the same time, jobs fled the central cities. U.S. industry began to look overseas for production. American plants closed. Cities across the Midwest and Northeast experienced the "rust belt" phenomenon of deindustrialization. Blue-collar workers of all backgrounds were hard-hit, but those in the central cities had the fewest alternatives. Retail and many office jobs left the cities for the suburbs, often relocating in areas beyond the reach of public transportation. The employers that remained in downtowns were often the so-called **FIRE enterprises:** finance, insurance, and real estate businesses. Employment with these firms required a college degree, professional presentation, and a long list of credentials, none of which low-income urban residents possess. The only employment these businesses offered central-city residents was an occasional position as custodian or security guard (if one had a completely clean police record). Privileged workers shuttled between the high-rise office buildings downtown and the new suburbs on new urban expressways, making as little contact as possible with city residents. Institutional ghettoes become jobless ghettoes. Even as employment opportunities grew, there remained a fundamental mismatch: Where jobs were located and what those jobs demanded of workers did not match up with where the unemployed were and what they had to offer. Inner-city residents turned to informal economic activities, which often included drug trafficking and other illegal enterprises. The central city became a place of danger and mystery for suburbanites and a place of danger and degradation for its residents.

In a book he ironically titles *Amazing Grace,* Jonathan Kozol (1995) describes the experience of children in the Mott Haven neighborhood of the South Bronx, one of the poorest urban areas in the United States:

> The houses in which these children live, two thirds of which are owned by the City of New York, are often as squalid as the houses of the poorest children I have visited in rural Mississippi, but there is none of the greenness and the healing sweetness of the Mississippi countryside outside their windows, which are often bolted against thieves. Some of these houses are freezing in the winter. In dangerously cold weather,

the city sometimes distributes electric blankets and space heaters to its tenants. In emergency conditions, if space heaters can't be used, because substandard wiring is overloaded, the city's practice, according to *Newsday*, is to pass out sleeping bags. . . . In humid summer weather, roaches crawl on virtually every surface of the houses in which many of the children live. Rats emerge from holes in bedroom walls, terrorizing infants in their cribs. In the streets outside, the restlessness and anger that are present in all seasons frequently intensify under the stress of heat. (Pp. 4–5)

The Making and Unmaking of the Postindustrial City

Deindustrialization, particularly coming on the heels of decades of housing discrimination, devastated many cities. This was not the final word on urban America, however. For one thing, newcomers have not stopped arriving. New York City was the ultimate immigrant metropolis at the beginning of the twentieth century, and it remains a magnet for newcomers. One reason industry hasn't left New York altogether is that newcomers, especially Asians, have continued to come seeking work, particularly in the textile industry. Other newcomers, especially from the Caribbean, continue to begin new small businesses in the city. Other immigrant metropolises continue to grow as well, including Miami, Los Angeles, and Washington, D.C. Some immigrant entrepreneurs prosper, but many immigrant workers struggle under conditions reminiscent of the sweatshops of Asia or of New York in the last century, with low wages, uncertain employment, no benefits, long hours, and vulnerability to exploitative employers (Waldinger and Lichter 2003; Nee, Sanders, and Sernau 1994). Nonetheless, the latest influx of labor and capital has brought new energy and growth to urban economies.

Inland cities such as Detroit and Chicago also continue to draw newcomers, including Latinos and groups from the Middle East, the Caribbean, and Asia, often "secondary migrants" from the two coasts. Many rust-belt cities have also made a determined comeback. They have invested in urban redevelopment projects such as Baltimore's Inner Harbor, Cleveland's Nautica, and Pittsburgh's Station Square. They have built new stadiums, new aquariums, new amusement parks, new "halls of fame," and new casinos. Overall investment in older urban areas is again up, but it is unclear how much this has benefited the urban poor. Often the cities' new enterprises have taken the form of upscale and high-priced attractions for the well-off, who consciously avoid the attractions' neighboring urban areas. For example, Detroit has three new casinos and a new sports stadium, but the latter is virtually walled off from the surrounding urban neighborhood, and few nearby residents

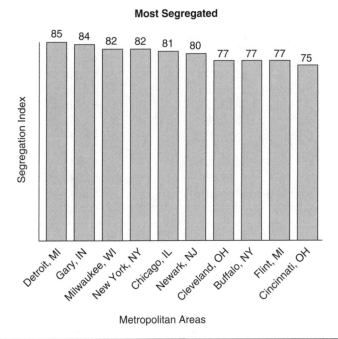

Exhibit 10.1   Urban Segregation: Ten Most Segregated Metropolitan Areas in the United States

Source: Based on data from U.S. Census Bureau, Current Population Surveys, 2000.

could afford tickets to games there. "Riverboat" casinos in Gary and Michigan City have brought new profits to investors such as billionaire Donald Trump and have generated some tax revenue but offer few jobs to residents and have little to do with the cities, staying safely anchored just offshore. Gary and Detroit remain the most segregated cities in the United States, followed closely by Milwaukee, New York, Chicago, and Newark, New Jersey (see Exhibit 10.1).

U.S. cities now are not only immersed in the global economy, they often mirror it in their contradictions: diverse interactions yet persistent segregation, islands of prosperity amid wastelands of poverty, job growth amid chronic underemployment, economic growth amid stagnant wages and declining fortunes. These problems are not unique to the United States; this is also a fair description of postindustrial London and Amsterdam, and of the new Prague of the Czech Republic. Increasingly, this description can be applied to world cities anywhere—Bangkok or Manila, Buenos Aires or Rio de Janeiro. As Pico Iyer (2000) writes:

> The place we reassuringly call a global village looks already a lot like a blown-up version of Los Angeles, its freeways choked, its skies polluted, its tribes settled into discontinuous pattern—the flames of South Central rising above the gated communities of Bel Air. (P. 76)

Rank	City	Avg Income Top 20% of Households	Avg Income Bottom 20% of Households	Ratio
1	Washington, DC	$186,830	$6,126	30.5
2	Atlanta, GA	$172,773	$5,858	29.5
3	Miami, FL	$125,934	$4,294	29.3
4	New York, NY	$159,631	$5,746	27.8
5	Newark, NJ	$93,680	$3,747	25.0
6	Boston, MA	$145,406	$5,832	24.9
7	Los Angeles, CA	$162,639	$7,124	22.8
8	Fort Lauderdale, FL	$176,053	$7,831	22.5
9	Cincinnati, OH	$117,086	$5,440	21.5
10	Oakland, CA	$163,931	$7,642	21.5

Exhibit 10.2a   Income Gaps in the Central Cities of the 10 Largest Metro Areas

Source: D.C. Fiscal Policy Institute, Washington, D.C., and U.S. Census Bureau, Current Population Surveys, 2000.

City	State	Persons Under 18 Total	Number Poor	Percent Poor	Rank (1 = lowest)
10 Worst					
Brownsville	Texas	48,000	21,732	45.3	245
Hartford	Connecticut	35,624	14,701	41.3	244
New Orleans	Louisiana	127,566	51,707	40.5	242 (tied)
Providence	Rhode Island	44,547	18,045	40.5	242 (tied)
Atlanta	Georgia	90,755	35,624	39.3	241
Buffalo	New York	75,530	29,200	38.7	240
Miami	Florida	77,285	29,770	38.5	239
Gary	Indiana	29,787	11,387	38.2	238
Cleveland	Ohio	133,335	50,629	38.0	236 (tied)
Laredo	Texas	62,205	23,660	38.0	236 (tied)
10 Best					
Sunnyvale	California	26,490	1,571	5.9	10 (tied)
Bellevue	Washington	22,838	1,353	5.9	10 (tied)
Scottsdale	Arizona	38,457	2,177	5.7	9
Pembroke Pines	Florida	34,448	1,883	5.5	8
Thousand Oaks	California	29,726	1,616	5.4	6
Westminster	Colorado	26,681	1,440	5.4	6
Plano	Texas	62,981	3,082	4.9	5
Overland Park	Kansas	37,993	1,209	3.2	3 (tied)
Livonia	Michigan	23,855	761	3.2	3 (tied)
Gilbert	Arizona	37,239	1,173	3.1	2
Naperville	Illinois	40,818	952	2.3	1

Exhibit 10.2b   Big Cities (100,000 or More Population) With Highest and Lowest Child Poverty Rates in 1999

Source: Data from the U.S. Census Bureau, Census 2000. Ranking by Children's Defense Fund.

We miss the truly sweeping effects of these contradictions, however, if we assume they are only the pathologies of the largest cities. Global economic trends have also changed the nature, and the faces, of small cities and urban regions with a similar mix of prosperity and poverty. Global change is also remaking the countryside. Some of the cities in the United States with the highest levels of inequality are those that are heavily dependent on commerce, government, and services in the new global economy. Their gaps between rich and poor rival those of the most unequal countries in the world (see Exhibit 10.2a). Poverty abounds in cities across the country, largely absent only from elite new "edge cities" at the periphery of major metropolitan areas (see Exhibit 10.2b).

Brain-Gain and Brain-Drain Cities

The loss of industrial jobs has been partially offset by gains in information technology jobs, jobs in research and development, and other high-technology employment. This shift gained momentum in the 1980s and has continued into the twenty-first century. Rarely, however, have job gains and losses taken place in the same locations, even though at times they have been in close proximity. The industrial small cities in the San Francisco Bay Area withered while nearby San Jose and the "Silicon Valley" boomed. Flint, Michigan, spiraled downward to take first place in *Money* magazine's annual listing of the worst places to live in the United States, as General Motors closed one plant after another and unemployment soared. At the same time, Ann Arbor, only a short drive down the highway, was booming. Similarly, the year that *Money* named Rockford, Illinois, as the worst place to live, the magazine placed Madison, Wisconsin, only a short drive to the north, on the top of its list of the best places to live. In Texas, Austin boomed while McAllen experienced record poverty. Albany, New York, thrived while declining urban fortunes followed the entire string of cities along the old Erie Canal all the way to Buffalo. The cities with declining fortunes all have one thing in common: They had all been centers of old industry and were not well situated to attract new enterprises. The growing cities are centers of government and education: Austin, Albany, and Madison are state capitals and also home to major universities. Ann Arbor is situated at a convenient exurban distance from Detroit and is home to the University of Michigan. In some cases, the cities experiencing growth with new technology and service industries are right next to struggling industrial cities. For example, Ann Arbor is near Ypsilanti, and Albany is near Schenectady. In these particular cases, a housing boom in one city ultimately began to benefit the other. Yet rarely can the unemployed in one city find work down the highway in the other, as the nature of the jobs has changed.

Richard Florida (2002, 2004) describes the rise of "brain-gain" cities— that is, cities with both the jobs and the cultural and entertainment offerings that attract college graduates, particularly bright, creative, multitalented people. His list of brain-gain cities includes those mentioned above as well Atlanta, Boston, Denver, Minneapolis, San Diego, San Francisco, Seattle,

Portland (Oregon), Raleigh and Durham (North Carolina), and Washington, DC. He notes that these cities are attractive because they look to the future rather than the past, perhaps because they have progressive leaders. Brain-gain cities also have very tangible advantages over other cities in attracting highly educated people: The government offices and corporate headquarters located in these cities seek well-educated employees, their universities produce research spin-offs, and their commercial hubs bring in new commerce, including the dot-com variety. Sometimes these cities can use their advantages to accelerate their gains. As Florida notes, some routinely poach talent from their struggling neighbors: Baltimore, Cleveland, Milwaukee, St. Louis, and San Antonio. Brain-drain cities lose jobs, and then they lose the people who are most likely to create new jobs. This is a problem for older big cities, but it is most severe for small to midsize cities that do not have broad economic bases to attract newcomers.

The expression *brain drain* was first coined to describe the plight of countries that educate their citizens only to watch them flee, with their educations, to places where they can enjoy higher returns on their education. India and the Philippines have been classic brain-drain countries. Their graduates in the health and technology fields have often found better, and maybe their only, options overseas. The growing trend in the United States toward moving white-collar jobs offshore may begin to offset this, at least in the technology sector (having an offshore doctor is still an inconvenience that hasn't been overcome). Cities in India such as Hyderabad and Mumbai (formerly known as Bombay) are seeing thriving and dynamic technology and marketing sectors while other parts of the country continue to struggle with massive unemployment and poverty.

In one sense, the brain-gain/brain-drain phenomenon is just the labor market at work. People are pushed from areas of low opportunity and pulled toward areas of high opportunity. But only certain types of workers can make these moves at certain times: young, college educated in the right fields, without constraining family ties during times of economic growth. For others, the prospects are dimmer. The unemployed steelworker in Allentown, Pennsylvania, or Gary, Indiana, may find little in Seattle or San Diego except a higher cost of living. Yet as the young and the educated leave older communities, both opportunities and services decline, leaving pockets of stagnation, unemployment, and poverty. The metropolitan counties with the highest poverty rates represent some of these struggling places (see Exhibit 10.3).

Suburban Poverty: The Urban Fringe

In Latin America, poor communities often surround large urban areas. Traditionally, elites have been concentrated near the city centers and the poor have been pressed out to the margins of metropolitan areas, living in shantytowns that climb steep mountainsides or are carved from steep ravines and other inhospitable land. In the United States, people have come to associate

Rank	County	Percent	Major City
1	Hidalgo County, TX	36.2	McAllen
2	Cameron County, TX	34.8	Brownsville
3	Bronx County, NY	29.8	New York (Bronx)
4	El Paso County, TX	26.7	El Paso
5	Tulare County, CA	25.0	Visalia
6	St. Louis City, MO	24.0	St. Louis
7	Nueces County, TX	23.1	Corpus Christi
8	Orleans Paris, LA	21.7	New Orleans
9	Philadelphia County, PA	21.2	Philadelphia
10	Baltimore City, MD	20.6	Baltimore
11	Kings County, NY	20.4	New York (Brooklyn)
12	Fresno County, CA	20.1	Fresno
13	Kern County, CA	19.8	Bakersfield
14	Mobile County, AL	19.6	Mobile
14	New York County, NY	19.6	New York (Manhattan)
16	Fayette County, KY	18.8	Lexington
16	Caddo Parish, LA	18.8	Shreveport
18	Shelby County, TN	18.5	Memphis
19	Miami-Dade County, FL	18.3	Miami
20	District of Columbia, DC	17.5	Washington, D.C.
21	Jefferson Parish, LA	17.4	
22	Suffolk County, MA	17.2	
23	Oklahoma County, OK	16.6	
24	Denver County, CO	16.4	

Exhibit 10.3   Percent of People Below Poverty Level Population in Metropolitan Counties

Source: Data from U.S. Bureau of the Census, 2002.

poverty with the central city. Increasingly, however, it is spilling into the urban fringe, amid and beyond the better-off suburbs. This movement is driven in part by the gentrification of some older neighborhoods. Well-off urbanites who are tired of long commutes move back into older central-city neighborhoods, buy up and fix up old houses, and drive up property values, often displacing the poor in the process. Public housing is also scarce in many large urban areas. Older projects are being dismantled at the same time newer projects are being delayed. For increasing numbers, this means that the only affordable housing is on the urban fringe: in poor satellite communities, in trailer parks, in adjoining areas where old industry, a concentration of railroad yards, or some other feature makes them undesirable for middle- and upper-class suburban development.

A major trend in the United States for more than the past decade has been the suburbanization of poverty. Despite stereotypes of poor inner-city ghettoes and remote mountain "hollers," the greatest growth in U.S. poverty is in the suburbs. In some ways, this simply mirrors the sweeping suburbanization of the entire country. It also reflects the fact that entire rings of U.S. suburbs are now more than 50 years old. Some of these suburbs are not aging gracefully. In many U.S. cities, people with the means to seek the good life have two options. Some are returning to the city center, buying up and gentrifying old brownstones, quaint flats, and townhouses. A new generation of upscale urban apartments and condominiums is also encouraging movement back to the city. The new urban gentry tend to be people who have no children: singles and young professional couples, childless two-career couples, and increasing numbers of "empty nesters" whose children have grown up and left home. For these people, urban life provides a more interesting setting than the suburbs and allows them to avoid long commutes on increasingly congested urban expressways, many of which are now also half a century old. These newcomers, with no young children at home, are not particularly concerned about the quality of urban schools or with finding safe places for children to play. Families with means who are concerned about their school districts and want to have more space are moving ever further out, beyond the ring of inner suburbs to "exurbia." New suburbs with houses built on larger lots, new exurban developments, and former quaint small towns with new development have all become part of this movement.

Left in between the city and exurbia are the older suburbs. Some of these attract new immigrants and Latino workers, as central cities once did. It doesn't hurt to be able to speak Spanish if one does business in Arlington, Illinois, just outside of Chicago. In some of these suburbs, business signs display messages in several alphabets, as immigrants and refugees from Asia, the Caucasus, and Eastern Europe offer a mix of services and wares. These are struggling suburbs, hoping they are on their way up.

Other older suburbs serve those who find themselves on their way down. Suburbs have always been places for families, and now this increasingly includes poor families, often single mothers and their children. Divorce and/or unemployment may have devastated their incomes, but they still feel most at home in the suburban environment. Their "fall from grace," as Katherine S. Newman calls it (1988), may have been dramatic. She tells the story of those who are down and out in Bel-Air (the elite area above Beverly Hills), sleeping in their Lincoln Town Cars or on their friends' $5,000 couches. More often, the shift is less dramatic: Those who always considered themselves middle-class find that their incomes have fallen below the poverty level for their family size. Urban fringe poverty can be quite apparent and concentrated, as in a trailer park on the city's edge, but more often it is dispersed and hidden. The poor residents of the urban fringe may seek employment in the city but face many of the other problems of the rural poor: lack of reliable transportation, lack of social services and accessible health care, isolation, and lack of political clout.

Rural Poverty: Forgotten Places

Around the world, the poorest communities with the least access to services are those that are rural. This is sometimes forgotten in the emphasis on urban poverty. Compared with urban poverty, rural poverty is more hidden, gets less policy attention, and makes duller settings for prime-time police shows. Rural poverty got U.S. attention in the 1930s as Appalachian coal mines went bankrupt and plains farmers were driven from their land by bank foreclosures and the Great Dust Bowl. Franklin Roosevelt's presidential administration provided some relief and sponsored programs for rural electrification, but then public attention to the issue subsided. It arose again in the 1960s with the publication of Michael Harrington's book *The Other America* (1962). Robert Kennedy toured poor counties in the Southeast, and Martin Luther King, Jr., launched his "poor people's campaign" across the same region. But with the urban unrest and riots of the later 1960s, America's attention was again riveted on the problems of cities. The farm crisis of the 1980s drew some attention to rural poverty, as farmers in the Midwest lost their land to low commodity prices, high interest rates, and agribusiness competition. The farm crisis also faded from public attention, however, even though small farms are still being lost at a rapid rate.

Although urban problems and urban poverty are again currently dominating the public policy agenda, the poorest places in the United States remain rural counties: the Pine Ridge and Rosebud Sioux reservation areas of South Dakota, the rural Hispanic communities of New Mexico, and the black farming communities of Mississippi and Louisiana (see Exhibit 10.4).

Rural poverty, especially rural nonfarm poverty, is less likely than urban poverty to gain public attention in part because by its very nature it tends to be hidden in out-of-the-way places such as bypassed small towns and hidden valleys and "hollows." Compared with the urban poor, the rural poor are more likely to be working poor and less likely to make use of public assistance, sometimes out of pride and sometimes simply because assistance is not offered nearby or is difficult to access. If chemical dependency is part of the problem, it is most likely to involve alcohol, or possibly marijuana, and less likely to involve drugs that get first attention from law enforcement. Likewise, crime among the rural poor is not of the sort to get much attention or make good television programming: petty theft, drunken driving, domestic disputes. Yet a map showing violent crime rates by county in the United States almost perfectly replicates the shaded areas in the map in Exhibit 10.4b. The rates are higher in rural counties with persistently high poverty and high unemployment than they are in any U.S. city.

Two-parent families are more common among the rural poor than among the urban poor, although both parents may be underemployed. The housing situations of the rural poor may raise an occasional eyebrow or comment, but they don't generate any great public outcry. Typical housing may be an aging trailer home with a plywood addition and porch on the front and a

County	State	Rural or Metropolitan	Total Persons Under 18	Number in Poverty	Percent in Poverty
Buffalo	South Dakota	Rural	840	519	61.8
Ziebach	South Dakota	Rural	1,021	625	61.2
Shannon	South Dakota	Rural	5,395	3,292	61.0
Starr	Texas	Rural	19,948	11,875	59.5
Todd	South Dakota	Rural	3,827	2,210	57.7
East Carroll Parish	Louisiana	Rural	2,830	1,607	56.8
Owsley	Kentucky	Rural	1,180	666	56.4
McDowell	West Virginia	Rural	6,279	3,325	53.0
Madison Parish	Louisiana	Rural	4,131	2,172	52.6
Holmes	Mississippi	Rural	6,839	3,583	52.4
Brooks	Texas	Rural	2,481	1,286	51.8
Wolfe	Kentucky	Rural	1,807	930	51.5
Humphreys	Mississippi	Rural	3,613	1,825	50.5
Sharkey	Mississippi	Rural	2,174	1,095	50.4
Perry	Alabama	Rural	3,469	1,708	49.2
Wilkinson	Mississippi	Rural	2,641	1,300	49.2
Zavala	Texas	Rural	3,902	1,911	49.0
Bennett	South Dakota	Rural	1,254	612	48.8
Corson	South Dakota	Rural	1,481	722	48.8
Wilcox	Alabama	Rural	3,975	1,928	48.5
Tensas Parish	Louisiana	Rural	1,734	838	48.3
Leflore	Mississippi	Rural	11,219	5,406	48.2
Allendale	South Carolina	Rural	2,949	1,417	48.1
Clay	Kentucky	Rural	5,961	2,852	47.8
Sumter	Alabama	Rural	4,256	2,028	47.7
Edwards	Texas	Rural	607	288	47.4
Luna	New Mexico	Rural	7,434	3,501	47.1
Jackson	South Dakota	Rural	1,038	481	46.3
Zapata	Texas	Rural	3,983	1,839	46.2
Coahoma	Mississippi	Rural	9,910	4,564	46.1
Magoffin	Kentucky	Rural	3,535	1,627	46.0
Jefferson	Mississippi	Rural	2,745	1,263	46.0
Mellette	South Dakota	Rural	719	331	46.0
Hidalgo	Texas	Metropolitan	198,744	90,831	45.7
Webster	West Virginia	Rural	2,205	1,008	45.7
Phillips	Arkansas	Rural	8,428	3,845	45.6
Hancock	Georgia	Rural	2,432	1,105	45.4
Martin	Kentucky	Rural	3,501	1,591	45.4

Exhibit 10.4a

Source: Data from the U.S. Census Bureau, Census 2000. Ranking by Children's Defense Fund.

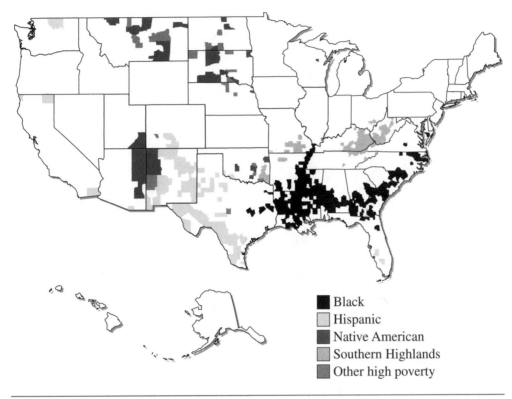

Black
Hispanic
Native American
Southern Highlands
Other high poverty

Exhibit 10.4b   38 Counties Having Higher Child Poverty Rates Than the Poorest Big Cities, 1999

Data from the U.S. Census Bureau, Census 2000. Calculations by the Children's Defense Fund. Map prepared by the Economic Research Service.

still older outbuilding behind. Home heating may be minimal and often dangerous, provided by old kerosene heaters and stoves. Natural hazards such as tornados are a real concern, but the biggest danger to such a home is usually fire from a hazardous heating system. Several cars may be scattered on the lot around the trailer, not out of carelessness but because of the need to keep at least one vehicle running by scavenging parts from others; professional repairs are expensive, and the rural poor cannot depend on public transportation. The residents may work in neighboring communities or seek the handful of remaining rural jobs, such as working on the county road crew. Many of these jobs are seasonal.

Many Americans tend to view rural poverty as something associated with certain locations and earlier times, such as the rural South of Faulkner's novels or Harper Lee's *To Kill a Mockingbird*. According to anthropologist Janet M. Fitchen (1981, 1991), however, who has devoted her career to studying rural poverty in the northeastern United States, rural poverty is growing and extending into new regions. Likewise, sociologist Cynthia Duncan (1992, 1999) has found that rural poverty is alive and well in Appalachia, in the U.S. South, among migrant workers and on American Indian reservations, and even in New England.

The Black Belt

The region of the United States often referred to as the black belt is a long, slightly curved region that stretches from the rural Carolinas across northern Georgia, Alabama, Mississippi, and Louisiana. The name refers not to the martial arts movies playing in the region's old movie theaters but to the fact that many of the counties there are majority black. The bulging buckle of this belt consists of low-lying counties on either side of the Mississippi River in Louisiana and Mississippi, an area long known simply as the Delta. In the past few decades, African American poverty in the United States has come to be viewed as an inner-city problem, but the poverty in the black belt is largely rural. This was cotton country: Many of the region's residents are descended from slaves. When slavery ended, many of their ancestors became sharecroppers—working the land for white landowners who took a share of what they produced. They borrowed money for planting from the landowners and so remained in a state of perpetual debt, living lives that were not much different from those they had under slavery. Over time, many left the area for the urban North and West, but others remained. Some still work as farm laborers, others work in sawmills or whatever industry may be present. One thriving industry in the area has been food processing (as described in Chapter 3).

Much of the South is changing rapidly. Coastal areas are filled with new retirement communities, resort developments, and beautiful seaside estates. Cities such as Atlanta pride themselves on their glittering modernity. Some highland cities with attractive locations and educated populations, such as Chapel Hill and Asheville, North Carolina, are attracting new high-technology industry and prosperous newcomers. The black belt stands between and largely apart from these developments, instead occupying the hottest, most depleted locations with enduring intergenerational poverty and little but the most arduous, most hazardous, and often lowest-wage employment in the country.

American Highlands

Standing above the black belt geographically if not always economically are the highland regions of long-standing poverty in Appalachia and the Ozarks. Highland poverty ranges over a vast area, but its greatest concentration lies in eastern Kentucky and parts of West Virginia. The population here is primarily white, although some African Americans and occasionally

Native Americans such as the Choctaw and Cherokee live here as well, along with those whose ancestry mixes these various groups. Farms have dotted this area for generations but have always been marginal, given the difficult terrain and soil. Coal was king for a while in the eastern portion of this region. Mining was always demanding, dangerous work that could shorten lives abruptly, through accidents, or slowly, through lung disease. Eastern high-sulfur coal is no longer favored by industry, and many of the mines are now limiting operations. Sawmills still operate, but they offer their own hazards and unpredictable employment. Certain locations within this region have boomed with tourist-driven economic growth, such as Pigeon Forge, Tennessee, and Branson, Missouri. For those who are not headline entertainers or well-heeled entrepreneurs, most jobs are in low-wage service activities (there are a lot of new motel beds to be made up). The prosperity of these locales has been slow to extend to outlying areas, just as in the southern lowlands, where, as Thomas A. Lyson (1989) has noted, there are "two sides to the Sunbelt" and places that are "high tech, low tech, and no tech" (Falk and Lyson 1988).

The Barrio Border and the Rural Southwest

Wooden one-room shacks cluster under the sparse shade of a few cottonwoods. They have no running water and no electricity. Their occupants work in fields and seek day jobs in the city in the off-season. Some think about leaving their families to try their luck in the North. Life on the border is hard. These people live not on the Mexico side of the border but on the Texas side. They have arrived in the United States, but they certainly have not "arrived" at economic success in U.S. society. Having traveled through the rural Southwest toward Mexico, Robert D. Kaplan (1998) contends that the real border he encountered runs through the middle of Tucson; it is not the one the Border Patrol tries to guard a bit further south. North Tucson is primarily Anglo and has prospered with tremendous growth, many newcomers from the North, and spin-off industries from the University of Arizona. South Tucson is primarily Latino, gains most of its newcomers from the South, and remains poor.

It shares this poverty with much of the rural Southwest. Hispanic Americans whose families have been in northern New Mexico since the 1500s (long before there was either a Mexico or a United States) and those who have been in the United States only a few months both remain at the bottom of the economic ladder in this growing region. Newcomers in Texas often work as migrant farm laborers, picking vegetables or melons on the large irrigated commercial farms and then heading north, sometimes as far north as Michigan, to pick vegetables and cherries during their limited seasons. Farms across the United States depend on these workers, and some farm owners have even begun to lobby for relaxed immigration rules or guest worker programs, such as the old Bracero Program, so that they can have access to more workers. In northern New Mexico, the farmers tend to work their own land. Yet the prosperity of high-tech Los Alamos (in the second wealthiest county in the nation), trendy Santa Fe, growing Albuquerque, and ski-resort-filled Taos escapes them as they work the dry, rugged land with limited irrigation and mechanization. This land also includes Native American reservations, including the largest in the United States, occupied by the Navajo. Miners covet some of this land and tourists visit other portions, but throughout most of it, rural poverty still prevails. Select international students who come to the well-endowed Armand Hammer United World College of the American West study in a U.S.-styled castle in Montezuma, New Mexico, but are often dismayed as they travel out into neighboring Mora County, the 15th poorest in the country, where they see rural poverty that looks much like that found in the developing countries from which they came.

The Central Plains

In *The Grapes of Wrath* (1939), his classic novel set against the 1930s Dust Bowl that swept the great swath of plowed-up grassland that is the Great Plains, John Steinbeck depicts a conversation between tenant farmers scratching at the exhausted land and the spokesmen for the landowners and banks that are evicting them. "Sure, cried the tenant men, but it's our land. We measured it and broke it up. We were born on it, and we got killed on it, died on it. Even if it's no good, it's still ours." But the spokesmen remind them of bigger structural forces:

> Well, it's too late. And the owner men explained the workings and the thinkings of the monster that was stronger than they were. . . . You see, a bank or a company . . . those creatures don't breathe air, don't eat side meat. They breathe profits: they eat the interest on money. If they don't get it, they die. . . . When the monster stops growing, it dies. . . . The bank is something more than men, I tell you. It's the monster. Men made it, but they can't control it. (P. 2)

Each of the past U.S. Censuses has shown that the portion of the country most consistently losing population comprises the farm states of the Great Plains. As farming consolidates into ever larger, ever more mechanized operations, there are fewer small farms and fewer farmers. This in turn affects the many small communities that grew up to serve the needs of small farms; businesses close and downtowns become all but deserted except for a café and a tavern where the remaining local residents can mourn their losses.

Small farmers have faced multiple crises over the past hundred years. The Dust Bowl crisis of the 1930s was brought on by drought and soil erosion but also by low prices and heavy debt. Similarly, bad weather contributed to the farm crisis of the 1970s and 1980s, but mostly the crisis arose because of farmers' heavy borrowing (which the government had encouraged) and falling commodity prices. Now another farm crisis is occurring, created by unpredictable weather patterns, a new blight affecting wheat crops, rising interest rates, and grain prices at all-time lows. Once again, many small farms are being lost and their equipment auctioned. Because modern farming requires huge inputs of land and materials yet offers modest per acre returns, small farmers have had to borrow heavily. Currently, the total debt of small farmers in the United States exceeds that of all the countries of Latin America put together. This debt burden, coupled with a dependence on distant markets that they cannot control or anticipate, leaves farmers continually vulnerable and often extremely income poor, even if their land itself amounts to considerable wealth.

The poorest people on the Great Plains are not grain farmers, however, but the scattered residents of several large Indian reservations. They live in areas too remote to be likely candidates for casinos and tourism, and their access to capital and resources is very limited. Rural Great Plains reservations have extraordinary unemployment rates, topping 30% even amid national economic growth, as well as the country's highest rates of poverty, alcoholism, and family disruption.

For all their differences, the rural poor groups described above are all faced with similar choices. They can leave the lands and ways of life their families have known for generations and stake their hopes on uncertain futures in distant cities that may or may not have a place for their skills, or they can remain where they are and hope that new opportunity will come to their communities—a hope that has yet to be realized in any of these places (see Exhibit 10.5).

Poverty in the Global Ghetto

Measuring poverty in a world of enormous income and lifestyle differences and divides is not an easy task. The United Nations Development Program (2004) estimates that at least one-third of the world is poor. By "poor," the UNDP means desperately poor. One way this organization measures poverty is by setting the arbitrary but revealing baseline of income of less than $1 per day. At least 1 billion people in the world currently fall below this level. Another 1 billion exist on less than $2 per day. A second, maybe more meaningful, measure of absolute poverty used by the UNDP is the inability to satisfy basic needs for nutrition, health, and education. At least 1.5 billion are poor by this standard, probably closer to 2 billion or more. The poorest continent is Africa, where as much as half the population meets the UNDP description of absolute poverty. Asia has become a land of rich and poor, but with its huge population it may hold more than two-thirds of the world's

Characteristic	Nonmetro	Metro
Median household income (dollars)	34,654	45,257
Poverty population (millions)	7.5	27.1
Prevalence of poverty (percent)	14.2	11.6
Poverty by region of U.S. (percent)		
Northeast	10.7	10.9
Midwest	10.7	10.1
South	17.5	12.7
West	14.3	12.1
Poverty rates for selected groups (percent)		
Non-Hispanic Blacks	33.2	22.7
Non-Hispanic Whites	11.0	7.2
Hispanics	26.7	21.4
Children (younger than age 18)	19.8	16.0
Elderly (age 65 and older)	11.9	10.0
Nonelderly adults (age 18-65)	12.4	10.2
Educational attainment of poor adults (percent)		
Less than high school education	44.5	40.0
No more than high school education	32.8	30.7
More than high school education	22.8	29.3
Income levels of poor adults (percent)		
Less than half the poverty line	36.4	41.9
Between half and 75 percent of the poverty line	28.2	25.5
Greater than 75 percent of the poverty line	35.4	32.6

Exhibit 10.5a   Selected Poverty Rates and Economic Indicators, 2002

Source: U.S. Department of Agriculture, 2004.

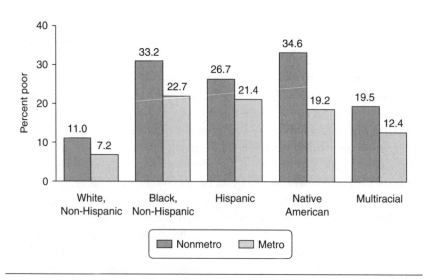

Exhibit 10.5b   Poverty by Region and Characteristics

Source: U.S. Department of Agriculture, 2004.

poor. Some gains have been made in absolute poverty, especially in China and India, but relative poverty is probably more acute than ever. There are now few places so remote that the inhabitants are unaware of the world of riches; billboards depict luxury in the poorest cities, and satellites broadcast American television programs to the most out-of-the-way villages.

The world's poorest people are rural. Worldwide, rural poverty may account for three-quarters of the planet's poor: small farmers, landless peasants and migrant laborers, charcoal gatherers, and others. These people often lack access to even the minimal health and housing facilities available to the poor of the cities; they often have no safe, drinkable water or adequate nutrition and have the world's shortest life expectancies. Yet urban poverty is increasing. At the beginning of this millennium, the world passed an important milestone: According to U.N. estimates, a majority of the planet is now urban. Desperate people are moving from rural to urban areas around the world in search of jobs, education, and basic services. Many remain very poor. In income terms they may indeed raise their standard of living by moving, but they also become acutely aware of relative poverty as they rush up to the windows of elegant cars to sell trinkets or wash windshields and as they huddle around televisions that show them more of what they do not have.

Many of the world's poor are confined to the margins—both the margins of the burgeoning cities and the margins of the new global economy. Some literally inhabit the garbage dumps of the great world cities. In the Philippines, they wade through the torn plastic bags and mounds of refuse outside their doors and stumble amid the smoke of ever-burning fires of rubber and rubbish on "Smoky Mountain," a great hill of trash, to gather the usable and salable refuse of Manila's urban explosion in a scene that evokes the biblical Gehenna, the New Testament's image of hell. Outside of Tijuana, dump dwellers slog through the black mud and toxic yellow ooze of another mountain of trash, fighting wild dogs for usable scraps. Next to this dump is a vast cemetery of buried and half-buried children. Here, amid one of the most horrific wastelands on the planet, they make their homes and their plans (Urrea 1996). At night these people can climb the great hills of trash and admire the bright lights of San Diego shimmering in the desert air just a few miles away.

Swedish economist Gunnar Myrdal (1970), observing trends in Europe, the United States, and the developing countries, wrote of an urban and rural "underclass" that is cut off from society, its members "lacking the education and skills and other personality traits they need in order to become effectively in demand in the modern economy" (p. 406). Poverty persists in advanced industrial economies, with the United States having the highest poverty rates among advanced industrial countries, just behind Great Britain and Ireland (see Exhibit 10.6). The trend for economic globalization to leave in its wake a segment that is excluded from full economic participation, political power, and social integration is growing. Latin American scholars have analyzed this trend as marginalization. The poor on the urban margins

High human development rank	Human poverty index (HPI-2)		Probability at birth of not surviving to age 60 (% of cohort) 2000–05	People lacking functional literacy skills (% ages 16-65) 1994–98	Long-term unemployment (% of labour force) 2002	Population below income poverty line (%)	
	Rank	Value (%)				50% of median income 1990–2000	$11 a day 1994–95
Sweden	1	7.1	8.3	8.5	0.2	6.4	4.3
Norway	2	6.5	7.3	7.5	1.1	6.5	6.3
Netherlands	3	12.9	8.8	17.0	1.3	14.3	17.6
Finland	4	12.2	8.7	16.6	0.7	12.8	7.4
Denmark	5	8.2	8.7	10.5	0.8	7.3	7.1
Germany	6	12.4	9.4	18.4	3.4	8.0	—
Luxembourg	7	15.8	12.6	20.7	0.5	17.0	13.6
France	8	11.1	7.5	—	1.7	11.8	—
Spain	9	15.3	9.3	22.6	1.2	12.3	—
Japan	10	14.8	8.9	21.8	1.2	12.5	15.7
Italy	11	8.4	10.2	10.4	2.2	5.4	4.8
Canada	12	10.5	9.7	—	0.7	6.0	0.3
Belgium	13	10.8	10.0	—	3.0	8.0	9.9

High human development rank	Human poverty index (HPI-2)		Probability at birth of not surviving to age 60 (% of cohort) 2000–05	People lacking functional literacy skills (% ages 16-65) 1994–98	Long-term unemployment (% of labour force) 2002	Population below income poverty line (%)	
	Rank	Value (%)				50% of median income 1990–2000	$11 a day 1994–95
14 Australia	14	9.1	11.0	9.6	0.8	9.2	—
15 United Kingdom	15	10.3	9.2	14.4	4.1	8.3	7.3
16 Ireland	16	11.0	8.8	—	4.6	10.1	—
17 United States	17	11.6	8.6	—	5.3	12.7	—

Exhibit 10.6 Advanced Industrial Countries

Source: United Nations Human Development Report, 2004.

are often also on the margins of the urban economy and the life of the society. In a book titled *Networks and Marginality* (1977), a study of Mexico City, Chilean-born anthropologist Larissa Adler Lomnitz terms this group "the hunters and gatherers of the new urban jungle." In North America, scholars such as sociologist William Julius Wilson have continued Myrdal's use of the term *underclass,* often with strong racial implications. While conceding that the term can be used as an attack on the supposed failing value system of the poor, Wilson (1987) uses it to refer to the extensive social isolation of the new urban poor. In Europe, the preferred term has been *social exclusion* (Bhalla and Lapeyre 2004), and the emphasis has been on the fraying of the social fabric of basic supports. The extent and composition of the excluded groups vary across the continents, but the problems in all cases are rooted in weak attachment to the formal labor market (unemployment in Europe, underemployment in North America and Latin America), which leads to, and is compounded by, residential and social isolation and political disempowerment. The global village contains a multinational, multiracial ghetto whose inhabitants have not been welcome to share in the village's riches, and the global economy has become an economy of exclusion (Sernau 1994) for this global underclass. One of the great policy challenges of the twenty-first century is to find ways to reincorporate the excluded into the vital life of national societies and the global economy.

In the developing world, the problems of the underclass are only magnified. Unemployment in Kinshasa, Congo, has topped 60%. In many Latin American cities, those making do the best they can in the informal economy may constitute half of the urban population (Portes, Castells, and Benton 1989). Further, government interventions are limited in developing countries by low incomes, national deficits, and international debt. When countries in financial crisis turn to the International Monetary Fund (IMF) for assistance, they are typically ordered to undergo "structural adjustment." This often means devaluing their currency, removing price controls, privatizing industry and banking, and cutting government spending. These measures are intended to promote the fiscal stability of a country and control inflation. Yet even the IMF and the World Bank have come to admit that the people most often hurt by these requirements are the poor. Food prices soar, government health and social services are cut way back, schools and clinics close. The IMF contends that this is necessary. The World Bank asks governments to give some attention to emergency relief for the poor, but it also contends that the pain is necessary. Critics assert that the national economies of poor countries are being balanced on the backs of the poor and that the least advantaged are being starved to pay interest on loans to wealthy countries and to secure new loans (see Exhibit 10.7).

Dennis Gilbert (2003) argues that what he terms the "age of growing inequality" in the United States is largely explained by the greater freedom given to market forces. This is not just a U.S. phenomenon but a global

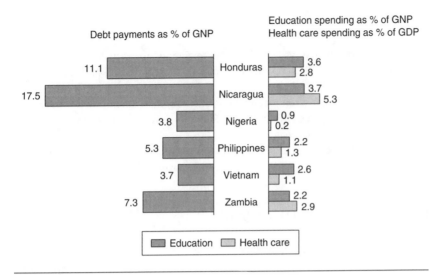

Exhibit 10.7   Debt's Burden

Source: World Bank, *World Development Report*, 1999/2000.

experiment: Around the world, governments are giving greater play to market forces, by design or in response to international demands. Gains have been made in controlling runaway inflation, bloated bureaucracies, and inefficient systems. Yet a common result has been a growing divide between the well-off and the very poor.

Reducing global poverty will certainly require international action and determination on the part of both rich and poor countries. One proposal for addressing global property emerged from the work of a variety of church and human rights groups and came to be called Jubilee 2000; it involved declaring the year 2000 a "jubilee" year to forgive debts, as described in the Old Testament. Wealthy creditor countries would agree to forgive the unre-payable debt of the poorest nations, and these poor countries in turn would agree to use the money they would have spent on debt service to address their people's needs in the areas of health care, education, poverty reduction, and basic survival. An earlier proposal that emerged from U.N. conferences on social development was called the 20/20 proposal. High-income donor countries would agree to focus at least 20% of their foreign assistance on basic needs provision for the poor (most now goes for military aid), and the poor countries themselves would agree to spend at least 20% of their domestic budgets on health care, education, and basic needs for their poorest citizens. This modest but far-reaching proposal has yet to be fully implemented by either rich or poor nations. The beginning of the millennium has come and gone, and wealthy donor countries are still undecided about participating. The world's 2 billion poor are still waiting.

KEY POINTS

- Urban poverty has grown with the migration of many low-income rural dwellers to the cities and the flight of industry, commerce, and well-off residents from city centers.
- Rural poverty has often received less attention than urban poverty, but many of the poorest places in the United States and around the world are isolated and exploited rural areas.
- Around the world, economic growth has been coupled with growing income divides and persistent, or even worsening, poverty.

FOR REVIEW AND DISCUSSION

1. What locations in the United States have experienced severe urban and rural poverty? What have been the major causes of this concentration of poverty? How does rural poverty differ from urban poverty?
2. What are the major causes and effects of poverty around the world? What are the elements in webs of social exclusion for the world's poor? What kinds of proposals have been offered to help alleviate the worst conditions of poverty around the world?

MAKING CONNECTIONS

U.S. Census

Look at the *County and City Data Book* published by the U.S. Bureau of the Census or check the county and city data available online at www.census.gov. You will find a great deal of information on the distribution of poverty. Look at the data on poverty for your state or region. Are there local concentrations? What accounts for the patterns of poverty you find? How is poverty correlated with patterns of race and ethnicity, the local labor market, and patterns of job growth and decline?

United Nations

See the U.N. Development Program's latest *Human Development Report,* available in hard copy and online at www.undp.org, to get an overview of global patterns of poverty and development. Where do you see signs of hope in alleviating poverty? Where do you see signs of decline? What seem to be the most important factors in reducing poverty?

Community Development

Many low-income individuals and families live in areas of concentrated poverty and limited opportunity. Where are these locations in your community? What efforts are being made toward positive change in these locations? Visit a community center, neighborhood association, nonprofit community development corporation, or other community-based organization working in a low-income community. Learn about the community demographics: Who lives there, in terms of income, race/ethnicity, family composition, age, and so forth? Have the area's demographics changed over time? Is the area undergoing neighborhood transition? What is the housing situation—many absentee landlords, subdivided large houses, apartments, public housing? Who, if any, are the area employers and retailers? What are the neighborhood's most pressing problems and what is being done to address them? Are local residents actively involved in these efforts? You may find opportunities to engage in community improvement activities, on your own or with a group. If you become involved in this way, note your observations.

Photo Essay: Deindustrialization

Elena Grupp

South Bend, Indiana, grew from a center of regional trade to a major manufacturing center with the growth of firms such as the Oliver Chilled Plow Works and the Studebaker Company, which went from making wagons to automobiles.

When these operations closed in the early 1960s, South Bend became the first major U.S. city to experience deindustrialization. Unionized manufacturing jobs were lost, and the city has struggled to revitalize the "Studebaker Corridor" of abandoned multistory factory buildings. Some buildings were used by the school corporation, and a few near the water became restaurants and apartments. Almost none, however, have found use in manufacturing, their multistory brick frames considered obsolete.

The Oliver Plow Works chimney now stands in isolation in an open field as a monument to an earlier time. The rest of the Works was torn down to make room for new industry that has yet to arrive.

This derelict Studebaker factory is hidden from public sight by newer buildings surrounding most of the property. Unused train tracks run through the long grass. To capture this shot, it was necessary to trek behind the building through a rarely frequented alleyway and shoot through a chain-link fence.

A "For Rent" sign beside a Studebaker warehouse indicates that there are still some functional areas inside of the building. The sign's location on busy Sample Street doesn't seem to have made up for the unfortunate condition of the building, however. This building is across from the recently constructed St. Joseph County Jail, which was built on the demolished Avanti & Standard Surplus sites.

Liberal quantities of razor-wire, barbed-wire, and chainlink fences coupled with heavy policing make this Studebaker factory, like most others, hard to get into. Because there have been few attempts at renovation or even meeting minimal safety requirements, it would be unwise, says the city, to attempt entrance.

The rise of Studebaker and Oliver gave rise to great mansions, converted now into a museum and a restaurant; and also to streets filled with small homes for workers. Some of these small buildings still stand, mostly as absentee-owned rentals in transitional neighbor-hoods.

One resident seems to seek security from rising local crime rates in both divine protection and the more worldly protection of ADT security services.

South Bend's economy is now built around the service sector: hospitals, universities, accounting firms, and others. This has dramatically changed both the face of the city and the structure of the labor market. These transitions, and the abandoned spaces, have been repeated in nearby Gary, Indiana (US Steel); Flint, Michigan (General Motors); and across the Northeast and North Central United States.

11

Reversing the Race to the Bottom

Poverty and Policy

The freest government cannot long endure when the tendency of the law is to create a rapid accumulation of property in the hands of a few, and to render the masses poor and dependent.

—Daniel Webster (1782–1852)

From Welfare to Work

In 1999, New York Mayor Rudolph Giuliani proposed that residents of the city's homeless shelters should be required to work. Some saw this as largely a punitive policy, citing as evidence the attitude of one of the mayor's advisers, who suggested that the poor, adults and children alike, should be given green suits similar to those issued to prisoners and sent out to clean the streets. Others, however, thought it was just good sense; the homeless would be helped by the rigors of steady work. Then came the surprise: At least one-third (or more, depending on how occasional work was measured) of the city's homeless were already working, some of them full-time. They were working at low-wage jobs that didn't provide enough income for them to afford New York's high rents and housing deposits, didn't cover medical emergencies and special needs, or otherwise were not enough to keep them from becoming homeless.

"**Welfare to work**" became a national slogan in the United States and the goal of a wide range of state programs in the 1990s. A question that policy makers had grappled with from the 1970s onward was, How can we get people off of welfare—a term that had gone from referring to well-being to referring to government support payments—and onto the work rolls? This was the essence and core of antipoverty policy. In the 1980s, President Ronald Reagan promised to end welfare as we knew it and made some funding cuts and other changes. In the 1990s, President Bill Clinton also promised to end welfare as we knew it and crafted a proposal that was never taken up in Congress. Finally, in 1996 Congress and the president agreed on a welfare reform law that gave states great latitude in changing their welfare systems. Many had already begun programs to move people from welfare to work. The programs seemed to work: Hunger and homelessness remained big problems but they did not soar, and people were not thrown onto the streets as some pessimists had predicted. Welfare rolls declined and many people went to work, helped by a continually growing economy. But the optimists were also to be disappointed: Poverty didn't go away. People got jobs but remained poor, and the concept of the working poor moved to center stage. Work alone is not the answer. Poor people are now more likely to work than in the past, but they work in low-wage, no-benefits, precarious employment that comes and goes. They look for help from family and friends. They make use of state and local programs to help with the burdens of child care. They go on, now working, but still poor. This shouldn't have come as a surprise to Americans, given that this pattern already existed across the border in Mexico and in much of Latin America, where generous government support programs have always been rare. Somehow, however, it did surprise many who had been convinced that work must be the answer.

It is clear that our analysis of poverty must also shift from welfare to work, and especially to wages. Some have argued that this has been the underlying problem all along (Wilson 1978; Zinn 1989). Before we examine antipoverty policy, we need to look at changing economies and at how the changing world of work has sculpted the face of poverty in different locations.

The Challenge of the Margins: Antipoverty Programs

The poor have always been with us, at least since the first kingdoms and stratified agrarian societies, and so have antipoverty proposals. Ancient kings often saw it as their duty to tend to the problems of the poor—especially the "deserving poor," such as widows and orphans—without radically challenging the overall system. Christian, Muslim, Jewish, and Confucian ethical systems all call on good rulers to defend the powerless and for good people to give alms (gifts of charity) to the poor. When obeyed, such injunctions provide a good beginning toward coping with poverty, but they are inadequate for addressing the needs of large, complex industrial societies. The first society to confront these problems was industrial Britain. Nineteenth-century reformers such as William Wilberforce first challenged the evils of slavery and then turned their attention to the masses of poor and displaced people who filled the streets of Britain's great industrial cities. Poorhouses and orphanages were set up for homeless adults and children. Government food distribution plans supplemented the soup kitchens operated by churches and civic and fraternal groups. Preacher-reformers of the time, such as John

Wesley, agreed with the prevailing opinion that drunkenness, idleness, and sin were at the root of poverty, but they also preached that the poor who were willing to change their ways needed to be given a second chance. The children of the poor, in particular, needed options, and Sunday schools were begun, not to teach religion, but rather to teach basic academic skills to those children who worked the rest of the week. Later, the reformers challenged child labor practices and proposed systems of public schooling.

These proposals and programs designed to provide the poor with housing, food, and basic education also found their way across the Atlantic, with many U.S. states adopting similar programs. Church, civic, and fraternal groups increased their social outreach in the 1800s and have been more active in social programs in the United States than almost anywhere else in the world. Yet government programs were limited and local, and very little national-level action was taken. By the beginning of the twentieth century, Progressive Era politicians such as "Fighting Bob" La Follette of Wisconsin and Theodore Roosevelt of New York promised poor farmers and workers a "square deal" and began to institute reforms in tax laws and monetary policies that were geared toward helping low-income workers. The prevailing opinion since colonial times, however, had been that poverty is the result of personal failings, and so government involvement in attempts to help the poor is unnecessary and unwise (Feagin 1972).

The New Deal

Attitudes toward helping the poor underwent a major change during the Great Depression of the 1930s. The United States had gone through many economic panics, often at 10- or 15-year intervals over the course of the previous 100 years, but the depression inaugurated by the stock market crash of 1929 dragged on longer than any previous downturn. This depression was also worldwide, with much of Europe, and Germany in particular, very hard-hit. U.S. President Herbert Hoover tried to continue the laissez-faire or "hands off" economic policies of his two predecessors. The view of the Republican Party of the time was that the country should be left alone to lift itself out of the Depression without government interference. But by 1931, breadlines had become commonplace and the homeless unemployed were living in shantytowns they called "Hoovervilles." When Franklin Delano Roosevelt came to office in 1933, unemployment was at 25%, and if discouraged workers and the underemployed had been counted, the figure would have been much higher. Suddenly it seemed that the old ethos no longer applied: Good, hardworking people were losing their jobs and their land, becoming indigent and homeless. If this was indeed the result of a societal failure rather than just personal failures, then concerted social action was needed.

Franklin Roosevelt called for just that. Adapting the "square deal" slogan of his distant cousin Theodore Roosevelt, FDR called for a "New Deal." This New Deal would be targeted at the unemployed, poor farmers, minorities, and anyone else who seemed to be getting a "raw deal." Roosevelt's challenge to the Great Depression called for relief, recovery, and reform. First, the hungry, homeless, and unemployed needed relief. Then the country needed to be led toward recovery. Finally, reforms were needed to prevent a recurrence of economic depression. The first relief programs operated essentially as "block grants" to states, which they could spend on emergency assistance as needed.

By 1935, with 20% of the U.S. population now receiving some form of relief, the emphasis shifted to getting people off of handouts and into employment ("welfare to work," although that expression wasn't used). Roosevelt began public works programs to employ men and youth in need of work. The Civilian Conservation Corps (CCC), formed during Roosevelt's famous "first hundred days" in 1933, had been building trails, bridges, and lodges in parks. To this was added the Works Progress Administration (WPA), which built roads and public buildings in major cities. These were make-work programs, and the WPA worker leaning idly on his shovel was a popular cartoon image, but many of the buildings constructed by the WPA, some with fine craftsmanship, still serve the public today.

Two other relief programs established in 1935 had even longer-lasting results. One of these was the program that became Aid to Families with Dependent Children (AFDC). Already during the Progressive Era of the beginning of the twentieth century, many states and localities had begun "widow's aid" or "mother's aid" programs. Up until that time, because she would likely have very limited income-earning options, a woman who lost her husband would often have to place her children with relatives or in an orphanage or workhouse. These aid programs were intended to help such women keep their children with them. The programs were generally accepted by the public because they targeted one group of poor people that most had long agreed were "deserving," widows and their children, and they served to keep families together, also a long-standing ideal. AFDC made these state programs part of a national effort. The program quickly expanded from providing aid to widows to covering single parents under a wide range of circumstances, but it remained largely a mother's program.

The other enduring program was really a cluster of programs that became known as Social Security with the Social Security Act of 1935. Retirement programs were established to ensure that the elderly would have income in their later years (and to encourage them to leave the workforce and make room for younger unemployed workers). At first these programs covered only certain categories of workers (some, such as railroad workers, had their own programs), but they gradually expanded and consolidated to cover all workers. Supplemental Security Income (SSI) programs provided income to disabled workers and their families, as well as to others who were not readily employable.

Roosevelt's recovery proposals included building the national infrastructure, especially in poor rural areas. The Tennessee Valley Authority

(TVA) used hydroelectric projects as the basis for wide-ranging rural development, as well as rural electrification, programs. Reforms included expanded worker protection and expanded consumer protection, especially in banking and finance.

The New Deal programs were radical for their time, at least in the eyes of critics, and set a new standard for activist federal government programs. Yet in many ways, they were built on values that would have accorded well with those held by the conservative welfare reformers of the 1980s and 1990s: They emphasized work, helping only the deserving, maintaining strict state and local control, providing temporary assistance to people who were en route to self-sufficiency, and family preservation. It was really the changing national society and economy that most changed these programs, and proved the undoing of some.

The most effective antipoverty program has always been broad-based prosperity and a tight labor market with high wages. These conditions began with the tremendous need for labor during World War II. Even before the United States entered the war, the nation's self-appointed role as what Roosevelt called "the arsenal of democracy" demanded huge industrial output and created far more jobs than the CCC or the WPA. The economic growth that followed in the 1950s and 1960s greatly reduced levels of absolute poverty, probably more so than either the civil rights movement or government programs of the time. Yet policies also help determine who does and does not benefit from economic expansion.

When Harry Truman became president on the death of Roosevelt, he tried to continue the reforms of his predecessor. He adopted the agenda of progressives such as Hubert Humphrey of Minnesota and Adlai Stevenson of Illinois, promising a "fair deal" that would address persistent poverty and expand the protection of industrial workers. Truman earned a reputation for being gutsy as he fought for reform, but the public's enthusiasm for such measures waned in the new conservative mood the 1950s. In the presidential election of 1952, Adlai Stevenson lost to General Dwight Eisenhower and Americans' renewed hope that a rising tide of economic prosperity would eventually lift all boats.

The War on Poverty

Antipoverty policies gained new attention during the Kennedy administration in the 1960s. After years of emphasis on the promotion of economic growth, it was becoming clear that not all Americans were sharing in postwar prosperity. In his book *The Affluent Society* (1958), economist John Kenneth Galbraith warned of the dangers of an economy driven by a consumer culture that divides the haves from the have-nots and fails to provide for the basic needs of all. Galbraith became an economic adviser to President John F. Kennedy. Michael Harrington's powerful book probing the topic of poverty in the United States and the many places left out of economic growth, *The Other America* (1962), gained nationwide attention. It particularly gained the attention of Robert Kennedy, the U.S. attorney general and President Kennedy's influential younger brother, who had been similarly dismayed by the extent of the rural poverty he saw as he campaigned for his brother in Appalachia. Somehow, the great new day for America that the Kennedys had promised must also include this other America. In 1960, they asked Mollie Orshansky, a government economist, to devise a measure of absolute poverty. Orshansky proposed calculating the cost of feeding a family (so that the income level needed varies by family size) and multiplying this figure by three. This is the measure the federal government still uses today. As a result, the year 1960 is often used as the benchmark against which U.S. trends in poverty are measured.

When Lyndon Johnson became president after John Kennedy's assassination, he brought with him a pledge to carry on the fight against poverty. Capturing the Kennedy vision, Johnson called for the United States to become a "Great Society," one free from poverty and deprivation. As part of his Great Society vision, Johnson began the War on Poverty. The food stamps program was established to attack the embarrassing paradox of hungry people in a land of supposed plenty. A program for mothers and young children, the Special Supplemental Nutrition Program for Women, Infants, and Children, known as WIC, targeted those most at risk from poor nutrition. Medicare was created to ensure health care for the elderly, who were still among the poorest Americans. Medicaid extended coverage to the poor on AFDC. Great Society programs were also aimed at going beyond relief to real reforms: better schools, job training and job creation both in the inner cities and in places of rural poverty, expanded health care for the poor and uninsured. With a landslide victory in the 1964 presidential election, Johnson pushed through more legislation for social programs than had been seen since Franklin Roosevelt's first 100 days in office. VISTA (Volunteers in Service to America) expanded the Peace Corps model to send U.S. volunteers into poor urban and rural settings in the United States. The Job Corps and the Neighborhood Youth Corps trained high-risk young people for employment. Upward Bound helped prepare low-income youth for college. Head Start focused on low-income and special-needs preschool children, seeking to give them the skills and confidence necessary for school success. "Hurry up, boys," Lyndon Johnson insisted as he prodded, charmed, and intimidated his Democratic majority into action, "before Landslide Lyndon becomes Lame Duck Lyndon."

Although many of the Great Society programs were later attacked as failures in the 1980s and some were discontinued, poverty did drop substantially in the 1960s, no doubt helped by a growing economy. The energy and money spent on the War on Poverty, however, was gradually drained by another conflict, the war in Vietnam. Government spending and attention moved from the domestic front to Southeast Asia. The enemy most feared was no longer rural and urban poverty (which was also a big part of the struggle in Southeast Asia) but global communist expansion, as it had been for much of the 1950s.

The War on Welfare

In 1968, the Nixon administration took over the struggle in Vietnam and also what was left of the struggle against poverty. Nixon distanced himself from the more conservative side of the Republican Party, represented by Ronald Reagan, in part by asserting that he would continue the fight against poverty. He explored the idea, touted by some progressives as well as some conservative economists such as Milton Friedman, of a negative income tax to replace the existing cluster of federally funded welfare programs. Instead of applying for AFDC, food stamps, WIC, and the help provided by many other programs, those earning below a certain level would just get a single check from the Internal Revenue Service—the reverse of paying income tax, hence the name. Proponents of this plan noted that it would require far less bureaucracy and paperwork than the existing programs, carry much less stigma for those receiving aid (the checks would look like income tax refunds), and give poor families discretion in how to use the money to better their situation.

Two grand experiments—maybe the biggest social science experiments ever funded—were set up in Seattle and Denver to explore the possible outcomes of a negative income tax program: the Seattle and Denver Income Maintenance Experiments, known as SIME and DIME. Welfare recipients who volunteered to participate were given fairly generous cash payments in place of the aid they would have received from the existing network of welfare programs. A team of social science analysts—sociologists, economists, and others—carefully monitored the results. First came the good news: People who participated showed no less interest in leaving the program for work than did those receiving other forms of aid; despite the program's more generous nature, recipients still were eager to become self-sufficient, and many did. Then came the bad news: Divorce and marital separation increased among those in the income maintenance group. It seemed that poor women who were guaranteed an income were now more likely to leave troubled marriages and unemployed, alcoholic, or abusive husbands. This trend alarmed members of Congress. Although the program passed one test, in that it did not destroy the work incentive, it failed another, that of family preservation. The experiments were scrapped and the idea of a negative income tax was shelved.

Antipoverty programs in the 1970s were always overshadowed by other concerns: Vietnam and Watergate for Richard Nixon, the Iran hostage problem and the war in Afghanistan for Jimmy Carter. The biggest limitations for both administrations, however, were the stagnant economy and high unemployment rates that limited revenues and provided few options for ambitious programs. The public's frustration with these ongoing problems revived interest in the conservative agenda of Ronald Reagan, who triumphed over the centrist Republican faction led by the man who became his running mate, George H. W. Bush, and the moderate Democratic policies of Carter.

For the first two years of the Reagan administration, the recession continued and poverty and unemployment increased. Some of the new homeless took to living in shantytowns they dubbed "Reaganvilles," which recalled the "Hoovervilles" of the early 1930s. Reagan continued to champion the hands-off government policies that had caused Herbert Hoover such grief, but Reagan was more fortunate. By late 1982, the economies of the world's advanced industrial societies were again beginning to grow, driven by new technologies and new global markets. The decade that was an economic disaster for much of the developing world was a major success for Japan, Western Europe, and North America. The Reagan administration contended that the way to keep this growth moving was to keep government interference as small as possible. It encouraged the privatization of government-owned or -controlled activities, as Prime Minister Margaret Thatcher did in an even bigger way in Great Britain. With few government-owned enterprises to begin with, the United States focused more on deregulation: giving a freer hand to industry and big business. Tax cuts were emphasized, especially for those in the upper income brackets. Because military spending was also increased, this meant that domestic spending on social programs needed to be cut drastically even as deficit spending rose markedly. The guiding philosophy was that of neoclassical economics, although it was nicknamed "trickle-down economics" or just "Reaganomics." The idea was that small government and a free hand to private enterprise would lead to ever greater growth, and the benefits of this growth—more jobs, more consumer goods, more private demand for workers' skills—would "trickle down" to everyone. In this view, growth was the answer to poverty.

Yet poverty remained and the number of poor Americans remained high. In general, the wealthier the family, the more it benefited from the growth of the 1980s, and the poorer the family, the less it benefited. The rich got richer, and the poor got nowhere. Conservatives blamed the influence of long-standing government programs such as AFDC for creating a subculture of dependency among the poor. Conservative analyst Charles Murray (1984) contended that this was why the United States was "losing ground" in the fight against poverty. Reagan adopted this philosophy and added his own concerns about abuse of the system by "welfare queens." According to conservatives, the system needed to be changed. It is interesting to note that when the Reagan administration–sponsored changes finally made it through Congress as the Family Support Act, the proposals were quite familiar: new education and job training, and a requirement that "welfare mothers" participate in job training and job search activities.

Once again, however, "reform" was difficult, and many Americans still seemed dependent on "relief"—relief that was disappearing. When older job creation programs were cut, many of the participants did not find private-sector jobs—instead, they became unemployed. When food stamps and nutrition programs were cut, hunger increased. The Physician Task Force (1985), which had been optimistic about ending hunger in the 1970s, found that hunger was back "with a vengeance" in the 1980s, with as many as 4.5 million American children hungry—hungry enough to be prone to chronic health problems and developmental disabilities. When housing programs were cut massively, homelessness increased. Treatment programs for alcohol and drug problems were cut, community programs for people with emotional problems (including Vietnam-era veterans) were slashed, and support for single women with children declined, all of which led to an increasingly visible homelessness problem. Hundreds of thousands of homeless were in shelters and on the streets, and many more were staying with friends and relatives as well as in low-price motels, at campgrounds, and wherever they could make other informal arrangements (Jencks 1994). Poverty was not "back"—it had never left—but extreme poverty was back in a way that had not been seen in the United States since the early 1960s.

That trickle-down economics was not working became more apparent as the country entered a mild recession during George H. W. Bush's presidency. Poverty had a new twist: The "intergenerational poor" were still as poor as they had ever been, and now they were joined by the "new poor." Many working-class people lost industrial employment and either became unemployed or took low-wage service-sector positions that turned them into working poor. Some white-collar workers also lost work to downsizing as U.S. firms tried to remain globally competitive, and those who were downsized fell quickly from sharing in prosperity to sharing in poverty. Bush had once derided Reagan's conservative policies as "voodoo economics," but he had now inherited these policies. In his campaign he called for a "kinder and gentler" America, in clear reference to the harsh realities of the growing income divide, but he also doubted that the federal government should play a large role in attempting to reduce that divide. Instead, in a now famous line, he called for Americans to combat poverty with "a thousand points of light"— that is, through local private and volunteer efforts. Bush administration officials suggested: "We concluded that there were no obvious things we should be doing that we weren't doing that would work. Keep playing with the same toys. But let's paint them a little shinier" (quoted in Gilbert 1998:251).

During the years of the Bush administration, increasing numbers of working and middle-class voters grew angry and impatient and turned against the policies that had won them over in 1980. They turned to the new activist agenda of Bill Clinton. This was not just a U.S. phenomenon. In Britain, Conservative Party leader John Major tried to continue what had once been popular policies under Margaret Thatcher, and soon lost the office of prime minister to Labour Party leader Tony Blair, who campaigned on economic strategies that sounded a lot like "Clintonomics": continued commitment to global trade and growth but new attention to the needs of workers who were losing in this process. Conservatives also were voted out of office in Canada. Throughout the early 1990s, concern was growing about the "new poverty" that was the dark shadow of the "new prosperity" of globalization.

The Clinton administration addressed this concern in several ways, beginning with a proposal for an economic stimulus package. Given that the recession was already easing, the package was probably unnecessary as economic "recovery," but it did contain important measures of relief for those who were not benefiting from the economy, along with reform for the future. In many ways, these measures involved returning to 1960s programs that had declined during the 1980s: restoring worker protections along with expanding Head Start, child vaccinations and prenatal care, and job training. VISTA, which had been dismantled, was now resurrected as AmeriCorps, a program that placed interns in social service settings. Hillary Clinton had worked for the Children's Defense Fund, and the president of that organization, child advocate Marian Wright Edelman, became an important adviser to the president, and children's issues were given new attention. Edelman had long noted that whereas the elderly were now at lower risk for poverty than in the past, the risk of poverty for children had been rising since the 1970s, making children the group most at risk for poverty. Along with his economic stimulus package, Clinton proposed both welfare reform and health care reform. He made a fateful decision to begin by addressing health care, arguing that lack of attention to this issue would otherwise undermine progress in any other area.

Health Care Reform

The need for health care reform was clear. Although the United States spends more on health care than any other nation in the world and has the greatest concentration of advanced medical technology, it lags behind much of the industrialized world in life expectancy, infant mortality, and child welfare. A recent World Health Organization study that ranked nations not just by years of life but by years of healthy life gave the top place to Japan; the United States appeared substantially further down the list (Associated Press 2000). Max Weber spoke of the "life chances" that people can expect to have based on their social position. In the United States, compared with those who are better off, the poor can expect shorter lives; greater chances of chronic illness, including heart disease, cancer, diabetes, and asthma; and far greater chances of late diagnosis and erratic treatment for such illness. Urban hospital emergency rooms treat, often after long waits, many poor who are suffering acute bouts of illness that could have been prevented by regular and routine medical care.

The problems in the U.S. health care system are rooted in the nature of that system and particularly in issues of inequality. The majority of the dollars spent on health care go to emergency intervention rather than basic prevention. A pregnant woman cannot afford prenatal care, and the result is that her premature, low-birth-weight, or otherwise high-risk infant spends time in a hugely expensive neonatal intensive care unit. Even if the newborn receives medical care, the results are often still not good. As Edelman has pointed out, although the United States does better with white infants, a black child born in the United States has less chance of seeing his or her first birthday than a child born in Botswana or Albania. In almost every category, it is the poor medical care given to low-income and nonwhite patients that lowers the U.S. average.

A big issue in health care is that of the uninsured, a group that has grown to include more than 45 million Americans—more if lapses in coverage and incomplete coverage are taken into consideration (American Medical Association 2004). Because most Americans with health insurance are covered through employer-provided group plans, changes in employment levels, particularly the loss of protected unionized jobs with full benefits, can have major effects on health coverage in the population. The uninsured often cannot afford to visit doctors for ongoing preventive care, and so may eventually need more expensive emergency care, perhaps in hospital emergency rooms, which cannot turn them away. This pattern of health care usage causes costs to climb, and the medical results are less satisfactory than those that could have been achieved with prevention.

This situation contrasts sharply with that just across the border in Canada, which has a single-payer health care system. Canadians choose their own private-practice health care providers, but a national insurance system pays the bills. In Canada, health care is considered to be a right of citizenship rather than a commodity for sale or a benefit of employment. The insurance does not cover purely elective procedures, and where services are limited, care is prioritized by need, in some cases leading to longer waits than in the United States. Most Canadians heartily endorse their system, and some in the United States have sought to use it as a model for reforming the American system. European health care systems vary in type, ranging from some that are similar to Canada's to systems (in the Scandinavian countries and Russia) in which most physicians are government employees, earning much less than their U.S. counterparts.

The Clinton administration's task force on health care reform, headed by Hillary Clinton, was aware of the problems with the existing system but was reluctant to propose the establishment of full nationalized health coverage, which would require tax hikes and displace private insurers. Instead, the task force proposed a system of health networks that would cover everyone, grouping individuals together to give them better bargaining power with private insurers. In addition, insurers' policy rates would be subject to national review. The proposal was designed to address as many of the problems of the U.S. system as possible without completely overturning it. As it grew increasingly complicated, however, suspicion and opposition grew, especially from private insurance companies and the powerful American Medical Association. The proposal died in Congress.

The Return of Welfare Reform

The backlash that occurred in response to the Clinton administration's failed health care proposal, which helped elect a Republican Congress in 1994, made it clear that it was now going to be much more difficult for the president to advance his welfare reform proposal. Clinton's welfare reform advisers were two highly respected social scientists, Mary Jo Bane and David Ellwood. In his 1988 book *Poor Support,* Ellwood had argued that although there are clear "helping conundrums," or contradictions, such as how to support poor children fully without encouraging people to have children they cannot support, a program could be crafted that would address basic U.S. values and still respond to current realities. He suggested a five-part plan:

1. *Ensure that everyone has medical protection.* Ellwood proposed that the government should be the insurer of last resort.

2. *Make work pay, so that working families are not poor.* Ellwood called for expanding the Earned Income Tax Credit, which gives tax "refunds" (even exceeding the amount paid in taxes) to working families with dependent children who fall below low-income thresholds. He also proposed raising the minimum wage and expanding other tax credits, such as for child care.

3. *Adopt a uniform child support assurance system.* Noncustodial parents would have any child support payments automatically deducted from their paychecks, as Social Security taxes are, and the government would assure minimum child support in cases such as unemployment of the noncustodial parent.

4. *Convert welfare into a transitional system designed to provide serious but short-term financial, educational, and social support for people who are trying to cope with temporary setbacks.* This transitional assistance would be available to both single-parent and two-parent families.

5. *Provide minimum wage jobs to persons who have exhausted their transitional support.* Others have also suggested that the government serve as employer of last resort (Wilson 1996), as in programs like the WPA, to ensure that there are jobs for those leaving assistance. This would be especially important during a recession or in a high-unemployment region.

The health care portion of this plan was already defeated by 1994, except for minor provisions such as the Family and Medical Leave Act, which protected the jobs of people who need to take time off of work to care for newborns or other family members. The other proposals were featured prominently in the Clinton reform plan. Compared with the administration, however, the majority of the members of Congress were less willing to spend money on child care and training, more concerned about enforcing work requirements, and more eager to place responsibility for welfare matters with the states. After succeeding in convincing Congress to remove some of its harsher proposals, Clinton signed the Welfare Reform Act of 1996, also known as the Personal Responsibility and Work Opportunity Reconciliation Act, or PRWORA. The act limited the receipt of welfare benefits to periods of two years without special extensions and to a five-year lifetime maximum. Rather than giving the money directly to individuals in a shared arrangement with the states, the federal government would give the money as block grants to the states for the states to use as they liked within these broad guidelines.

Passage of the Welfare Reform Act stirred intense controversy. When Health and Human Services Secretary Donna Shalala presented the terms to the annual meeting of the American Sociological Association in New York City, she was met with murmurs and polite but icy silences. Outside, the streets were filled with angry protesters decrying the plan. Critics called it "welfare deform" and asserted that it blamed and penalized the poor, that it targeted children, and that it would lead to streets filled with begging mothers and homeless centers filled with young children. Others worried that states would engage in a new race to the bottom, with each offering less and less in welfare payments to encourage the poor to go somewhere else more generous. Shalala acknowledged that the act wasn't perfect, but she argued that it was the best that could be achieved with the current Congress.

In fact, most of the critics' dire predictions have not been borne out. Homelessness and hunger remain national problems, but these preceded the Welfare Reform Act. States have not been overly generous, but many have been somewhat more creative than critics expected. Many have experimented with job training programs, child-care provision, child support collection plans, modest educational supports, and a variety of welfare-to-work programs similar to what Ellwood proposed for the national government. One of the harshest provisions that had remained in the bill as it was signed into law was the denial of benefits to legal immigrants; this was struck down by the courts because the U.S. Constitution gives full rights and protections to all legal residents. Since passage of the act, many states have reported large and dramatic declines in their welfare rolls. Wisconsin leads the list with more than 90% of its former recipients now off of state payments. Again, however, the biggest factor in these changes seems to have been the unprecedented growth of the economy. Because of a tight labor market, many employers were willing to hire people they might otherwise have shunned, including single mothers with spotty work histories. The critics' fears might have been realized if the plan had been implemented during a recession, and many still wonder what might happen during a national recession: Would these new, low-skill, limited-experience workers who have used up their welfare benefits be the first fired? Where then would they turn for help?

An immediate issue raised by welfare reform concerns those who have been termed the *multiproblem poor.* Wisconsin moved 90% of welfare recipients from its rolls quickly but has had little success with the remaining 10%. Other states have had similar experiences: good initial success with the most employable but then stalled progress with the remainder. Those who are difficult to move out of state welfare programs often have multiple deficits in the job market: physical disabilities, learning disabilities, multiple young children, arrest records, chemical dependency, emotional problems, and/or big gaps in education and job skills. Moving them into the workforce takes a much larger investment.

Further, although the nightmare visions of the critics have not come to pass, neither have the optimistic visions of reform's proponents. More people are working, but this has only increased the ranks of the working poor. Just as the needs have grown, funding for programs such as food stamps has been declining, and there has been no decisive movement on health care. The Earned Income Tax Credit for working poor families is not increasing to match increasing need. The minimum wage has risen significantly in recent years, but it is still, in real dollars, below the level of value it had in the 1970s. The minimum wage remained at $3.35 per hour for much of the 1980s while inflation continued. Hikes to $5.35 have made up some of that loss, and a minimum wage of $6.25 would go further toward restoring the wage floor. Yet estimates of what is needed for true self-sufficiency are much higher. Some analysts have suggested that a true "living wage" would be closer to $10 an hour. Perhaps the term should be *family wage,* given that the issue of what is enough depends heavily on how many dependents the wage earner has to support. A study of costs in the part of the midwestern United States where I live, an area where the cost of living is relatively low, found that a single mother with one dependent would need close to $10 an hour to be truly self-sufficient (see Exhibit 11.1). Unemployment is low here and jobs abound, but the average starting wage is in the range of $6 to $7 an hour. This means that for many parents and others with dependents, work, even full-time work, does not necessarily mean self-sufficiency. Many people who work still need some charitable or government assistance.

Thus although poverty rates slipped slightly during the growth of the 1990s, it became clear that growth would not cure poverty (see Exhibit 11.2). It took the economic growth of the entire duration of the Clinton administration just to allow the poor to recover what they had lost in the 1990–92 recession, and they were still below 1970 levels in real dollars. Although the poor fared better in the 1990s than they did in 1980s, by far the largest share of economic growth went to the upper income groups. Thus absolute poverty was not increasing in the United States, but **relative poverty** continued to increase, and the poor were left further behind. When recession returned in the first years of George W. Bush's presidency, poverty rates again rose, almost wiping out the gains of the Clinton years. With tax cuts favoring the wealthy—cuts for high income earners, for large estates, and for capital gains on investments—the inequality gap and relative poverty continue to grow.

Further, we find enduring patterns in who is poor. Despite the attention often focused on welfare dependency and long-term and intergenerational poverty,

Making it in St. Joseph County, Indiana

Here's a breakdown of the monthly costs of living in St. Joseph County for various family sizes as estimated by the Self-Sufficiency Standard for Indiana. "Housing" includes utilities except for phone, which is included in "miscellaneous." "Transportation" is based on owning and operating an 8-year-old car. "Health care" assumes that the employer provides health insurance in which the parent helps to pay.

Montly costs	Adult	Adult + infant	Adult + preschooler	Adult + infant + preschooler	Adult + schoolage + teenager	Adult + infant + schoolage + teenager	Two adults + infant + preschooler	Two adults + preschooler + schoolage
Housing	415	546	546	546	546	682	546	546
Child care	0	374	440	814	440	1254	814	880
Food	162	237	245	318	420	428	457	501
Transportation	147	147	147	147	147	147	290	290
Health care	90	188	167	208	213	228	261	240
Miscellaneous	81	149	155	203	177	274	237	246
Taxes	194	329	354	486	365	702	554	590
Earned income tax credit (–)	0	(60)	(44)	0	(83)	0	0	0
Child care tax credit (–)	0	(48)	(46)	(80)	(44)	(80)	(80)	(80)
Child tax credit (–)	0	(33)	(33)	(67)	(67)	(100)	(67)	(67)
Monthly self-sufficiency wage	**$1,090**	**$1,828**	**$1,931**	**$2,576**	**$2,114**	**$3,535**	**$3,012**	**$3,146**
Hourly self-sufficiency wage	**$6.19**	**$10.39**	**$10.97**	**$14.64**	**$12.01**	**$20.08**	**$8.56***	**$8.94***

Exhibit 11.1 Living Costs and Wages in St. Joseph County, Indiana

*Per adult

Source: From Indiana Coalition on Housing and Homeless Issues, reported in *South Bend Tribune*, January 23, 2000. Reprinted with permission.

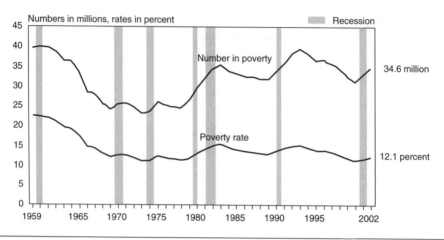

Exhibit 11.2   Poverty Rates

Source: U.S. Census Bureau, Current Population Survey, 1960–2003 Annual Social and Economic Supplements.

most poor people remain poor for fairly short periods, the majority for less than two years. Yet certain groups are at much higher risk of poverty and of recurring spells of poverty. Race matters: Most poor people are white, but blacks and Latinos are at significantly higher risk of poverty than whites (see Exhibit 11.3). Gender and family status matter: Single parents, and single mothers in particular, show the greatest risk of poverty and have the hardest time escaping poverty (see Exhibit 11.4). As large numbers of American children now live in single-parent families as well as in two-parent working poor families, children constitute a group that is particularly at risk. The national poverty rate hovers around 12%, but more than 20% of children live in poor families. Compounding these factors increases the risk: 50% of black children are poor, and two-thirds of children born to single black mothers are poor.

These figures also show how intertwined social and economic situations are, and how economic policy is bound up in social policy. Changes in family structure have had big effects on poverty as well. In one study, 90% of divorced mothers in the research sample, both black and white, fell into poverty after divorce (Arendell 1986). Many states have attempted to cope with this problem, and save welfare dollars, by searching out nonsupporting fathers. Wisconsin led the nation in this attempt, and, as with the state's welfare reform efforts, the results were mixed. The state had reasonable success with middle-class fathers who had secure employment and could be forced or embarrassed into paying child support. As for other fathers, however, many were in prison, out of work, chemically dependent, erratically employed, or otherwise poor prospects for providing child support. Even in social changes, analysts have noted the important role of wages and employment. A major factor contributing to the large number of low-income single black mothers, for instance, seems to be the dismal employment prospects of large proportion of young black men. In William Wilson's phrase, young black men with limited skills and work experience are "unmarriageable," and ultimately black women say "to hell with them" and go on with their families without their children's fathers.

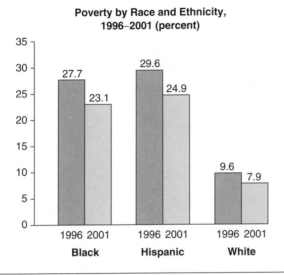

Exhibit 11.3   Poverty Rates by Race and Ethnicity

Source: 1997 and 2002 National Survey of America's Families

Notes: "White" and "black" include non-Hispanics only; "Hispanic" includes all races. Estimates for 1996 use weights based on the 2000 Census and may differ from previously published estimates using weights based on the 1990 Census. All estimates for blacks and Hispanics are significantly different from estimates for whites at the 0.10 level.
*Difference from the 1996 percentage is significant at the 0.10 level

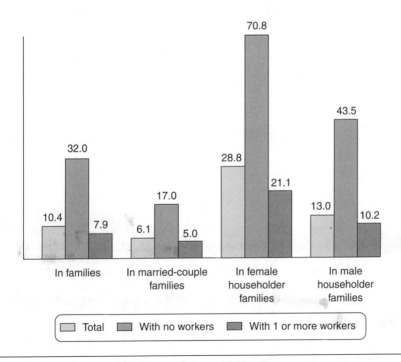

Exhibit 11.4   Poverty Rates by Family Type and Employment

Source: U.S. Census Bureau, Current Population Survey, 2003 Annual Social and Economic Supplement

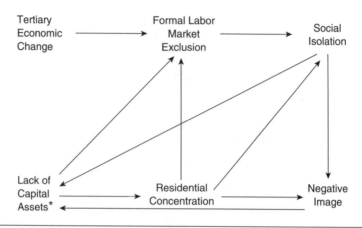

Exhibit 11.5   Web of Exclusion

Source: Sernau, 1996.

*Social, political, and market.

The interplay of social and economic factors related to poverty has been termed the "cycle of poverty" (Myrdal 1944), or sometimes just the vicious circle of poverty. In fact, these factors more often form something that resembles a tangled web rather than a simple circle (see Exhibit 11.5). Economic changes displace workers, and displaced workers are often concentrated in clusters of poverty. The resulting social exclusion further isolates and stigmatizes the poor, frustrating their attempts at upward mobility (Bhalla and Lapeyre 2004). Perhaps the surprising part of this web of poverty is not that people remain entangled in it but that as many as do manage to free themselves and move up. The desire to move up is strong, but for the economically vulnerable, the downward pull is also strong. Although many escape permanently, others do so only briefly, returning to the tangle of poverty several times.

Extreme Poverty: Homelessness and Hunger _____

Problems of Extreme Poverty

Even while overall poverty declined a bit in the late 1990s (only to start upward again in 2001, as shown in Exhibit 11.2), extreme poverty—defined as income less than half the poverty level—continued to rise. In 2000, 39% of Americans living in poverty met the standard of extreme poverty. The National Coalition for the Homeless (2004) attributes this situation to a combination of eroding employment opportunities and retrenchment in social services. The minimum wage has continued to lose value relative to a rising cost of living. In every state, the minimum wage is insufficient to allow a worker to afford even a one- or two-bedroom apartment without help from some kind of subsidy. In a typical state, a worker earning the minimum wage would have to work 89 hours per week to afford a two-bedroom apartment without spending more than 30% of income on housing (as recommended by affordable housing guidelines). As a result, fully one-quarter of the homeless in major cities are employed but still cannot afford housing. The picture is bleaker if the worker has dependents. Half of the new jobs created since 1994 pay less than the poverty threshold for a family of four ($18,811 in 2003).

Wage growth for the lower-income end of the labor force has been lagging behind consumer prices in general and housing costs in particular. An interesting irony takes hold in a globalized economy: The costs of nonnecessities fall—electronic gadgets, using new technologies and made by low-paid foreign workers, fill store shelves and are often cheaper every year. Necessities, however—such as decent housing, high-quality health care, high-quality education, heating, and transportation—cannot be outsourced or readily imported. The result is that the poor may be able to buy more gadgets and diversions than their predecessors could have dreamed of, but they can't afford the things that really matter to their families.

Fraying in the Safety Net

Even before its elimination in the Welfare Reform Act of 1996, AFDC had been declining in value. The typical state's benefits had lost almost half their value to inflation between 1970 and the mid-1990s. When AFDC was replaced with Temporary Assistance for Needy Families (TANF), the situation worsened for many families that were able to leave the welfare rolls quickly. The median TANF benefit for a family of three is about one-third of the poverty level. Even combining TANF benefits and food stamps leaves a family below the poverty line.

The face of poverty continues to get younger. More and more children are in extreme poverty, as fewer families are lifted to even half the poverty line through cash assistance. Often when families leave the welfare rolls they also leave their eligibility for subsidized housing. Even for those receiving TANF, help with housing is severely limited: Only one TANF family in four receives housing subsidies. Parents and children fleeing domestic violence face a particular risk of homelessness if the custodial parent, typically the mother, does not have strong job prospects. A woman and children in such a situation may face the choice between an abusive home and no home at all.

Persons without families have considerably fewer expenses but now often are ineligible for any assistance. AFDC was often paralleled by General Assistance, a small benefit available to impoverished singles. In many locations this has been severely restricted or eliminated, also giving rise to increased homelessness. Persons with disabilities qualify for SSI, but they typically need to spend more than two-thirds of their SSI income to rent a one-bedroom apartment.

Homelessness

Homelessness often begins with a crisis: divorce or desertion, accident or major illness, drug or alcohol addiction, sudden job loss, fire, or eviction. Some slowly sink into homelessness as they exhaust their resources. Many people without homes of their own are not found on park benches but in the homes of family members and friends on a short- or long-term, or sometimes rotating, basis. Those who do not have such social support networks, or who exhaust their welcome, are most at risk of landing in shelters or on the streets. It is impossible to know exactly how many displaced people without true homes of their own there are in the United States, but we can measure how many use shelters and other facilities for the homeless. These individuals come from backgrounds that are quite diverse, reflecting a variety of risk factors for homelessness.

Homelessness is not just a result of personal crises and changes in relationships; it also reflects the structure of both the labor market and the housing market. Affordable housing has been disappearing over time. In the 20 years following 1973, more than 2 million low-rent units disappeared in the United States: torn down to make way for other uses or remodeled into higher-priced housing. By 1995, the deficit in affordable housing was almost 4.5 million units behind demand. Due to very low interest rates, the recession of 2001 through 2003 was accompanied by a housing boom, but most of the new homes constructed were priced in the range of $150,000 to $350,000, far beyond the means of many.

Renters have not been faring much better than home owners. A person who is forced to change residence is often faced with having to pay a large "security" deposit as well as first and last months' rent before he or she can move in to a rental property. The total amount required is often well over $1,000—much more in some areas of the country. In major urban areas such as New York and Los Angeles, the amount may be more than $3,000, even for a small apartment. Renters can sometimes avoid having to pay so much up front by moving into old and poorly maintained housing, but then they face health and security risks, and often astronomical utility bills. Residents of old buildings with erratic rental histories can find that they need to pay deposits of as much as 4 months' worth of service even to get their heat turned on—an expense of close to $1,000 in cold northern climates.

Families with children are now the fastest-growing homeless group. Children now make up 39% of the homeless in the United States (see Exhibit 11.6), a figure that represents a 5% increase in just two years. In small communities and rural areas, children and their families now constitute the largest group of homeless. The needs of these families have led to a proliferation of homeless shelters and service providers for the homeless, most of which are largely locally funded. Some of these are little more than flophouses, providing those in need with a cot or a floor to sleep on on a cold night. Others, however, are increasingly offering comprehensive services.

	U.S. Adult Population (1996)	All Homeless Clients (N = 2938)	Clients in Homeless Families (N = 465)	Single Homeless Clients (N = 2473)
Sex[a]				
Male	48(%)	68(%)	16(%)	77(%)
Female	52	32	84	23
Race/Ethnicity[b]				
White non-Hispanic	75	41	38	41
Black non-Hispanic	11	40	43	40
Hispanic	10	11	15	10
Native American	1	8	3	8
Other	3	1	1	1
Age[c]				
Under 25	13	12	26	10
25 to 54 yrs.	59	80	75	81
55 or more yrs.	28	8	*	9
Educational Attainment[d]				
Less than High School	25	38	53	37
High School Graduate/G.E.D.	30	34	21	36
More than High School	45	28	27	28
Veteran[e]	13	23	5	26

Exhibit 11.6   Diversity of the Homeless Population

Source: Urban Institute analysis of weighted 1996 client data from the U.S. Census Bureau National Survey of Homeless Assistance Providers and Clients.

Note: Numbers do not sum to 100% due to rounding.

*Denotes percentage less than 0.5 percent but greater than 0.

Sources for U.S. adult population data:

[a]Bureau of the Census (1997a), data for 1996; table 14, N = 200 million. Age range is 18 to 64.
[b]Ibid., Table 23, N = 196.2 million.
[c]Ibid., Table 58, N = 193.2 million.
[d]Ibid., Table 245, N = 168.3 million persons 25 and older.
[e]Department of Veterans Affairs, data for 1995.

Centers for the homeless in San Diego, South Bend, and now in a variety of large and midsize cities offer a full "continuum of care," from addiction treatment through self-esteem programs to job training and placement, even housing assistance. Among the various programs aimed at dealing with the problem of homelessness, these comprehensive programs are the most effective at restoring self-sufficiency to their clients, but because they require long-term commitments, they are limited in the numbers of homeless they can serve.

Hunger

The growth in the numbers of homeless shelters has been accompanied by a similarly rapid growth in the numbers of soup kitchens and day facilities that provide meals to those in need, as well as food banks that donate bags of basic foods to poor clients. The main federal program to fight hunger has been the food stamp program, in which eligible recipients receive coupons that are good for the purchase of food at their local stores. This program was intended as an emergency supplement to other programs. The program's use of coupons, which recipients cannot use to purchase tobacco products, alcohol, or nonessentials, was in part instituted out of mistrust: It is a way to ensure that the assistance goes to food needs. Like many programs, this one has grown far beyond its original vision. According to researchers Hirschl and Rank, whose findings from a study of food stamp use were reported recently in the *Washington Post* (2004), fully one-half of all Americans have used food stamps at some point in their adult lives. Five out of six African Americans have received food stamps as adults at least once. Hirschl and Rank found that temporary bouts of poverty are common occurrences for many Americans. In fact, given that only half of the people who qualify for food stamps actually apply, the number of those temporarily in need could be even higher. In 2003, 21 million Americans received food stamps, but many more had previously received help at one point or another. Although attention continues to focus on chronic and even intergenerational poverty, this points to the frequency of acute, short-term poverty.

The growing presence of food banks and soup kitchens coupled with food stamps and other supplemental nutrition programs, such as WIC, means that few in the United States have no food resources or are at risk of dying from starvation. Chronic malnutrition, however, has been implicated in a wide range of health disorders and in children's educational delays.

Poverty Programs among Advanced Industrial Nations

The world's advanced industrial societies have less extensive poverty than do other societies, but just as every nation has its elites, so too does every nation have its poor. Of the world's industrial countries, the United States, with the largest economy and one of the highest average standards of living, also has one of the highest poverty rates. Who is poor also varies by society. With a system of social security programs similar to those in place in most of the wealthier countries, the elderly in the United States are no longer more likely to be poor than the general population. As Social Security payments are not very generous, however, those elderly who do not have pensions and personal savings are often "near poor," living just above the poverty line, and must budget very carefully. Health care costs not covered by Medicare, such as many medicines, can be a huge financial burden.

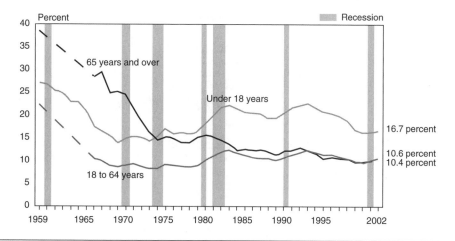

Exhibit 11.7   Trends in Poverty by Age

Source: U.S. Bureau of the Census, Current Population Survey, 1960–2003 Annual Social and Economic Supplements.

Echoing a concern about widows that goes back to ancient times, very elderly women who have outlived their husbands and their savings are particularly at risk, especially if they themselves do not have extensive work records. The greatest concern, however, may be for the youngest Americans (see Exhibit 11.7). Given the high proportion of young, single-parent households, the relatively high proportion of poor minorities, and the limited family support programs in the United States, American children are more likely to be poor than are their European and Pacific Rim counterparts (see Exhibit 11.8).

In general, most European countries are far more generous than the United States, and somewhat more so than Canada and Australia, in social supports for their residents. Their minimum wages are higher, unemployment compensation more generous, maternity leave far more generous, and aid to poor families and individuals less restricted. This kind of combination of national social supports is often called the **welfare state**—a term that sounds unappealing to many Americans who have come to associate the very word *welfare* with handouts and dependency. Europeans, however, understand the term to refer to a state that is actively concerned about promoting the welfare of its citizens. Such programs are costly. Effective tax rates (total taxes after deductions) in Europe are often 40–50% of total income, whereas in the United States they are typically closer to 25–35%. These programs do, however, make a big difference in the level of absolute poverty. France, for example, which has a quite wide gap between rich and poor, would have as large a proportion of its children in poverty as the United States if it were not for very generous government supports. The government encourages French citizens to have children (the birthrate is actually below replacement level) and supports the care of children with generous tax deductions, health care provisions, maternity leave, and heavily subsidized day-care arrangements. The combined effect is a much lower rate of child poverty than that found in the United States.

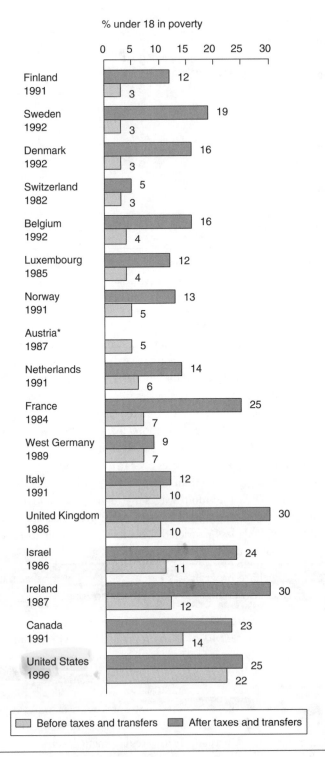

Exhibit 11.8   Child Poverty before and after Government Intervention

Source: UNICEF, 1996.

Despite their successes in lowering poverty rates, many European countries have begun to reexamine, and in some cases reduce, their welfare state policies. Arguing that such changes are needed for these countries to remain globally competitive and reduce government spending, and in particular to match the economic growth of the United States (and, to a lesser extent, a rebound in Great Britain, where such programs are also less generous), some have called for a less controlled and government-based redistributive economy: if not Reaganomics, at least Clintonomics. Some have admired Tony Blair's British "third way" between Thatcher's conservative free hand to markets and the controlling hand of the traditional welfare state, a position advocated by prominent British sociologist Anthony Giddens (1999). Others—particularly in France, the Netherlands, and the Scandinavian countries—however, are loath to give up their progressive image and sense of national solidarity for the more individualistic, sink-or-swim economic policies that have dominated in Great Britain and the United States. Many are still seeking a true "third way" between the high-wage/high-unemployment and low-poverty/low-growth economies of the welfare state and the low-wage/low-unemployment and high-growth/high-poverty-level model of the United States in the past two decades. The welfare state has also been strongest only in the area of relief, helping those hurt by the global economy, rather than in the area of reform, curbing the tendency of that economy to create big gaps between rich and poor. In Eastern Europe those gaps have grown markedly with the demise of state socialism, and although a few countries, such as Poland, are seeing economic growth, they are also seeing growing inequality and poverty. As Europe unites its economy under the European Union, the EU member nations must also consider whether they are willing to undertake a continent-wide welfare state. Funds have been traveling from the more prosperous north and west to the poorer south and east, but many voters in Europe are not eager to extend national fiscal solidarity to an international welfare system.

Japan has a vigorous national education and job training system, but with its more homogeneous population, prevailing traditional family structure (with few single parents), and smaller income gap between rich and poor, it has not had to offer welfare programs as extensive as those of other nations. Yet Japanese society is also changing: Workers are no longer assured of lifetime employment with companies that will provide all needed benefits, and spouses are no longer assured of lifetime marriage cemented by the social stigma of divorce. If these trends continue, Japan could find that it must deal with growing divides and dislocation as its Western industrial counterparts have.

The Private World of Poverty

In the 1980s, the United States under Ronald Reagan and Great Britain under Margaret Thatcher led the world into an era of privatization. Supporters of privatization asserted that the free market was the best way to create a tide that would "raise all boats." The free-market ideology included cuts in government funding for social services, reductions in labor protections, and an emphasis on private market delivery of health care and even education. These measures, coupled with a recession, sent poverty levels soaring back to heights not seen since the 1960s, particularly in the United States. Economic growth in the mid-1980s brought the poverty level back down somewhat, but never back to its lowest previous levels. The crucial shift is more subtle, however: In a privatized and marketized society, the life chances of the poor decline. When everything—housing, health care, even education—is for sale in the marketplace to the highest bidder, those with little income, who have little market power, have ever fewer options. The poor do not become poorer in dollar terms but in the quality of their lives and in what they can expect for the future.

Even our ways of thinking about and discussing poverty and well-being seem to have become privatized. Interest in individual volunteering and private charities is up, while trust in government or public policy to make lasting changes in our society is down. The word *welfare* once carried the connotation of well-being, as in the phrase *the common welfare,* something to be attained. Now it has come to connote bureaucratic programs and handouts, something to be avoided. The term *public* has largely lost its sense of "common to all citizens," as in *the public good* or *an informed public.* For many, it has come to mean only what is left over for the less fortunate: Public housing is for those who can't afford any better place to live, public transportation is for those who can't afford a car, a public defender is for those who can't afford an attorney, and public schools are for those who can't afford a private education.

Around the world, we are seeing the privatization of poverty. Poor communities still abound, but increasingly poverty is a private pain for individuals to cope with as best they can in an impersonal market. Workers' wages and benefits are eroded by international competitive pressure in what has been termed a race to the bottom. Latin America left the 1980s with a host of privatization schemes. In Chile, dictator Augusto Pinochet promoted growth by slashing labor protections and social services. Democracy has now been restored in Chile, but most of the social programs that Pinochet ended have never been restored. Mexico embraced privatization under U.S.-educated presidents Carlos Salinas and Ernesto Zedillo, and then in a major way under President Vicente Fox, a business executive turned politician. Big government and big labor were notoriously corrupt and inefficient in Mexico under previous administrations, and important reforms have taken place under Fox. Yet in the past, many categories of workers were protected by labor guarantees, health care benefits, and social security programs. These are all being cut. Uneven growth means that some workers have higher incomes and others simply have fewer protections and options (Sernau 1994).

This pattern has also reached the postsocialist world. China has been growing impressively, but it has also been dismantling many of its social programs. China was once famous for its "barefoot doctors," who brought basic preventive health care to the villages. As modern hospitals become more common, these health care workers are disappearing. Those who can afford to pay and live near medical centers get ever better care, but those without this access have lost their preventive care (Rosenthal 2001). To serve its billion people, India created a vast system of public clinics, but many of these are now "skeletons," starved for resources and trained doctors (Dugger 2004). The same has happened in Russia, where the people's health care choices are now between expensive private facilities and woefully inadequate public ones. Russian centrally controlled production was notoriously inefficient, but under that system workers did have a variety of assurances. Increasingly, successful Russian entrepreneurs and workers have access to previously unimagined luxuries while poor workers have few protections, few benefits, and diminishing hopes. The have-nots in Russia tell a joke more ironic than funny: "Everything the communists told us about the glories of communism was a lie, but, unfortunately, everything they told us about the horrors of capitalism was true."

Elsewhere in Europe, the welfare state is also under assault from new demands of the European Union, the World Trade Organization, and growth-oriented governments (Bhalla and Lapeyre 2004). It was one thing for Swedes or Danes to decide the full range of rights and benefits they would give their own citizens, but now a uniting Europe must decide what rights and benefits it will give all of its members. Some prefer the Anglo-American model of private markets as the means to growth. Others contend that this "savage capitalism" is fundamentally undemocratic and un-European. While the debate continues, around the world more people can glimpse affluence next door as they face the reality that they are on their own in the pursuit of their dreams. Many have been laboring literally under the billboards of multinational corporations promising the luxuries of private affluence for a long time already:

> We have glimpses of a new world economic order. I suspect the best glimpses will not be found primarily in the high rises of Tokyo or Manhattan, nor in the desolation of Mogadishu, Somalia and Kinshasa, Zaire (Congo). A more likely glimpse of the future can be seen in Bangkok, Thailand where economic miracles and social nightmares coexist daily. We might include Juárez and Tijuana in Mexico, São Paulo and Rio de Janeiro in Brazil, Kuala Lumpur and Jakarta in Southeast Asia, perhaps Lagos and Nairobi in Africa, probably Los Angeles and Miami in the United States—all those hot, noisy, hazy mixtures of hope and despair. They provide the products and services for a growing global middle class, along with the prospect of a door into that class for the otherwise excluded. Without fundamental changes, however, for many it will be a rapidly revolving door, offering an invitation, a quick glimpse of air conditioned affluence, and a rapid return to the street. (Sernau 1994:137)

KEY POINTS

- For half a century, controversies over poverty-related social policies in the United States have focused on federally funded welfare payments and unemployment, but new attention is now being given to low wages and the problems of the working poor.
- Large-scale federal antipoverty programs date from Franklin Roosevelt's New Deal of the 1930s and include the War on Poverty of the 1960s. Policies of the 1980s and 1990s emphasized economic growth and welfare reform, but poverty—including extreme poverty—persists.
- Two measures of extreme poverty, hunger and homelessness, have been on the rise in the United States in recent years, and an increasing proportion of the homeless population consists of families with children.
- The U.S. government tends to do less to alleviate poverty and inequality than do European governments, particularly in regard to poor children.
- Around the world, policies promoting growth and efficiency have benefited some, but many workers have lost benefits and an economic safety net to privatization and reductions in social supports.

FOR REVIEW AND DISCUSSION

1. Why is increasing attention being given to the plight of the working poor? Why are there so many working poor at a time of low unemployment? What do families need in terms of wages and benefits to be self-sufficient?
2. Which areas of the United States have experienced severe urban and rural poverty? What have been the major causes of this concentration of poverty? How does rural poverty differ from urban poverty?
3. What has the U.S. government done in its attempts to reduce the effects and extent of poverty? Which policies and programs have proven successful and which have not? Why is it difficult for governments to implement effective antipoverty measures?
4. What are the major causes and effects of poverty around the world? What are the elements in webs of social exclusion for the world's poor? What kinds of programs have been proposed to help alleviate the worst conditions of poverty around the world?

MAKING CONNECTIONS

Poverty Budget

What is it like to have to budget for a household on a poverty income? To find out, start by listing all of the living expenses for a family of three or four for one month (you might consider making lists for a two-parent family with two children and a single-parent family with two children). Be sure to include the costs of food, housing, transportation, health care, child care, telephone and utilities, clothing, and incidentals. To make sure the costs you list are realistic, check the prices on food at a small neighborhood grocery store (the kind of place where most low-income families shop), and check the local newspaper rental listings for the cost of housing in low-income neighborhoods. To arrive at the amount of the family's income for a month, do one of the following: Find out what the federal poverty guideline is for annual income for your family size and divide that figure by 12, use your state's TANF monthly welfare payment (this information is available at your local welfare office), or multiply the amount of the current minimum wage by 160 hours (representing a month of full-time employment). You might find it interesting to do all three of these and compare the resulting amounts. (If this is a group project, different groups may choose to take differing profiles of family size and working versus nonworking poor.) Are you able to meet all of your family's needs and stay within your "budget"? What types of families would find this most difficult?

How much did you budget for food? Find out how much assistance the family you are representing would qualify for in food stamps (see Edin and Lein 1997), then visit a neighborhood grocery store with a calculator and see how much food you are able to buy for a week with this amount. Try to plan a week's meals with the food you can buy without assuming that you also have a pantry full of basic supplies such as flour. Are you able to do it? Will you need to look for outside help from school lunch and breakfast programs, local food banks, or extended family members?

Homelessness and Hunger

Homelessness and hunger are two manifestations of extreme poverty. Where do the homeless in your community live? Many are probably with friends and relatives and in informal living arrangements. Are there shelters to which they can turn? Find out what the options are in your community or one nearby: homeless centers, rescue missions, Salvation Army houses, YWCA programs, and others. Any community resource agency should have this information available. Otherwise, begin at the Web site for the National Coalition for the Homeless, at www.nationalhomeless.org. First, note the site's "Facts" section, then go to "Directories" for state-by-state directories of homeless centers as well as links to missions run by religious organizations and Salvation Army sites. Select a local center and volunteer to work in the kitchen, help serve a meal, work at the front desk, help with children's programs, or fill other needs. What is the center's guest profile and how has it changed over time: men, women, children of differing ages? What programs does the center offer? Is it merely a "roof and meal" center, or does it operate programs for chemical dependency counseling, educational preparation, and job and housing assistance?

Many homeless centers also provide meals for those who are poor but not homeless. Where else can the hungry in your community turn? Find out about local food pantries or food banks, soup kitchens, and similar programs. Select one and take some food there or volunteer to serve a meal (this is a good group project). Who are the guests? How many are single individuals, and how many come with their families? Talk with the program's coordinators. Has demand for its services grown or shifted over time? Are the meals linked to other services aimed at helping guests cope with and recover from extreme poverty?

12 Challenging the System

Social Movements

When I give food to the poor, they call me a saint. When I ask why the poor have no food, they call me a communist.

—Dom Helder Camara, Brazilian archbishop (1909–99)

M ost people seem to accept their lot in life. They tend to their daily affairs without much sense of the course of history or their place in it. Yet every now and then an idea catches fire and ignites people's hopes and passions across distant places. Such an idea took hold in 1989 as democratic movements swept across Eastern Europe, tearing down the Berlin Wall and remaking the map of a continent. The tremors from that mass movement shook China, ultimately toppled the Soviet Union, and were echoed as far away as Latin America. That was not the first time a social/political movement has had wide-ranging consequences. In 1968, student uprisings and movements shook the United States, France, Mexico, and China. Some thought international revolution was at hand, not the least in the United States, where antiwar demonstrations came on the heels of race riots as the civil rights movement turned "militant." The events of 1968 did not radically change the world as some had hoped, but some ideas ignited five years before that time are still making their way around the world. In a single year, 1962–63, Rachel Carson's *Silent Spring* launched the environmental movement, Betty Friedan's *The Feminine Mystique* launched the new women's movement, Michael Harrington's *The Other America* launched the anti-poverty movement, Ralph Nader's attacks on General Motors launched the consumer movement, and Martin Luther King, Jr.'s "I have a dream" speech galvanized the civil rights movement. These two women and three men presented their ideas with such power that the effects are still being felt.

A similar convergence of ideas happened at least once before in "modern" times. The year 1848 offered all the democratic aspirations of 1989, all the violence and anger of 1968, and as many world-changing ideas as 1962–63. It may not have figured prominently in your high school history book, but the events of 1848 shook the political and social world. European cities were in flames as workers' movements, food riots, and democratic movements burst into the streets to do battle with the forces of the state. Some only wanted enough to eat, but others wanted to change the world. Karl Marx and Friedrich Engels captured the sentiments of some of the latter in *The Communist Manifesto.* In the United States, the Mexican-American war was countered with an antiwar movement whose ideas found form in Thoreau's *Civil Disobedience.* Frederick Douglass began the abolitionist newspaper the *North Star,* and the antislavery Free-Soil Party was created. And women in New York wrote that "men *and women* were created equal" and listened to Elizabeth Cady Stanton call for the extension of the right to vote to women (women did win the right to own land for the first time in New York).

In time, these various ideas intermingled. Engels went on to write about women's oppression. White feminist Lucy Stone rallied to black abolitionist Frederick Douglass's cause, and later Douglass joined Stone in working for women's suffrage. These are all examples of interrelated **social movements,** or ongoing efforts by groups to promote (or prevent) social change. Social movements work beyond the confines of the formal political process and often attract those who find themselves excluded from that process. Social movements gain force through **collective action,** people coming together in both planned and spontaneous actions and demonstrations. Many social movements have long and vigorous histories, but a pattern of **new social movements** appears to be emerging in postindustrial global society. These new social movements are typically global in scope, stressing international action that encompasses social, economic, and environmental issues. Given their broad nature, their supporters often span a wide variety of interest and age groups. When protesters took to the streets of Seattle in the spring of 2000 to protest meetings of the World Trade Organization, they were continuing in a long line of action that stretches back at least as far as 1848. Yet these demonstrators comprised an eclectic group typical of new movements; they included aging labor leaders in gray suits, young environmentalists in whale suits, and self-proclaimed "raging grannies" in pantsuits.

We can only guess what the activists of 1848 would have thought if they could have witnessed the demonstrations in Seattle in 2000. Would Marx have reminded labor leaders of his prediction that free trade would only serve to speed the revolution? Would Thoreau have donned a whale suit? Would Lucy Stone and Elizabeth Cady Stanton have joined the raging grannies? Would Frederick Douglass have demanded a face-to-face meeting with President Clinton or joined the Green Party?

We know that follow-up demonstrations occurred in Washington, D.C., and in London, and then in Spain, Italy, India, and Brazil. It remains to be seen what additional ideas and actions will emerge from the new social movements, but it is clear that they are likely to be grounded in more than 150 years of intertwined struggles over class, gender, and race.

_____ The Enduring Struggle: The Labor Movement

Karl Marx put it simply: History is the history of class conflict. The struggle between rulers and ruled, between owners and owned, has been fought with bronze spears, iron lances, muskets, and automatic weapons as well as with ballots, laws, and court orders. For the past 150 years, much of this class struggle has been waged between labor unions and big business.

In January 1848, Marx gave a speech about the workers' struggles that was soon to shake the world. In February of that year, he and Engels published *The Communist Manifesto,* outlining their view of the issues at stake (see McLellan 1988). By spring, Europe was in flames, literally and figuratively. Worker revolts, political uprisings, food riots, and street battles raged. As full-scale industrialization transformed the societies of Europe and North America, the world was in turmoil.

In the latter half of the 1800s, the United States came as close as it has ever come to Marx's vision of worker's hell. New industries sprang up rapidly, many falling into bankruptcy during frequent economic crises and "panics," and a few grew exponentially, making great fortunes for their fortunate founders. The difference between spectacular success and competitive collapse often depended on an employer's ability to secure a supply of cheap labor and then work those laborers to their limits. The workers came, some driven by desperation in their homelands, others pulled by recruiters' promises of wonderful opportunities. First were the Irish and the Germans, then the Poles, the Hungarians, the Italians, the Greeks, and the Russian Jews. From the 1840s until World War II, great streams of immigrant workers poured into the growing American industrial colossus and into an epic competitive struggle. Working conditions often amounted to 12 hours of pure misery. Wrote one visitor:

> [The workers] crouched over their work in a fetid air which an iron stove made still more stifling and in what dirt; hunger hollowed faces; shoulders narrowed with consumption, girls of fifteen as old as grandmothers, who had never eaten a bit of meat in their lives. (Quoted in Allen 1952:52)

Children worked in mills and mines, and young women labored in textile sweatshops. The worst abuses caught the attention of the press. The 1911 Triangle Shirtwaist Factory fire killed more than 150 young women and trapped some 700 workers in a building with locked stairwells. Stunned observers reported seeing young women leaping from the burning building and "hitting the pavement like hail."

Most often, however, challenges to working conditions came from early attempts to unionize workers. Union organizers saw themselves as part of a national and international workers' struggle. They organized workers broadly across industrial lines and often espoused a strong progressive, and occasionally socialist, ideology. These were the Knights of Labor, the International Workers of the World (IWW, derisively dubbed "Wobblies"), and the Western Federation of Miners. Employers fought back with lock-outs, with armed "security police," and with pressure for intervention from state militias. Bloodshed often resulted. Demands for an 8-hour workday brought 80,000 workers into the streets of Chicago and stirred worries of a workers' revolution. In the days following that demonstration, police attacked workers and picketers, resulting in a gun battle that left 11 dead and led to the declaration of martial law across the country. In 1877, wage cuts for railroad workers who were working 15- to 18-hour days resulted in bitter strikes that culminated in a battle between striking workers and 300 Pinkerton "detectives" that led to 10 deaths. A mill workers' strike in 1912 in Lawrence, Massachusetts, also saw the imposition of martial law and pro-voked an attack by police on a group of women and children from striking families (Brooks 1971).

Less radical and less threatening than the unions noted above was the American Federation of Labor (AFL), which organized "locals" around par-ticular craft specialties in a manner that at first was more reminiscent of medieval craft guilds than the all-inclusive workers' unions. Gradually these locals grew, and the AFL's power superseded that of many of the early unions. The AFL was not revolutionary, but as it grew, it too embraced a progressive political agenda. Samuel Gompers, the AFL's early and dynamic president, contended:

> What does labor want? We want more schools and less jails; more books and less arsenals; more learning and less vice; more leisure and less greed; more justice and less revenge; in fact, more opportunities to cultivate our better natures, to make manhood more noble, woman-hood more beautiful and childhood more happy and bright. (Quoted in AFL-CIO 2000:4)

Yet, because the AFL organized primarily skilled craft workers, many factory workers still had little collective voice or representation.

A growing nationwide progressive movement at the beginning of the twentieth century brought gradual workplace reforms. An increasingly sym-pathetic press moved from berating "radical" workers to publishing exposés of the worst working conditions. Journalists who became known as muck-rakers gained the public's attention with articles on the abuses that workers suffered, and novels focused on working conditions as well, such as Upton Sinclair's *The Jungle* and Frank Norris's *The Octopus*. First at the state level and then nationally, laws were enacted that banned child labor, limited working hours, set minimum wages, and established guidelines for workplace safety. Seeing the advantages of accommodation over bloody confrontation, employers themselves also gradually agreed to improve conditions. The progressive sympathies of the Woodrow Wilson administration and the tremendous need for labor during World War I gave new strength to the union movement, and membership in the AFL grew to more than 4 million. At the close of World War I, trade unionists from around the world met to create the International Labor Organization (ILO) in 1919. Yet the pendulum in this struggle began to swing again, and the 1920s brought sharp reactions against unions.

World War I had helped to create the new Soviet Union, and the United States was gripped by its first "red scare." Anyone with socialist sympathies—which were often seen as the motivation behind any type of union organizing—was suspect. Conservative presidential administrations saw prosperity in an ever-rising stock market and in the strict regulation of "disruptive" worker activity. Employers were quick to make use of their growing power and position in the government and often contracted with "union-busting" services to break strikes and intimidate organizers. Although the 1920s saw a sharp drop in immigration, employers also had a new population from which to recruit strikebreakers: poor African Americans who were leaving the rural poverty and Jim Crow segregation of the American South in search of urban industrial jobs in the North. With little public or political support and sharply divided by race, workers once more lost bargaining power.

The situation changed yet again in the 1930s. The stock market crash of 1929 and the subsequent Great Depression showed the need for economic reforms. Hardships extended now to many workers, and public support for "relief, recovery, and reform" grew. Franklin Roosevelt was elected president, largely because he built a coalition of workers that was both rural and urban, black and white, and offered all of them a "new deal." The National Labor Relations Act of 1935 protected workers' rights to organize and to strike without fear of immediate replacement. The northward flow of African Americans continued, but now protected from strikebreaking, unions began to include nonwhite members, for greater numbers meant greater power, both in bargaining and in elections.

World War II enormously increased the demand for workers. Women, in particular, moved into the industrial labor force in previously unheard-of numbers. Unions grew, but workers forwent strikes and allowed concessions as part of their show of support for the war effort. Following the war, the unions, now larger and more diverse than ever, with many female and nonwhite members, again began to assert demands. They faced another postwar red scare, however, and a conservative political agenda held unions in suspicion. The Taft-Hartley Act (1946), passed over Harry Truman's veto, sought to reduce union power and make union leaders liable for strike-related damages. Yet a growing economy needed more workers, and unions continued to attempt to consolidate their power. In 1955, the craft-organized AFL merged with the industry-organized Congress of Industrial Organizations (CIO) to create the AFL-CIO. Growing union power was key to the 1960 election of John Kennedy. At the same time, many Americans viewed unions with suspicion, in part because of charges of corruption and the alleged ties of union leaders to organized crime, such as Jimmy Hoffa of the Teamsters (a union that was not part of the AFL-CIO).

Other changes were occurring in the U.S. and world economy by the mid-1960s, however, the impacts of which few national leaders could have predicted. In 1964, the United States and Mexico began the Border Industrialization Program, which allowed for the opening of U.S. plants on the Mexican side of the border in a duty-free export zone. American industries were beginning to seek labor, especially labor for relatively unskilled assembly work, overseas. The deindustrialization of the American economy had begun. New jobs created during the 1970s were often in service sectors, which traditionally had not been unionized.

Global economic change was accompanied by major political changes in the United States as well as in Britain and Canada. Conservative governments with strong ties to business interests were growing in popularity. As Ronald Reagan entered the American presidency, one of his first cabinet appointments, to the post of secretary of labor, was Ray Donovan, a man who had been charged with union-busting activities. One of Reagan's first official acts as president was to fire the striking air traffic controllers and to seek replacements. Reagan defended this action on the grounds that the air traffic controllers were involved in an illegal work stoppage, yet the symbolism of the president firing hundreds of striking federal employees was powerful. Striking workers now faced a double threat: They could be easily replaced by the many workers left unemployed by the recessions and deindustrialization of the 1970s and early 1980s with little government interference, or they could be "replaced" by overseas workers eager for jobs at a fraction of U.S. wages. The first threat broke strikes at the Hormel meatpacking plant, Caterpillar equipment, and others. The second threat not only limited workers' ability to strike but was often used to force worker concessions (Fantasia 1988). Workers were told that if they did not "give back" some of their wage and benefit gains, their companies would no longer be competitive and would need to move operations offshore. Unions found themselves negotiating the size of wage *decreases* rather than increases. The government-business alliance of the 1980s was in many ways similar to those established in the 1920s and the 1950s: It was staunchly anticommunist and antiregulation, and it equated patriotism with compliant labor. In the 1980s, however, global economics further weakened the power of labor.

Organized labor's success in using electoral politics was also greatly diminished. Strong union backing for Walter Mondale's challenge to Ronald Reagan in the 1984 presidential election did little to fire up the Mondale campaign. Union membership, as a proportion of the total labor force (and the total electorate), was as low as it had been since the 1920s. The campaign contributions made by labor unions were considered suspect, and labor could not match the largesse of big business.

Finally, in 1992 organized labor believed it had reformed and united enough to again flex political muscle, and it found its candidate in Bill Clinton. Twice endorsed as "the best friend that laboring people in this country have," Clinton nonetheless had a contentious relationship with union leaders. He adopted much of their agenda: an increase in minimum wages, expanded health care, regulations protecting the right to organize and strike, and job security measures. But Clinton was also a strong proponent of free trade. One of his first presidential efforts was to push through the North American Free Trade Agreement (NAFTA), which had been negotiated by the Bush administration. Clinton insisted on "side agreements" to NAFTA that were intended to protect labor and the environment, but these proved difficult to enforce.

Labor endorsements were important to both Al Gore and John Kerry when they were presidential candidates, but in neither of their two campaigns were they sufficient to prevent the election of George W. Bush. Support from traditional labor strongholds may have helped to secure Pennsylvania and Michigan for the Democratic ticket, but it was insufficient to deliver the once heavily industrialized state of Ohio.

Labor has found some new strength in recent years. Union memberships are again rising, largely because of increased organizing among professionals and service workers. Creating a new working coalition that includes the remaining industrial workers, the many new low-wage service workers, and professionals—including many teachers, a growing number of professors, and soon perhaps even physicians—is a difficult task. The United States remains one of the least unionized countries in the world: Less than 15% of the members of the U.S. workforce belong to unions, compared with 30% in Germany, 40% in Ireland, and close to 90% in Sweden. Even Canada has three times the unionization of the United States.

Compared with unions in other countries, U.S. unions also appear to have less power to enforce work stoppages and to make effective demands. This is because the power struggle in the United States is not just a two-way fight between labor and business. Here, the power struggle also involves government at several levels as well as media and public opinion. As many Americans express suspicion of big business, big labor, and big government at the same time, building a winning political coalition can be very difficult. Firms engaged in fierce competitive struggles may be driven to bring down wages, and huge enterprises may have enormous resources on which to draw. In his 1979 book *Manufacturing Consent,* Michael Burawoy suggested that Marx's competitive capitalism was giving way to monopoly capitalism, which could offer a wide range of carrots and sticks to maneuver workers into willingly, if unwittingly, submitting to their own exploitation. Even when they are not so willing, many workers are unsure of how to mount a winning challenge to corporate power.

Given the global nature of capital, there is also new interest in mounting a global labor movement. However, it is very hard to organize workers in different countries with very different wage rates. Further, unions in many developing countries are even weaker than unions in the United States. In Mexico, for instance, unions have long been part of the governing coalition, but they remain so only by avoiding confrontational methods. In Mexico, university students strike; workers endure. In Poland, the workers' union Solidarity struggled against a socialist, supposedly pro-labor regime and saw its president, Lech Walesa, move from the shipyards to the presidential residence. Yet the new Poland has continued its economic growth only by embracing free-market economics and discouraging union activism. American unions, meanwhile, recruit college students as organizers in hopes of building a new generation of union sympathy and solidarity.

Gender and Power: The Women's Movement

The political battles of the past 150 years have been drawn along gender and color lines as well as class lines. Many view these as "new" struggles, but a quick look back shows that they are not really very new. In 1848, as Marx and Engels handed out the *Manifesto* and workers in Europe revolted, a remarkable women's movement was convening a conference in Seneca Falls, New York. This was the Victorian era, when the ideal woman was widely considered to be a devoted mother of many children, just like the beloved Queen Victoria. Somehow, the fact that this British queen also ruled over a quarter of the planet's population and was one of the most powerful women in history did not enter into the U.S. and British cultural ideal. Women were to tend the home fires, nurture children, and soothe their weary husbands, who were the ones out battling daily in the "modern" commercial marketplace. Men were to return home in the evenings to be lords of their ornate "castles," an image that Victorian architecture tried to convey. In shocking language for the time, the declaration issued by the women who attended that Seneca Falls conference denounced these cultural norms as "domestic slavery" (*slavery* was not a word used lightly in the United States in 1848). Queen Victoria reigned right up until the turn of the century, and so did Victorian ideals, yet from 1848 these ideals were continually challenged by early "feminists."

Women began to demand broader roles and more opportunities than those afforded by the Victorian ideal. Charlotte Perkins Gilman (1910) argued that "ideal" Victorian women, who were supposed to be devoted mothers yet largely free from sexual passion, were in fact "over-sexed." Such a woman was "compelled to live within a constricted, feminine world of domesticity in which her dress, manners, education, work, and social relations all were defined by her sexual function" (p. 12).

The concerns of the early feminists also moved beyond strictly gender issues. They noted that many families were not able to live the Victorian dream inside the family castle. The homes that survive from this period and define it in our imagination are the well-built structures of the wealthy. Most of the rest we have never seen, because they burned down in great fires in the nineteenth century or were demolished as urban blight in the twentieth. Stephanie Coontz reminds us of the lives of the have-nots of this era in her book *The Way We Never Were* (1992):

> For every nineteenth-century middle-class family that protected its wife and child within the family circle, then, there was an Irish or a German girl scrubbing floors in that middle-class home, a Welsh boy mining coal to keep the home-baked goodies warm, a black girl doing the family laundry, a black mother and child picking cotton to be made into clothes for the family, and a Jewish or an Italian daughter in a sweatshop making "ladies" dresses or artificial flowers for the family to purchase. (P. 11)

Concern for the poor and the powerless developed into a range of sweeping reforms that became known as the Progressive Movement, a period starting sometime in the 1890s and continuing to about 1920. The politicians associated with this period, like all politicians in the United States and elsewhere at the time, were men: William Jennings Bryan, Theodore Roosevelt, "Fighting Bob" La Follette, and Woodrow Wilson, representing Democrats, Republicans, and Independents. Yet much of the local- and national-level reform came from women.

At Hull House, a settlement house in Chicago, social reformer Jane Addams served poor immigrants in surrounding neighborhoods but also pulled together one of the most remarkable gatherings of women intellectuals and activists ever (see Addams [1910] 1998). Unable to vote or hold elective office, they nonetheless transformed their communities. Many were women from wealthy backgrounds who were not interested in maintaining their husbands' private "castles." Instead, they started schools, trade schools, museums, and projects to preserve immigrant culture. They provided options and a new start for "unwed mothers," and they fought for children's rights and worker's rights. They demanded universal education and health care, and they fought political corruption. When corrupt city governments failed, they took over trash collection to improve city sanitation. They earned the admiration of even "macho" politicians such as Theodore Roosevelt, but they also engendered a huge amount of opposition for having stepped far out of their place.

Although these women shared a concern for the well-being of families, children, and workers, two crystallizing issues drove their activism: temperance and women's suffrage. *Temperance* in this context referred to abstinence from alcohol; this was the movement that eventually brought national prohibition to the United States. Often lampooned as drum-beating eccentrics, the women in the temperance movement were actually concerned about the large numbers of working-class men, many from immigrant backgrounds, whose attempts to escape the harsh realities of the day led to alcoholism and the subsequent neglect and abuse of their families. Few Americans were yet worried about the cocaine and morphine in patent medicines of the time; alcoholism represented the drug crisis of the day.

Prohibition did not last, but the changes the women brought about through their other crusade did. Electoral democracy was spreading around the world, but everywhere it was men's democracy. At the time of the writing of the U.S. Constitution, the debate was whether the vote should go to all men or only to property-owning "landed gentlemen." The Fifteenth Amendment to the Constitution, ratified in 1870, extended the right to vote to former slaves but didn't mention the "domestic slaves" of the Seneca Falls declaration. Women in both the United States and Great Britain insisted that progressive political reform must include the rights of all citizens to vote and hold office—that is, full suffrage. Susan B. Anthony and Elizabeth Cady Stanton fought for the Thirteenth Amendment as part of the abolitionist movement, then tried to include women's suffrage in the Fifteenth Amendment. They failed. The Fourteenth Amendment, on citizenship rights, became the first to include specifically only males as citizens, and the Fifteenth said that the right to vote could not be denied due to race, color, or previous condition of servitude, but Anthony was unable to get Congress to include the word *sex* in that amendment.

The women's suffrage movement gained ground slowly. Women first voted in national elections in New Zealand in the 1880s. They voted in a few local and territorial elections in the United States, first in the Wyoming Territory in 1869, a fact that almost prevented Wyoming from becoming a state in 1890. National support gradually came from Woodrow Wilson and other reformers. Finally, in 1920 the Nineteenth Amendment to the U.S. Constitution gave American women the right to vote. Great Britain followed a few years later. Gradually the idea moved around the world: Women in Mexico first gained the vote in 1953, women in El Salvador couldn't vote until 1972, and in Switzerland women couldn't vote in national elections until 1974.

The women's vote did not change the power structures of the world's democracies overnight. Surveying the death and destruction of World War I, some hoped that female leaders would bring peace and greater humanity to the world. Those hopes were dashed as electoral democracy itself was threatened by a new wave dictatorial governments in the 1930s and 1940s. Some countries did turn to women leaders who proved popular and capable and led their nations toward progressive reforms and a commitment to peace: New Zealand, Norway, and Iceland among them. But other female leaders led their countries into wars and other conflicts: Golda Meir in Israel's Six-Day War, Indira Gandhi in one of India's struggles with Pakistan, and Margaret Thatcher in Great Britain's war with Argentina over islands in the south Atlantic.

Whether in electoral office or not, women have added important voices to the political process. Franklin Roosevelt promised poor and struggling Americans a "new deal," but it was his wife, Eleanor, who made sure he kept his promises. She traveled the country meeting with the poor and dispossessed and spoke out on behalf of women, children, workers, and other groups whose agendas might be forgotten. Internationally, she helped build the United Nations into both a force for peaceful dispute settlement and a champion of the rights of women, children, and the poor around the world. Her hesitant, serious manner was a marked contrast to the jaunty optimism and deal-making of her husband, yet her work and its influence outlived the president and helped set the global agenda for the latter half of the twentieth century.

During the years of World War II, millions of American women went to work in heavy industry, both keeping the country running and providing the huge amounts of industrial might that Franklin Roosevelt called for to win the war. Yet as the male veterans returned home, looking for jobs, education, and families, the cultural pendulum again swung back to glorify the domestic woman. Her hopes were to be contained in her "hope chest": a box filled with the silverware and china she would need for her marriage and her domestic role supporting her husband and children. Women's social role was again redefined and their political role greatly diminished. Yet a handful remained active in political struggles, especially around civil rights, just as they had a century earlier. Rosa Parks helped launch, and Bella Baker helped organize, what became the civil rights movement.

What most know today as the women's movement emerged alongside the demands for racial civil rights in the 1960s. Much of the spark came from Betty Friedan's 1963 book *The Feminine Mystique*. Basing her arguments in a strong sense of history, Friedan asserted that the emphasis placed on a "feminine mystique" in the 1950s was an intentional attempt to undermine the 100 years of social change that had taken place since the Seneca Falls conference in 1848. Indeed, Friedan noted, writers' in *Newsweek* had even suggested women should leave college (it was only making them discontented) and perhaps even give up the vote (it was too much for them)! Friedan wrote with subtlety and wit, but her publisher was anything but subtle in its promotion of her book; one blurb read: "Today American women are awakening to the fact that they have been sold into virtual slavery by a lie invented and marketed by men. One book has named that lie and told women what to do about it." A renewed women's movement was launched.

The women's movement gained full strength in the 1970s with the prominence of groups such as the National Organization for Women (NOW). *Feminism* and *women's liberation* were now familiar terms. The women's movement has faced pendulum swings similar to those experienced by the workers' movement, and sometimes these two movements have risen and fallen together. Both gained in the Progressive Era of the early 1920s and in the Roosevelt New Deal era of the mid-1930s to mid-1940s. Both lost ground in the backlashes of the 1950s and the 1980s. Susan Faludi (1991) suggests that the backlash against the women's movement is still in progress.

Interestingly, although most American women endorse key elements of the women's movement's agenda—wage equality between men and women, for instance—many are reluctant to call themselves feminists. The term has taken on negative connotations for some, who associate it with a "radical" or "antifamily" agenda. Some women believe that they can't be "true" feminists because they "aren't lesbians" or "don't hate men," although sexual orientation and gender animosity have never been central issues to the movement. There are many varieties of feminism, of course, just as those who support workers' rights include Marxists, radicals, and reformers who range from moderate to liberal in their political views. Marxist feminists, such as longtime activist Angela Davis, see capitalism, both national and global, as the ultimate oppressor of women. Radical feminists, more influential in academic writing than in national politics, question the very construction of the American family and whether it can ever be a good place for women. Liberal or reformist feminists are less eager to utterly remake the economy or society. Like labor reformers, they seek more limited tangible changes: pay equity, expanded child care, a more family-friendly work environment, greater opportunities for women, and expanded health care. Senator Hillary Rodham Clinton has been a vocal proponent of this reform approach.

A fairly recent development—emerging in the past two decades—has been the globalization of the women's movement, a trend akin to attempts to globalize the labor movement. In one sense this is nothing new: American feminists lectured in Europe in the 1900s and took notes on changes as far away as New Zealand in forming their own ideas. Yet as global interchange increases markedly, changes in one country can quickly spark calls for change in others. The global character of women's issues was expressed dramatically at the United Nations' Fourth World Conference on Women, which was convened in Beijing in 1995. This event provided a remarkable reminder of the growing global awareness of the common cause of women and the continuing risks and burdens facing girls and women around the world. Some who attended the conference undoubtedly considered themselves feminists, whereas others had likely never heard the term. Yet this highly publicized gathering was a definitive demonstration of both the multiplicity of women's problems and the commonality of women's hopes and struggles. Hillary Rodham Clinton (1995) addressed these commonalities in her speech at the conference:

> However different we may be, there is far more that unites us than divides us. We share a common future. And we are here to find common ground so that we may help bring new dignity and respect to women and girls all over the world—and in so doing, bring new strength and stability to families as well.
>
> By gathering in Beijing, we are focusing world attention on issues that matter most in the lives of women and their families: access to education, health care, jobs and credit, the chance to enjoy basic legal and human rights and participate fully in the political life of their countries. (P. 1)

Women's roles are changing rapidly around the world, yet the world remains a dangerous place to be female. In addition to highlighting the particular vulnerability of women and children, Clinton stressed the need for broader transnational awareness of human rights struggles: "If there is one message that echoes forth from this conference, it is that human rights are women's rights—and women's rights are human rights" (p. 6).

For activists who seek to improve the lives of women and children, it is natural to focus on the home and family. Yet around the world and across societies, we have seen that economic power is often the key to social power. The greatest burdens on women and children are no different from those on men: economic uncertainty and the drain of ill-compensated, degrading, and dangerous work. Improvements in home and family situations must be paralleled by changes in education and the workplace. Women's rights are intertwined with workers' rights. They are also intertwined with the full range of human rights. In the United States, women's rights issues have not only paralleled workers' rights issues, they have also been intertwined with what have become known as rights of citizenship, civil rights.

_____ Race and Power: The Civil Rights Movement

The third great challenge to existing power structures also has a long history, one at times intertwined with the other two. In 1848, the United States was just ending a war. With the marines still standing at attention in the "halls of Montezuma," having captured Mexico City, the Mexican government agreed to the Treaty of Guadalupe Hidalgo. In return for $15 million (a relatively modest sum even in those days), Mexico would cede more than one-half of its territory to the United States: the land that would become the states of California, Arizona, New Mexico, Nevada, Utah, and most of Colorado in addition to the previous acquisition of Texas. The United States gained a lot of land and a new "minority" group: Mexican Americans. These new Mexican Americans had not come to the United States, the United States had come to them.

The war with Mexico had been bitterly opposed in parts of the northern United States. Henry David Thoreau went to jail rather than pay his war tax and was inspired there to write his treatise on civil disobedience, a work that would later guide Mohandas Gandhi and Martin Luther King, Jr., among others. A tall, lanky congressman by the name of Abraham Lincoln rose in the U.S. Congress to denounce "Polk's War" as an unjust attack by a more powerful nation on the newly independent Mexico. A surly army captain by the name of Ulysses Grant wrote home to decry the foolishness and cruelty of this war. Northerners who opposed President Polk's war were not just concerned about international peace and justice. They understood that Polk, a southern Democrat, was interested in expanding the power of the South by adding a string of new southern slave states.

Battles over slavery had divided the nation since its founding. The drafters of the U.S. Constitution settled on the very odd "three-fifths compromise," in which three-fifths of the enslaved population would be counted for purposes of congressional representation. This was not, as some have alleged, a statement about the worth of individual black slaves, for northern abolitionists wanted none of the slaves counted and southern slaveholders wanted them all counted. The issue was power: whether the northern states or the southern states would have a majority in the new House of Representatives hinged on whether the southern states could count their slave population. With neither side willing to allow the other a majority, they settled on the strange fraction that would initially give North and South equal representation. The addition of each new state to the Union marked a battle over whether it would be slave or free and how it would tip the delicate balance of power. As the compromise of 1850 unraveled into new fighting along the border states, Lincoln became convinced that the nation could not stand "half slave and half free."

By this time, many had already come to the same conclusion, and the abolitionist movement was challenging the existing order by whatever means it could. John Brown took up arms and went down fighting at Harper's Ferry. Harriet Tubman worked the Underground Railroad, which helped escaped slaves travel to the North. The claim to freedom of one such slave, Dred Scott, after he had arrived in a free state went all the way to the U.S. Supreme Court in 1856. Chief Justice Roger Taney, writing for the majority, dashed Scott's hopes in the strongest language possible:

> [African Americans] are . . . not included, and were not intended to be included, under the word "citizens" in the Constitution, and can therefore claim none of the rights and privileges which that instrument provides for and secures to citizens of the United States . . . [African Americans] had for more than a century before been regarded as beings of an inferior order, and altogether unfit to associate with the white race, either in social or political relations; and so far inferior that they had no rights which the white man was bound to respect; and that the negro might justly and lawfully be reduced to slavery for his benefit.

The majority in Dred Scott concluded that "[T]he right of property in a slave is distinctly and expressly affirmed in the Constitution." Escaped slaves were to be returned. This decision, and the language Taney used, outrageous even for 1856, only further galvanized the abolitionist movement.

Half slave and half free was indeed an impossible way for the United States to go on. Finally, the members of the newly created Republican Party, split over who would lead them, gave their nomination for the presidency to Abraham Lincoln, who was by this time best known for his eloquence in denouncing slavery in his debates with Stephen Douglas during their race for the U.S. Senate (Lincoln lost that race). The Democrats were divided between northerners supporting Douglas and southerners who would not tolerate a northern candidate, even one with such a strong pro-slavery stance as Douglas. With the opposition divided, Lincoln won the presidency, even though he had to sneak into town for his inauguration for fear of assassination and, upon taking the oath of office, immediately faced southern secession. While fighting to save the Union, Lincoln was reluctant to grant freedom to all slaves for fear of antagonizing those border states still in the Union that nonetheless had slave populations. Many northerners, fearing black competition for jobs, were also strongly antiabolition. Finally, in 1863, with the new hope of battlefield victories, Lincoln delivered his Emancipation Proclamation. Slaves in the states rebelling against the Union were freed. As nobody in the states rebelling against the Union was listening to Lincoln, the proclamation immediately freed no one, but as the Union troops moved through the South, they could now offer freedom to the slave population, some of whom joined the ranks of the Union army themselves. In his 1876 "Oration in Memory of Abraham Lincoln," the great black abolitionist Frederick Douglass said of him:

> Viewed from the genuine abolition ground, Mr. Lincoln seemed tardy, cold, dull, and indifferent; but measuring him by the sentiment of his country, a sentiment he was bound as a statesman to consult, he was swift, zealous, radical, and determined.

The Thirteenth Amendment to the U.S. Constitution, ratified in 1865, made Lincoln's proclamation the law of the land. Slavery was over, but the struggle for empowerment was only beginning. The Fourteenth Amendment, most famous for its "due process of law" clause, guaranteed all citizens "equal protection of the laws." The Fifteenth Amendment promised full voting rights, regardless of a citizen's "previous condition of servitude." With the death of Lincoln and his promise of "malice toward none," in the years immediately following the Civil War white southerners faced humiliations at the hands of vengeful northerners. Yet black southerners found a renaissance. The first black members were elected to the U.S. Senate and the House of Representatives. For the first time, black children could legally attend school and learn to read. A few fortunate blacks gained title to land for the first time.

Then the tide turned. Returned to power by northerners more concerned about economic recovery than continued policing of the South, southern politicians began to implement the multitude of restrictions on black social and political participation that became known as Jim Crow laws. Black schools closed. Black farmers lost their land and were reduced to sharecropping, with an unpayable debt burden that kept them working as virtual slaves to their landowning former slave masters. Southern politics again became a white man's club, and the pattern for the next hundred years was set. In the case of *Plessy v. Ferguson* in 1896, the U.S. Supreme Court endorsed segregation, stating that "separate but equal" facilities for blacks, whether in rail cars or schools, did not violate the U.S. Constitution. Not until the 1950s would "equal protection," voting rights, and integration again be on the national agenda, with activists promoting reforms that many in the 1860s thought had been achieved.

Many black southerners responded to the heavy hand of Jim Crow by fleeing to the North. As the immigrant labor supply was shut off in the 1920s, it was replaced by a supply of African American labor migrants moving from South to North, rural to urban. In New York City, Harlem, long the home of new immigrants, became Black Harlem. In the 1920s this enclave experienced its own renaissance as its residents included black writers, such as Langston Hughes, black professionals and businesspeople, and black musicians who created the sound of jazz. The 1920s was the "Jazz Age" in large measure because a new sound was moving up the Mississippi River and the Atlantic coast. The sound that most black workers heard most often, however, was the slamming of doors of opportunity. Many lived in tiny "kitchenettes" in segregated neighborhoods and competed for the lowest-paying jobs, most often without union representation. Of all Americans, blacks were hit hardest by the Great Depression of the 1930s.

Still the struggle continued. Blacks converted in large numbers from Lincoln Republicans to Roosevelt Democrats. They gradually gained access to unions. And they fought in record numbers against Hitler's version of white supremacy, although they had to do it in segregated army units. As they returned home from serving in that war with new confidence and new demands, the sentiments that would become the civil rights movement were kindled (Takaki 1993). President Truman first integrated the military, an act that, like Lincoln's proclamation, could be done by executive order. In 1954, the U.S. Supreme Court reversed *Plessy v. Ferguson,* declaring in *Brown v. Board of Education* that "separate is inherently unequal." The Court found that segregation in schools is unconstitutional, although it would take a decade of intervention by federal troops to enforce this decision.

Most of the impetus for change, however, came from the actions of African Americans themselves. In 1956, Rosa Parks refused to give up her seat on a bus to a white passenger and the Montgomery, Alabama, bus boycott was in place. This ultimately successful boycott brought to the forefront an eloquent national spokesperson for the civil rights movement, Martin Luther King, Jr. King was not only the person reading Thoreau on civil disobedience and Gandhi on nonviolence, however. The Student Nonviolent Coordinating Committee (SNCC), a group led by college students such as Diane Nash and dozens of others, staged sit-ins for civil rights in Nashville and across the South. The Southern Christian Leadership Conference, with King and Ralph Abernathy and dozens of others, led marches, registered voters, faced the fire hoses turned on them by police, and went to jail in collective protests. The nobility and tenacity of the marchers and the bitter hatefulness of some of those who opposed them stirred many in the American public (predominantly in northern and western states) who were able to watch the events thanks to a new innovation: national television news coverage. Changing public sentiments forced federal government action, and in 1964, the U.S. Congress passed the Civil Rights Act, finally implementing the intentions of the Fourteenth Amendment. In 1965, Congress passed the Voting Rights Act, implementing the goals of the Fifteenth Amendment.

From 1965 on, however, the civil rights struggle grew more difficult. The laws had changed but the society had not, and violence escalated. James Meredith, who had been the first black student admitted to the University of Mississippi in 1962, was shot and wounded in 1966 as he was on his way to Jackson, Mississippi, to register black voters. In the turmoil that ensued, one of the leaders of SNCC, Stokely Carmichael, first used publicly the explosive term *black power*. Many African Americans saw the call to black power as a new call to awareness and black consciousness, a call to self-reliance rather than trusting in federal government actions, and a new call to solidarity. In South Africa, Steve Biko launched his "black consciousness" movement on similar principles: awareness of oppression, self-confidence in place of an instilled sense of inferiority, and a sense of solidarity in the struggle. This appealed to many African Americans, especially those in the urban North.

As whites listened to the speeches of Carmichael and of Malcolm X, and saw the creation of groups such as the Black Panthers (also organized in 1966), many of them understood the call to black power as a call to militant revolution that included a violent antiwhite agenda. The lines of a new divide between the races were being drawn, lines of suspicion and misunderstanding that persist to the present.

King himself preferred the less inflammatory phrase "Freedom now" to the term *black power,* knowing that freedom was something white Americans could not oppose without betraying their most basic political ideals. Yet King also knew that the movement needed to shift:

> When a people are mired in oppression, they realize deliverance only when they have accumulated the power to enforce change. . . . This is where the civil rights movement stands today. Now we must take the next major step of examining the levers of power which Negroes must grasp to influence the course of events. (King [1967] 1992:154)

King also spoke out against the war in Vietnam, which was draining resources from the War on Poverty and was disproportionately drafting and killing poor and nonwhite soldiers. The civil rights movement became intertwined with the antiwar movement, and these were eventually joined by other social movements concerned with rights. Young Native Americans from many backgrounds but with a common eagerness for change formed the American Indian Movement, a group that gained wide media attention when it seized the unused prison on Alcatraz Island in San Francisco Bay (an old treaty said that unused federal property should be returned to Native use). At the same time, "brown pride" spawned "La Raza," a movement of Mexican Americans in the southwestern United States devoted to "Chicano power." The intermingling of groups concerned with race, rights, and political mobilization was crystallizing into a larger movement.

Yet the undying common denominator remained poverty. King's last great initiative was his "poor people's campaign." While Michael Harrington (1962) wrote about "the other America" of rural and urban poor, and

Robert Kennedy toured this other America, King tried to organize the poor into a common movement. Increasingly, this became intertwined with the labor movement. King challenged so-called right-to-work laws, which made union organizing difficult:

> In our glorious fight for civil rights, we must guard against being fooled by false slogans, as 'right to work.' It provides no rights and no work. Its purpose is to destroy labor unions and the freedom of collective bargaining. We demand this fraud be stopped. (Quoted in AFL-CIO 2000:154)

King's final act was to go to Memphis, Tennessee, to march in support of striking sanitation workers. The strike, he told supporters, was about the fundamentals of human dignity. The following day, he was shot and killed.

New Challenges to Old Structures

Backlash: The First Challenge from the Right

The 1970s brought new opportunities for those who had finally been admitted to schools, universities, professions, and other places of power and opportunity. Yet economic stagnation and entrenched resistance often counterbalanced whatever progress had been made. Across the United States, a new generation of black mayors was elected, yet these mayors governed cities with declining resources and growing problems. Economic resources left the cities along with middle-class white flight to the suburbs, and the new mayors were forced to act as salesmen, trying to lure new investment; as beggars, trying to win contributions from suburban governments; and as lobbyists, trying to gain aid and support from increasingly reluctant state governments. African Americans had gained offices and titles, but real power remained elusive.

At the same time, the loss of industrial jobs and declining incomes for the working class created new fears of competition and struggles between workers. Working-class whites felt abandoned by the loss of the progressive New Deal agenda. In particular, they directed their anger and frustration at welfare programs, which they believed rewarded people for not working, and at affirmative action programs, which they feared would limit their own opportunities (Faludi 1991).

The unfortunate fallout of the affirmative action debate has often been to increase the divisions that separate those who share common vulnerabilities: workers of color, female workers, and working-class whites. Just as race, class, and gender are intertwined, so are the struggles over power and privilege. In delivering a speech at Indiana University at South Bend in 1997, liberal black philosopher Cornel West smiled but was earnest when he spoke of ultraconservative radio talk-show host Rush Limbaugh, calling him "Brother Rush, who speaks for his people and is right about the problems even if he is wrong about the solutions. He is still dividing when he needs to unite." One struggle is tied to another. In the nineteenth century, former slave and black abolitionist Sojourner Truth became an equally powerful advocate for women's rights. Hearing a man talk about how women should give up their "silly" struggle for power and equality and be glad to allow men to help them into carriages, she could only note, "Nobody ever helped me into no carriages, and ain't I a woman?"

In a memorable reading at President Clinton's first presidential inauguration, poet Maya Angelou called on those gathered to have the grace to look across racial, ethnic, historic, and geographic divides "into your sister's face" and "into your brother's eyes" and to say with simple hope, "Good morning." That hope has proved elusive. In Europe and the United States it took decades of labor struggle for workers in different factories to see their common cause and several more decades for workers in differing industries to see their common cause. In the United States, it took still longer for native-born and immigrant workers, and then for black and white workers, to see themselves as fellow strugglers rather than competitors.

We still live in a time in which workers in different countries are likely to see one another only as competitors, and maybe unfair ones, and in which men and women see themselves as competitors in a zero-sum game— that is, a game in which someone's gain must be someone else's loss. Yet in new ways men and women around the world are being pulled together to see their common stance as colaborers, coproviders, and coactivists in reform.

In his book *Race* (1992), Studs Terkel includes an amazing interview with C. P. Ellis, a former Ku Klux Klan leader turned labor activist. The son of a poor laborer who could never make ends meet, Ellis grew up angry and turned to the Ku Klux Klan, where he felt welcomed and at home. He rose to prominence as a Klansman, but he realized that he was still often snubbed by white elites and that he seemed to be worrying about the same issues as working-class blacks. As he grew active in city politics in Durham, North Carolina, he was called on to head a committee with Ann Atwater, a black civil rights activist. At first horrified by the prospect of working together, the two soon found how similar their lives, struggles, and hopes for their children were. Ellis left the Klan and was subsequently elected to lead a union that is 70% black. He now has a new message:

> I tell people there's a tremendous possibility to stop wars, the battles, the struggles, the fights between people. People say, "That's an impossible dream. You sound like Martin Luther King." An ex-Klansman who sounds like Martin Luther King! [He laughs.] I don't think it's an impossible dream. It's happened in my life. It's happened in other people's lives in America. (Quoted in Terkel 1992:278)

Durham, which has seen both civil rights and labor struggles, is now more often cited as an example of progressive change. Says Atwater, "At the end, C. P. and I could lock hands and say we won" (quoted in Terkel 1992:278).

Moral Values: The Second Challenge from the Right

Whereas many social movements are *progressive,* seeking social reform, social justice, and new forms of social organization, others are *conservative,* seeking to retain tradition, or even *reactionary,* seeking to bring back the customs of an earlier time that movement supporters perceive as more moral, pure, or satisfying than the present. The activities of the "religious right" can be seen in this light. During the 1980s, the Moral Majority, a group led by the Reverend Jerry Falwell, promoted a church-based social movement that would call the United States to national repentance and greater godliness. In choosing its opponents, the Moral Majority followed a pattern of backlash against movements its members saw as destructive. Particular targets included the National Organization for Women and the feminist movement, the gay rights movement, and the American Civil Liberties Union.

Although the Moral Majority garnered a lot of media attention, it never had the political clout that the Christian Coalition achieved in the 1990s. This group, led by conservative political activist and lobbyist Ralph Reed and featuring as its best-known spokesperson televangelist and presidential candidate Pat Robertson, sought to mobilize the votes of millions of conservative Christians and to influence elections at all levels, from school board to U.S. president. Other conservative groups, such as James Dobson's Focus on the Family, although less obviously engaged in electoral politics, also became strong supporters of this media- as well as church-based movement, which galvanized followers around the theme of "family values." This movement gained new prominence when presidential candidate George W. Bush, who was not originally the candidate favored by the movement's leaders, embraced many parts of its agenda, including restrictions on abortion and gay rights, greater government support for religious schools and organizations, and a national emphasis on supporting marriage and traditional family structures. A concern over "moral values" also appears to have been a major factor in the support Bush received from key groups in closely contested states in the 2004 presidential election.

The issue of "moral values," as defined by religious conservatives, has clearly tapped deeply felt concerns on the part of many. Although strongest in the American "heartland" of the Great Plains and rural South, this movement's agenda is now being echoed around the world. Conservative religious parties—Jewish (in Israel), Islamic, and Hindu (in India and Nepal)—are growing increasingly powerful across the Middle East and South Asia. Even where these parties have been restrained or repressed by national governments, such as in Algeria, Pakistan, and Turkey, they remain vigorous movements at the local level. Although different from the conservative Christian movement in their theology and methods, they have some concerns in common with the religious right in the United States. Although Jerry Falwell infamously suggested on Pat Robertson's television program *The 700 Club* that the terrorist attacks on New York City and Washington, D.C., of September 11, 2001, were indirectly the fault of feminists, gays, and civil libertarians, the truth is that both Falwell and Osama Bin Laden are worried about the purity of their faith and the moral state of their homelands in the face of challenges by women's movements, "vulgar" media, civil liberties activists, "secular humanists" and others (Falwell does not advocate terrorism as a response to these threats, of course).

Global trends suggest that conservative religious movements must also be igniting deep-seated feelings. Some observers see them as reactions against an increasingly secular, consumer-oriented, and media-dominated globalized culture. Supporters of these movements long for a more "moral" world that emphasizes faith and family. Some have suggested that parties and governments ignore this longing at their peril, and that they would be wise to bring moral values back into their agendas. Others suggest that the language of "morality" needs to be broadened: Do not growing numbers of hungry, homeless children or the excesses of the suddenly wealthy and powerful coupled with the displacement of those who have worked hard their entire lives also constitute a moral outrage? Christian columnist Leonard Pitts, Jr. (2004) asks:

> So I look at the success conservatives on the so-called Christian right have had in claiming [Jesus] as their exclusive property and I wonder, where in the heck is the Christian left? Where are the people who preach—and live—the biblical values of inclusion, service, humility, sacrifice, and why haven't they coalesced into an alternative political force?

If the moral values movement does broaden its base, as seems to be happening in a variety of places, such as some African American churches, it may become harder to categorize the movement as "right" or "left." At this point, however, it appears that the issue of "values" will continue to have prominence in U.S. political debate for some time to come.

No Sweat: International Labor and Consumer Activism

In the balance of power between capital and labor, capital has increasingly gone international, yet most labor organization remains local and national. Analysts of global industry, such as Mexico's Jorge Bustamante, contend that the international power of capital will have to be met by the international organization of labor. A model for how this might be done can be seen in the current organization of many labor unions in the United States: "Locals" are in turn organized into larger national unions, such as the United Auto

Workers, and these in turn cooperate in national labor organizations, such as the AFL-CIO. The next step is for national organizations to cooperate in international workers' groups. In many ways this idea is as old as the "Wobblies," the International Workers of the World, and other early labor organizations. It is difficult to implement, however, given not only the distances between locations but also the great disparity in wages and conditions among workers in differing locations. Could Mexican workers earning less than the equivalent of $10 a day rally behind their fellow workers in the United States who are striking for more than $10 per hour? Will U.S. workers rally to the cause of Mexican workers when they see their Mexican counterparts mostly as low-wage competitors?

International labor organization will be a difficult task, but forces of globalization may make it easier. Certainly, communication among workers across national borders is becoming easier than ever before. Further, falling industrial wages in wealthier countries are bringing these workers closer to their fellow workers in middle-income countries such as those in Latin America. Industrial plants themselves are becoming more similar internationally, and movement is no longer one-way. For example, Ford operates nearly identical plants in locations across North America, Europe, and Latin America, shifting production among them depending on consumer demand and the profitability of each. Engineers regularly travel among these plants, and a new generation of multicultural, multinational labor organizers may soon find themselves making the same circuit.

Labor organizing has been very difficult in some locations where political pressures, police intervention, local customs, and social structure weigh against organizing particular groups, such as young women in electronics plants. More and more, however, multinational corporate policy has been swayed by growing consumer activism. Large companies, along with their celebrity promoters, have been embarrassed into investigating charges of sweatshop conditions in their plants and those of their suppliers. A few years ago, television personality Kathie Lee Gifford faced accusations that the line of clothing carrying her name was largely sweatshop produced. The Nike sports apparel company has long been the target of protests for the low wages it pays the workers in its East Asian factories. Nike executives at first contended that the company couldn't be held responsible for conditions in plants that were owned by local suppliers, but as sales of Nike shoes slipped, they have slowly agreed to more inspections and controls. Jeans maker Levi Strauss has attempted to bolster both its image and its sales by publicizing the company's aggressive stance on refusing to work with sweatshop suppliers.

College students have been particularly active in the effort to force manufacturers to improve working conditions. In 1999, students around the United States began a boycott of Reebok sports shoes in response to conditions at Reebok's Indonesian factories. The company eventually admitted that the factory conditions were "subpar" and promised to remedy the problem. The boycott was coordinated by an organization called United Students

Against Sweatshops, which was bolstered by international human rights organizations and the Worker Rights Consortium. Students then went on to tackle the multimillion-dollar clothing business that is operated from their own campuses. Universities, especially those with well-known sports teams—such as Notre Dame and the Big Ten schools, including Michigan, Wisconsin, and Indiana—bring in large amounts of revenue through the sale of clothing with school colors and team logos. These clothes are imported from hundreds of suppliers, who in turn often outsource the manufacturing process to thousands of low-wage facilities. In recent years, students on many university campuses have successfully lobbied their administrations to set tougher standards for the conditions under which such clothing is made, including agreeing to work only with manufacturers inspected and approved by the Worker Rights Consortium. The very nature of the outsourcing process makes this difficult, as it is no small task to inspect thousands of small "needle shops," but the principle that sweatshop conditions must not be tolerated has been established. The financial risks to a university for refusing to accept sweatshop-produced goods are also substantial. For instance, in response to campus antisweatshop activism in the spring of 2000, Nike announced that it would withhold millions of dollars from Michigan and Brown Universities and $30 million it had promised to Nike owner Phil Knight's alma matter, the University of Oregon (Greenhouse 2000). Just as boycotts and calls for university divestment of funds from companies doing business in South Africa were the hallmark of student activism in the 1980s—also a frustrating task, given the complexity of international business, but one that was ultimately successful—boycotts and inspections of sweatshops have become central to student activism in the twenty-first century.

Consumer activism has also targeted one of the worst abuses perpetrated by industry in our time: the use of forced child labor. Children 6 years old and even younger are sent to work in, or even sold to, sweatshops, where they hook rugs, work looms, or stitch soccer balls that will eventually be stamped Nike, Dunlop, and Adidas. As Schanberg (1996) reports, some earn nothing and are lucky to get a meal; others work for as little as 6 cents an hour:

> Silgli is only three, barely able to hold a needle, but she has started to help her mother and four sisters in India; together they can earn 75 cents a day. Sonu spends his days cutting chicken feathers for shuttlecocks in a workers' slum riddled with tuberculosis. Amir left school after the third grade and now spends his days sitting on a concrete floor using his feet as a vise while sharpening scissors and tools for a Pakistani metal shop for 2 dollars a week. (P. 41)

Just as in the United States, Europe, and Japan a century ago, these brutal conditions often prevail amid growing wealth and luxury. Schanberg goes on:

In India the vistas of child labor are much the same: eight-year-olds pushing wheelbarrows of heavy clay across brickyards; four-year-olds stitching soccer balls with needles longer than their fingers; fragile-looking girls carrying baskets of dung on their heads; little boys hacking up sputum as they squat before their looms, trying to do good enough work to avoid the masters' blows. All this takes place against a backdrop of rising affluence enjoyed by the privileged classes—lavish villas with high walls topped by iron spikes and satellite dishes on the roofs, luxury cars driven by liveried chauffeurs. (P. 45)

Officials in countries such as Pakistan and India argue that Westerners misunderstand the complexity of the issue of child labor, the role that child labor has played historically in these countries, and the real needs of desperately poor families. It is true that child labor has played a major role in the early industrialization of many wealthy countries. Children worked in the mines and operated the looms of early industrial Great Britain, just as children worked in the mines and mills of the United States in its early industrial development. They were the preferred workers wherever wages were the lowest and working conditions the most cramped. Likewise, in Japan, children slaving in the mines and factories of Battleship Island helped launch Japan quickly into the industrial era.

Yet it is also true that concerted activism brought down these practices. Progressives such as Jane Addams denounced the "theft of childhood." Labor leaders fought to outlaw child labor, both in response to moral outrage and because the low wages of children undermined the wages of all workers. Journalists took up the call for reform, seeking to embarrass the great industrialists into changing their ways. Popular sayings in the early twentieth century contrasted the new worlds of work and leisure; for example: "The factory is next to the [golf] links so that the children can watch the men at play." Social activist Lewis Hine traveled the United States for four years, from 1908 to 1912, photographing children working in cotton mills, mines, and fields. His work inspired the 1916 Keating-Owen Act, which restricted child labor; however, this act was in force for only two years before it was declared unconstitutional. In 1924, a proposed constitutional amendment that would have banned child labor failed to win ratification. Finally, in 1938 President Roosevelt signed the Fair Labor Act, limiting hours and conditions under which children can work and giving a priority to schooling. Still, a 1997 study by the U.S. General Accounting Office found a 250% increase in child labor law violations between 1983 and 1990 in the United States.

A similar struggle against child labor is now taking place internationally, aided by the growing availability of fax and e-mail communication, and now joined by the voices of children themselves. One of the most recognized anti–child labor crusaders, and one of the most successful, is Craig Kielburger, a young Canadian who began his work in 1995 at age 12 after reading a story about child workers in Pakistan. He has since traveled the world and met with human rights notables from South African Archbishop Desmond Tutu and Pope John Paul II to Mother Teresa and the Dalai Lama. He has met with the Canadian prime minister and with the representatives of international organizations and has visited working children around the world, including those who live in the Manila dump I described in Chapter 10. His organization, Free the Children, now has more than 100,000 young members in 20 countries and has successfully freed scores of child laborers while working for more sweeping international changes (Rowe 1999). Craig's mother looks back in astonishment: "It's amazing because it just started with this group of 12-year-olds. I can't believe it myself. It just shows how much children want to be involved. It's amazing what they can do, just amazing." In contrast to stereotypes of his generation as apathetic and materialistic, Craig claims that "young people are longing for something more meaningful in their lives, something more challenging, something that allows them to prove themselves" (quoted in Rowe 1999:72).

Human Rights and Civil Rights

Across the planet, new attention is being given to human rights abuses. When President Jimmy Carter tried to make human rights a pillar of his foreign policy in the 1970s, he faced apathy and Cold War suspicions. At the beginning of the twenty-first century, in contrast, human rights investigations are taking place on all continents. Many of them involve struggles over inequality—issues of race and class as well as workers' rights and poverty.

In postapartheid South Africa, a body known as the Truth and Reconciliation Commission has been assigned the task of investigating human rights abuses that took place during the struggles to maintain and end apartheid. Archbishop Desmond Tutu, internationally recognized for his nonviolent efforts to bring down apartheid, has presided over the commission's proceedings, which have included investigations of the actions of both blacks and whites. Many issues remain unresolved, but there have been rare moments when both truth and reconciliation have been achieved. In one case, an elderly black woman, now blind, whose son was brutally murdered and who was made to watch the torture and killing of her husband by South African security forces, asked for only three things from the white sergeant who did the killing: that he would help her give her husband a proper burial, that he would come to visit her once a month so she could be "a mother to him," and that someone lead her across the courtroom to him so that he would know her forgiveness was real. On hearing her requests, many in the courtroom wept, the accused fainted, and someone broke into a chorus of "Amazing Grace."

In West Africa, the country of Nigeria now has a democratically elected government after years of military rule, yet huge inequalities and heavy-handed military actions continue. The irony that the poorest part of the country is also the location of the vast oil deposits that fuel the country's economy is not lost on residents. Long ignored by national governments with leaders from the north, who have a different ethnic and religious background, the southerners along the oil-rich Niger delta have tried everything from siphoning oil from the pipelines to taking Shell Oil Company workers hostage. Whether President Obasanjo, formerly one of the ruling generals, can move Nigeria toward stability, democracy, and a more equitable distribution of resources is still unknown; his ability to do so may determine whether Nigeria, Africa's largest country, will enter a new era of progress or a new round of civil war (Onishi 2000).

In Latin America, the Chilean Senate stripped General Augusto Pinochet of his immunity from prosecution. Pinochet came to power in the 1970s in a coup that toppled the elected socialist government with covert aid from the United States and several large corporations. Pinochet launched the "Chilean miracle" of export-driven, free-market-led growth. To make sure this growth was unimpeded, however, he had thousands of dissidents arrested, detained, tortured, and apparently murdered. These included not only socialist leaders, but all types of labor organizers or sympathizers; unions and student dissent were crushed. Finally, in 1988, Pinochet allowed a referendum on the continuance of his regime, and Chileans voted "*no mas*" in overwhelming numbers. Some credit Pinochet with stabilizing and globalizing the country's economy, but most were eager to return to the freedoms and democracy that had been lost in the process. The aging Pinochet was arrested in Britain in 1999, threatened with trial in Spain, and then returned to Chile, where the general and "senator for life" may now face trial for the brutality of his regime.

Argentina and Uruguay have already held investigations into the excesses of their generals during "dirty wars" against labor leaders and social activists, although with few convictions resulting and a general amnesty for the generals. Guatemala has ended almost a half century of fighting between leftist guerillas and peasant groups and right-wing military governments with on-and-off U.S. backing. Its fragile democratic government must now find ways to reestablish peace and security along with respect for human rights in what may be the most unequal country in the world, with its vast gap between rich and poor as well as a racial/ethnic divide between the Spanish-speaking and Mayan-speaking halves of the society.

In Asia, supporters of Tibetan autonomy, including a wide range of international celebrities, have rallied behind the Dalai Lama, himself now an international celebrity, to challenge China's often brutal dominance. Less fashionable but equally determined are the efforts of Daw Aung San Suu Kyi, winner of the Nobel Peace Prize and daughter of a former leader, to end the brutal military rule of the country that was once known as Burma and is now officially Myanmar. In Europe, international tribunals struggle to bring to trial the perpetrators of war crimes in Bosnia and Kosovo.

International organizations such as Amnesty International and Human Rights Watch have taken on key roles as observers, reporters, and pressure groups in each of these situations. Amnesty International, in particular, has become an organization whose mission is resonating with students. It has established many campus chapters that have taken up the causes described above as well as many lesser-known causes. Sometimes the students in these chapters have met brick walls of frustration, but sometimes they have helped to win the release of prisoners and detainees.

Amnesty International has also taken up the cause of North American prisoners, noting the problems of overcrowding and unsafe conditions in the U.S. prisons as well as opposing the U.S. death penalty. Issues of race and class converge with those of human rights in the United States as well. The prisoners waiting on the death rows of U.S. state prisons are rarely the criminal masterminds or kingpins depicted in popular movies; rather, they are disproportionately poor, ill educated, poorly defended, and black. Critics of the U.S. legal system have noted that although perpetrators of corporate fraud, embezzlement, and financial manipulations cost the U.S. public billions of dollars as well as risk lives through unsafe products and practices, they rarely receive prison sentences and often pay fines that are only a fraction of their total gains. So-called suite crime is often sweet indeed for those who get away with it. In his book *The Rich Get Richer and the Poor Get Prison* (1998), Jeffrey Reiman notes that the perpetrators of petty street crime, whose total take often doesn't even raise them out of poverty (many muggers barely earn minimum wage!), often receive harsh sentences despite the widespread perception that the criminal justice system is overly lenient. In 1990, New York Governor Mario Cuomo commented on this divergent "justice":

> If you're a kid from [a poor neighborhood] and you get caught stealing a loaf of bread, they send you to Rikers Island [prison] and you'll be sodomized the first night you're there. But if you're a businessman ripping us off for billions, they'll go out and play golf with you. (Quoted in *Washington Post* 1990)

Global human rights issues encompass not only class, race, and ethnicity but also gender (see Exhibit 12.1). The use of rape and other sexual abuse as a weapon of intimidation caught the world's attention during the Yugoslavian wars, but it has also featured prominently in the repression in Tibet and Myanmar, in West Africa, and in South America. In many places around the world, women remain not only the poorest of the population but also the most vulnerable. The U.N. Conference on Women held in Beijing in 1995 highlighted what has been a growing internationalization of the women's movement. Across national borders, women are coming to realize that they have common concerns and are starting to tie these issues to the question of

- Freedom from discrimination—by gender, race, ethnicity, national origin, or religion.
- Freedom from want—to enjoy a decent standard of living.
- Freedom to develop and realize one's human potential.
- Freedom from fear—of threats to personal security, from torture, arbitrary arrest, and other violent acts.
- Freedom from injustice and violations of the rule of law.
- Freedom of thought and speech and to participate in decision-making and form associations.
- Freedom for decent work—without exploitation.

Exhibit 12.1   The Seven Freedoms

From United Nations, Human Development Report (2000).

human rights. At the Beijing conference, Hillary Rodham Clinton (1995) summarized the transnational issues:

> It is a violation of human rights when babies are denied food, or drowned, or suffocated, or their spines broken, simply because they are born girls. It is a violation of human rights when women and girls are sold into the slavery of prostitution. It is a violation of human rights when women are doused with gasoline, set on fire and burned to death because their marriage dowries are deemed too small. It is a violation of human rights when individual women are raped in their own communities and when thousands of women are subjected to rape as a tactic or prize of war. It is a violation of human rights when a leading cause of death worldwide among women ages 14 to 44 is the violence they are subjected to in their own homes. It is a violation of human rights when young girls are brutalized by the painful and degrading practice of genital mutilation. It is a violation of human rights when women are denied the right to plan their own families, and that includes being forced to have abortions or being sterilized against their will. . . . Women must enjoy the right to participate fully in the social and political lives of their countries if we want freedom and democracy to thrive and endure. (Pp. 5–6)

No Dumping: Environmental Racism and Classism

When the residents of the Dudley Street area, one of the poorest in the Roxbury section of Boston, came together to try to revitalize their neighborhood, they had to begin with an environmental issue. Their neighborhood was being used as a dump: Vacant lots were filled illegally and overnight with the discards of more affluent areas, and everything from cast-off appliances to potentially toxic waste was landing in their backyards. The primarily black, Latino, and Cape Verdean (immigrants from islands off the western coast of Africa) residents came together under the banner of "Don't Dump on Us." The story of how this wasteland (in more ways than one) became a scene of revitalization is presented in a fascinating book by Peter Medoff and Holly Sklar titled *Streets of Hope* (1994) and by the documentary film *Holding Ground: The Rebirth of Dudley Street* (Lipman and Mahan 1996).

Many of the world's poor find themselves the unwanted recipients of the castoffs of more affluent neighbors, whether those across town or across the globe. Observers have noted that people of color with limited political voice are especially vulnerable to such dumping and have dubbed the practice "environmental racism." In his book *Dumping in Dixie* (1990), African American sociologist Robert D. Bullard documents the concentration of sites for toxic and unsightly waste in areas where most of the residents are poor and black. In a follow-up titled *Confronting Environmental Racism* (1993), he describes the efforts of local groups to challenge the use of their neighborhoods as dumping grounds for the by-products of municipal landfills, incinerators, polluting industries, and hazardous waste storage.

These problems have now taken on an international dimension, as the world's wealthier nations seek places to dump wastes they can't dispose of at home without creating an outcry. A favorite site is India, where it is not called "dumping" but rather "recycling." This recycling involves teenage boys tearing open car batteries from Australia without protection, breathing toxic fumes, suffering burns from battery acid, and risking brain damage while they pull lead plates from the batteries. In other sites, workers melt used plastic bags from the United States into material for new bottles, toys, and other products, earning as little as 30 cents a day while breathing fumes that can cause cancer, birth defects, and damage to the kidneys and nervous system (Guruswamy 1995). Greenpeace and other international environmental activist groups have taken up the cause of these workers, recognizing that environmental action divorced from workers' rights can turn into cruel ironies.

In the past, environmentalism has often been viewed as a pastime of the rich, the concern of those who are well enough off to worry about aesthetics and not their next meal. Yet it is often the poor who are most vulnerable to the effects of environmental degradation. For a time, at least, the rich can avoid the heat brought on by global warming by buying bigger air conditioners; they can avoid breathing polluted urban air and drinking contaminated water by driving cars with cabin air filters and buying prestigious brands of bottled water. They can hope to import dwindling natural resources from as far afield as necessary and to export their waste products to locations far away. The well-off can hope to reserve condominiums or time-shares in ever more remote and "pristine" locations. The poor, however, must deal with their world as it is: They must labor in the polluted air, carry the contaminated water to their homes, and live amid toxic landscapes. The notion that there is a trade-off between protecting the environment and promoting economic development is a false one. Truly broad-based, sustainable development that will benefit many into the future must take into account both social and environmental concerns (Weaver, Rock, and Kusterer 1997). Longtime civil rights activist and Rainbow Coalition founder Jesse Jackson stresses these links:

> The promise of the new world order is threatened by the twin injustices of poverty and environmental destruction. Environmental destruction falls most heavily on the shoulders of the disadvantaged, and this is a threat to democracy itself. What is democracy, after all, if your air is too polluted to breathe? What is economic development if your land is too poisoned to farm? What is international law if rich countries dump their toxic wastes on the shores of poorer nations? Environmental justice is fundamental to the new world order. The United States is proud of Liberty's promise to the world: "give me your tired, your poor, your huddled masses, who yearn to breathe free." We must now extend the right to breathe free to every nation and every individual, for the right to breathe free is the most basic human right of all. (Quoted in Porritt 1991:38)

The environmental movement has begun to find much in common with other social movements. The best protectors of the land are often indigenous populations who have learned to work sustainably with their local resources. Many sustainable development and restoration projects have begun to work closely with local women's groups. It is often the women who tend both the land and their families closely, and so understand the intimate health connection between the two. A successful environmental movement for this century cannot take the form of ecological imperialism, in which wealthy nations dictate to poor nations how they should use their land. Rather, it must empower local communities to protect the land they love and value as they share ideas with umbrella movements that encompass the globe.

One World After All

People who have a great deal of money to spend on a vacation and who also prefer rugged adventure travel to relaxation on sedate beaches (Bobos, perhaps?) can book passage on a Russian icebreaker to travel to the North Pole. The Russian navy gets extra cash, and the adventurers get to make a claim that few of us can: "I've stood on the North Pole." A group of people who made that journey in August 2000 cannot make that claim, however. It's not that their icebreaker didn't get to the North Pole—it did so easily. Rather, when they got there they were the first people ever to see open water at the pole. The ship floated in the open water, then took the passengers some six miles away to an ice floe so they could at least stand near the pole. No one knows for sure why this occurred or whether it was a sign of global warming, but climatologists estimated that the North Pole had not been ice free for the previous 60 million years ("No Ice at Pole," *New York Times* 2000). The ship's passengers thought of their low-lying home cities—New York, Los Angeles, London, Tokyo—and wondered whether the threat of rising oceans as a result of global warming is real. Whether or not global warming was the cause of their odd experience, it is a powerful reminder that we all, rich and poor alike, do inhabit one world after all. The poor encounter global problems first, but eventually they reach us all.

Soon it may be possible for the adventurous and wealthy to move beyond the poles to commercial space travel. Microsoft cofounder Paul Allen is currently helping to finance attempts to build a space plane that could eventually take paying passengers into space at the cost of something like $200,000 per ticket. But what sort of planet will these passengers look down on? One to which they are glad to return, or one of social and environmental devastation? No matter how wealthy, they will still need to come back down to our common home. This is also a strong reminder about the limits to the promise of endless growth. We cannot promise all the poor of the world that they too can have a Lincoln Navigator. Even if we could manage to deliver all those SUVs, the planet wouldn't be able to handle them. We must begin serious discussion, and take concerted action, regarding just how much is enough, what lifestyles the planet can sustain, and how we can best ensure that everyone has the opportunities necessary to lead a rewarding, meaningful life. This is a discussion in which Marx, Douglass, Stanton, and Thoreau would be very much at home.

The debate concerning who can best lead us toward a better world has often focused on two sectors: the private profit-making sector (big business) and the public, regulatory sector (big government). Truly big government was tried with state socialism, and faith in the productive power of big business has fostered global free-market neoliberalism. The former brought gross inefficiencies, and the latter seems to be bringing gross inequalities. One solution is to find an optimal balance between the productive power of the market at its best and the democratic concerns of the public sector at its best. Many have also called for renewed attention to the "third leg" of a stable system (Giddens 2000), **civil society**, or the social sector. This is the arena of voluntary action, citizens' groups, and action coalitions, as well as neighbors gathered on the front porch or in the village plaza to discuss common concerns. Civil society is not a replacement for responsive government—for one thing, it often lacks large resources—but it can often find and meet needs that are overlooked by the other two sectors.

Ironically, the best response to global pressures is often local action, exactly because it often finds the overlooked people and their potential, revealed in the details of their daily lives. Localism is not isolationism. Often the insights that help in the struggles of local people come by fax, e-mail, or satellite dish from distant sources where people are struggling with similar issues. Even as the world is ever more divided by gaps between the prosperous and the poor, the privileged and the oppressed, people are finding new ways of making connections across borders and barriers. The global economy is enriching a few, often at the expense of the many. But new global connections across old barriers can enrich our understanding, empower our actions, and embolden our spirit.

I am aware that the danger in reviewing so many problems at once, as I've done in this book, is that it may engender a feeling of "What's the use?" I offer here a word of encouragement: Don't underestimate the power of concerted action. A tired seamstress on a public bus helped to launch the U.S. civil rights movement. A former Klansman and a low-income neighborhood organizer changed the racial climate of an entire community. An old woman brought truth and reconciliation to a bitter situation in South Africa. A 12-year-old Canadian boy began a movement to end child labor. These people didn't act alone, but they did start small. You may often feel powerless, but if you are a student at a college or university of any kind, you are already on the way to having more privilege, prestige, and power than most of the people in the world. How are you going to use that position? A divided, troubled world is awaiting your answer.

KEY POINTS

- Social movements, using various forms of collective action, have been crucial to social change over the past 150 years.
- The labor movement, which has its roots in workers' struggles during the early industrialization of Europe and the United States, continues to address global issues of low wages and lack of job security.
- The women's movement also began in early industrial societies, as women called for greater social, economic, and political participation, including the rights to own land and to vote.
- The civil rights movement has its origins in the abolitionist movement, which fought against slavery in the 1800s. Civil rights activists grappled with legal issues, such as voting rights and desegregation, in the 1950s and early 1960s, and have gone on to address persistent race-related economic inequalities.
- New social movements have emerged to continue the work of earlier movements and to address global problems. These new movements include efforts to close down sweatshops and end child labor, to protect human rights, and to confront dangers to the global environment.
- The private sector (business), the public sector (government), and the social sector (voluntary action and civil society) all need to be involved if we hope to meet the worldwide challenge of creating equitable, sustainable economic development.

FOR REVIEW AND DISCUSSION

1. How have the labor movement, the women's movement, and the civil rights movement been intertwined during the past 150 years of U.S. history?
2. In what ways have social movements moved from the national to the international level? Give examples of some international movements.
3. How are the issues of gender inequality, poverty, racism, human rights, and the environment intertwined? What social movements have arisen to address these issues, and how are they interrelated?
4. What are some ways in which students have been involved in social change movements?

MAKING CONNECTIONS

You can explore the topic of environmental activism by visiting the Web sites of organizations such as the Sierra Club (www.sierra.org), the National Wildlife Federation (www.nwf.org), and the Nature Conservancy (www.tnc.org). If you are particularly interested in the subject of human rights, visit the Web site of Amnesty International (www.amnesty.org) and that of the U.S. chapter of this organization, Amnesty-USA (www.amnesty-usa.org). These sites provide a lot of current information on human rights issues and suggest ways in which individuals can become involved in promoting respect for human rights around the world. If you are particularly interested in social movements related to gender issues, the National Organization for Women maintains a Web site (www.now.org) that is filled with information on women's issues as well as the history and current state of the women's movement. You can find the union perspective on the labor movement, workers' rights, executive pay, and globalization at the AFL-CIO's Web site (www.afl-cio.org). This site also has links to a number of international and local labor organizations. At the Web site of the elder statesman of civil rights groups, the NAACP (www.naacp.org), you can find information on the civil rights movement, current issues and controversies related to the movement, and links to other organizations. Another good civil rights–related site is that of the Southern Poverty Law Center (www.splcenter.org), which provides information on the center's "Teaching Tolerance" campaign.

Explore some of the Web sites of organizations related to a social movement of particular interest to you. How do these sites present themselves to the public? How are they organized? What are the most pressing issues they emphasize, and what tactics do they use to advance their cause? How do they relate to other groups and organizations—local, national, and international?

Many of these organizations have local chapters, some of which may be in your community; some may have student chapters on your campus. Check your local phone book for community groups or your campus office of student life/campus organizations for information. Many of the Web sites mentioned above, including those for Amnesty International, the NAACP, the Sierra Club, and the Nature Conservancy, also have listings of local groups. Amnesty International's site also has information on starting student chapters. Visit a local meeting of one of these groups and talk with chapter officers and active members. How is this group organized, and how does it relate to national and international organizations and movements? What are the issues currently of particular interest to the group, and how is the group addressing them locally?

References

Addams, Jane. [1910] 1998. *Twenty Years at Hull House.* New York: Penguin.

AFL-CIO. 2000. *100 Years of Struggle and Success.* Washington, DC: AFL-CIO Public Affairs Department.

Allen, Frederick Lewis. 1952. *The Big Change.* New York: Harper & Row.

American Medical Association. 2004. "Number of Uninsured Americans Rises." Retrieved from http://www.ama-assn.org/ama/pub/category/13793.html

Anderson, Elijah. 1990. *Streetwise: Race, Class, and Change in an Urban Community.* Chicago: University of Chicago Press.

Anderson, Sarah and John Cavanagh. 2000. *The Rise of Corporate Global Power.* Washington, DC: Institute for Policy Studies.

Arendell, Terry. 1986. *Mothers and Divorce: Legal, Economic, and Social Dilemmas.* Berkeley: University of California Press.

Associated Press. 2000. "Study: Japan Has Healthiest Lives while Some U.S. Poor Rate Very Low." *Chicago Tribune,* June 5.

Astin, Alexander W. 1992. "Educational 'Choice': Its Appeal May Be Illusionary." *Sociology of Education* 65:255–60.

Banfield, Edward C. 1970. *The Unheavenly City.* Boston: Little, Brown.

Barnet, Richard J. and John Cavanagh. 1994. *Global Dreams: Imperial Corporations and the New World Order.* New York: Simon & Schuster.

Becker, Gary. 1964. *Human Capital.* New York: National Bureau of Economic Research.

Bell, Daniel. 1973. *The Coming of Post-Industrial Society.* New York: Basic Books.

Bernard, Jessie. 1981. "The Good Provider Role: Its Rise and Fall." *American Psychologist* 36:1–12.

Bhalla, A. S. and Frédéric Lapeyre. 2004. *Poverty and Exclusion in a Global World.* 2d rev. ed. London: Macmillan.

Blalock, Hubert M., Jr. 1967. *Toward a Theory of Minority-Group Relations.* New York: John Wiley.

Blau, Peter and Otis Dudley Duncan. 1967. *The American Occupational Structure.* New York: John Wiley.

Blauner, Robert. 1972. *Racial Oppression in America.* New York: Harper.

Bluestone, Barry and Bennett Harrison. 1982. *The Deindustrialization of America.* New York: Basic Books.

Bodley, John H. 1990. *Victims of Progress.* 3d ed. Mountain View, CA: Mayfield.

Bonacich, Edna. 1973. "A Theory of Middleman Minorities." *American Sociological Review* 38:583–94.

Bordewich, Fergus M. 1996. *Killing the White Man's Indian: Reinventing Native Americans at the End of the Twentieth Century.* New York: Doubleday.

Bourdieu, Pierre. 1984. *Distinction: A Social Critique of the Judgement of Taste.* Translated by R. Nice. Cambridge, MA: Harvard University Press.

———. 1986. "The Forms of Capital." In *Handbook of Theory and Research for the Sociology of Education,* edited by John G. Richardson. Westport, CT: Greenwood.

Bradshaw, York W. and Michael Wallace. 1996. *Global Inequalities.* Thousand Oaks, CA: Pine Forge.

Brodkin, Karen. 1998. *How Jews Became White Folks and What That Says about Race in America.* New Brunswick, NJ: Rutgers University Press.

Brooks, David. 2000. *Bobos in Paradise: The New Upper Class and How They Got There.* New York: Simon & Schuster.

Brooks, Thomas. 1971. *Toil and Trouble: A History of American Labor.* 2d ed. New York: Dell.

Bukharin, Nikolay. [1921] 1924. *Historical Materialism: A System of Sociology.* New York: International.

———. [1917] 1973. *Imperialism and the World Economy.* New York: Monthly Review Press.

Bullard, Robert D. 1990. *Dumping in Dixie: Race, Class, and Environmental Quality.* Boulder, CO: Westview.

———. 1993. *Confronting Environmental Racism: Voices from the Grassroots.* Boston: South End.

Burawoy, Michael. 1979. *Manufacturing Consent: Changes in the Labor Process under Monopoly Capitalism.* Chicago: University of Chicago.

Burch, Philip H., Jr. 1980. *Elites in American History: The New Deal to the Carter Administration.* New York: Holmes & Meier.

Cardoso, Fernando Henrique and Enzo Falletto. 1979. *Dependency and Development in Latin America.* Berkeley: University of California Press.

Cassidy, John. 1997. "The Return of Karl Marx." *New Yorker,* October 20–27, pp. 248–59.

Center for Responsive Politics. 2000. "The Big Picture." Retrieved from http://www.opensecrets.org

Chambliss, William. 1973. "The Saints and the Roughnecks." *Society* 11:24–31.

Clinton, Hillary Rodham. 1995. *Remarks for United Nations Fourth World Conference on Women.* New York: United Nations.

CNN Election 2004. Retrieved from www.cnn.com/ELECTION/pages/results/states/US/P/00/epolls.0.html, November 4, 2004.

Cohen, Mark Nathan. 1997. "Culture, Rank and IQ: The Bell Curve Phenomenon." Pp. 252-58 in *Conformity and Conflict,* 9th ed., edited by James Spradley and David McCurdy. New York: Longman.

Coleman, James S. 1992. "Some Points on Choice in Education." *Sociology of Education* 65:260–62.

Coleman, Richard P. and Lee Rainwater (with Kent A. McClelland). 1978. *Social Standing in America: New Dimensions of Class.* New York: Basic Books.

Collins, Randall. 1977. "Some Comparative Principles of Educational Stratification." *Harvard Educational Review* 47:1–27.

Coontz, Stephanie. 1992. *The Way We Never Were: American Families and the Nostalgia Trap.* New York: Basic Books.

Cose, Ellis. 1993. *The Rage of a Privileged Class.* New York: HarperCollins.

Dahl, Robert. 1961. *Who Governs? Democracy and Power in an American City.* New Haven, CT: Yale University Press.

Dahrendorf, Ralf. 1959. *Class and Class Conflict in Industrial Society.* Stanford, CA: Stanford University Press.

Davis, Allison, Burleigh B. Gardner, and Mary R. Gardner. 1941. *Deep South: A Socio-Anthropological Study of Caste and Class.* Chicago: University of Chicago Press.

Davis, Kingsley and Wilbert Moore. 1945. "Some Principles of Stratification." *American Sociological Review* 10:242–49.

Diamond, Jared. 1997. *Guns, Germs, and Steel: The Fates of Human Societies.* New York: W. W. Norton.

Dillon, Sam. 2004. "Chicago Has a Nonunion Plan for Poor Schools." *New York Times,* July 28. Retrieved from http://www.nytimes.com/2004/07/28/education/28chicago.html

Domhoff, G. William. 1967. *Who Rules America?* Englewood Cliffs, NJ: Prentice Hall.

———. 1974. *The Bohemian Grove and Other Retreats: A Study in Ruling-Class Cohesiveness.* New York: Harper & Row.

———. 1978. *Who Really Rules? New Haven and Community Power Reexamined.* New Brunswick, NJ: Transaction.

Dore, Ronald. 1976. *The Diploma Disease: Education, Qualification, and Development.* Berkeley: University of California Press.

Douglass, Frederick. 1876. "Oration in Memory of Abraham Lincoln." Retrieved from http://teachingamericanhistory.org/library/index.asp?documentprint=39

Du Bois, W. E. B. 1903. *The Souls of Black Folk.* New York: Dover.

Dugger, Celia. 2004. "Deserted by Doctors, India's Poor Turn to Quacks." *New York Times,* March 25.

Duncan, Cynthia. 1992. *Rural Poverty in America.* Westport, CT: Auburn House.

———. 1999. *Worlds Apart: Why Poverty Persists in Rural America.* New Haven, CT: Yale University Press.

Duneier, Mitch. 1992. *Slim's Table: Race, Respectability, and Masculinity.* Chicago: University of Chicago Press.

Durkheim, Émile. [1895] 1964. *The Division of Labor in Society.* New York: Free Press.

Edin, Kathryn and Laura Lein. 1997. *Making Ends Meet: How Single Mothers Survive Welfare and Low-Wage Work.* New York: Russell Sage.

Education Commission of the States. 2004. "No Child Left Behind." Retrieved from http://nclb2.ecs.org/Projects_Centers.htm

Edwards, Mike. 1997. "Boom Times on the Gold Coast of China." *National Geographic,* March.

Ehrenreich, Barbara and Annette Fuentes. 1981. "Life on the Global Assembly Line." *Ms.,* January.

Ellwood, David. 1988. *Poor Support: Poverty in the American Family.* New York: Basic Books.

Falk, William W. and Thomas A. Lyson. 1988. *High Tech, Low Tech, and No Tech: Recent Industrial and Occupational Change in the South.* Albany: State University of New York Press.

Faludi, Susan. 1991. *Backlash: The Undeclared War against American Women.* New York: Crown.

Fantasia, Rick. 1988. *Cultures of Solidarity: Consciousness, Action, and Contemporary American Workers.* Berkeley: University of California Press.

Feagin, Joe R. 1972. "Poverty: We Still Believe That God Helps Those Who Help Themselves." *Psychology Today,* November, pp. 101–29.

Featherman, David L. and Robert M. Hauser. 1978. *Opportunity and Change.* New York: Academic Press.

Fernandez-Kelly, Patricia. 1983. *For We Are Sold, I and My People: Women and Industry in Mexico's Frontier.* Albany: State University of New York Press.

Firebaugh, Glenn. 2003. *The New Geography of Global Income Inequality.* Cambridge, MA: Harvard University Press.

Fitchen, Janet M. 1981. *Poverty in Rural America: A Case Study.* Boulder, CO: Westview.

———. 1991. *Endangered Spaces, Enduring Places: Change, Identity, and Survival in Rural America.* Boulder, CO: Westview.

Florida, Richard. 2002. *The Rise of the Creative Class: And How It's Transforming Work, Leisure, Community, and Everyday Life.* New York: Basic Books.

———. 2004. *Cities and the Creative Class.* New York: Routledge.

Foner, Eric and John A. Garraty, eds. 1991. *The Reader's Companion to American History.* Boston: Houghton Mifflin. Retrieved from http://college.hmco.com/history/readerscomp/rcah/html

Food and Agriculture Organization of the United Nations. 2000. Data retrieved August 20, 2000, from http://www.fao.org

Forbes. 1996. "The Forbes 400." October.

———. 1999. "The Forbes 400." October.

———. 2004. "The Forbes 400." October.

Frank, Andre Gunder. 1967. *Capitalism and Development in Latin America.* New York: Monthly Review Press.

Friedan, Betty. 1963. *The Feminine Mystique.* New York: W. W. Norton.

Galbraith, John Kenneth. 1958. *The Affluent Society.* Boston: Houghton Mifflin.

Gans, Herbert J. 1990. "Deconstructing the Underclass: The Term's Danger as a Planning Concept." *Journal of the American Planning Association* 56:271–77.

———. 1995. *The War against the Poor: The Underclass and Antipoverty Policy.* New York: Basic Books.

General Accounting Office, Report 01-128. "Gender Equity in Higher Education. Retrieved from www.gao.gov/new.items/d01128.pdf

Gerth, Hans H. and C. Wright Mills. 1946. *From Max Weber: Essays in Sociology.* New York: Oxford University Press.

Giddens, Anthony. 1999. *The Third Way: The Renewal of Social Democracy.* Oxford: Blackwell.

———. 2000. *Runaway World: How Globalization Is Reshaping Our Lives.* London: Routledge.

Gilbert, Dennis. 1998. *The American Class Structure in an Age of Growing Inequality.* 5th ed. Belmont, CA: Wadsworth.

———. 2003. *The American Class Structure in an Age of Growing Inequality.* 6th ed. Belmont, CA: Wadsworth.

Gilbert, Dennis and Joseph Kahl. 1982. *The American Class Structure: A New Synthesis.* Homewood, IL: Dorsey.

Gilman, Charlotte Perkins. 1910. "Our Androcentric Culture." *Forerunner* 1(3).

Goertzel, Ted. 1997. "President Fernando Cardoso Reflects on Brazil and Sociology." *Footnotes* (publication of the American Sociological Association), November.

Goffman, Erving. 1979. *Gender advertisements.* Cambridge, MA: Harvard University Press.

Goode, William. 1992. "Why Men Resist." In *Rethinking the Family: Some Feminist Questions,* rev. ed., edited by Barrie Thorne and Marilyn Yalom. Boston: Northeastern University Press.

Greeley, Andrew. 1972. *That Most Distressful Nation.* Chicago: Quadrangle.

Greenhouse, Steven. 2000. "Nike Chair Cancels Gift to Alma Mater: Sweatshop Position Irks Shoe Billionaire." *San Francisco Chronicle,* April 25.

Guruswamy, Krishnan. 1995. "World's Waste Is Piling Up in India." *South Bend Tribune,* June 4.

Gutmann, Matthew C. 1996. *The Meanings of Macho: Being a Man in Mexico City.* Berkeley: University of California Press.

Hallinan, Maureen. T. 1994. "Tracking from Theory to Practice." *Sociology of Education* 67:79–84.

Harrington, Michael. 1962. *The Other America: Poverty in the United States.* New York: Macmillan.

Harris, Marvin. 1989. *Our Kind.* New York: HarperCollins.

Hayden, Jeffrey. 1996. *Children in America's Schools* [Film]. Los Angeles: Saint/Hayden; Columbia: South Carolina ETV.

Hechter, Michael. 1975. *Internal Colonialism.* Berkeley: University of California Press.

Hennig, Margaret and Anne Jardim. 1977. *The Managerial Woman.* New York: Doubleday.

Herrnstein, Richard J. and Charles Murray. 1994. *The Bell Curve: Intelligence and Class Structure in American Life.* New York: Free Press.

Hochschild, Arlie R. (with Anne Machung). 1989. *The Second Shift: Working Parents and the Revolution at Home.* New York: Penguin.

Hochschild, Arlie R. 1997. *The Time Bind: When Work Becomes Home and Home Becomes Work.* New York: Metropolitan.

Hoover Institution. 2004. "Campaign Finance Reform Legislation." Retrieved from www.campaignfinancesite.org/legislation/mccain.html

Ibarra, H. 1992. "Homophily and Differential Returns: Sex Differences in Network Structure and Access in an Advertising Firm." *Administrative Science Quarterly* 37:422–47.

Inglehart, Ronald and Wayne E. Baker. 2000. "Modernization, Cultural Change, and the Persistence of Traditional Values." *American Sociological Review* 65:19–51.

Isbister, John. 1998. *Promises Not Kept: The Betrayal of Social Change in the Third World.* West Hartford, CT: Kumarian.

Iyer, Pico. 2000. "Citizen Nowhere." *Civilization,* February/March.

Jaffee, David. 1998. *Levels of Socio-Economic Development Theory.* West Hartford, CT: Praeger.

James Madison Center. 2004. "Benjamin Franklin on Native Americans." Retrieved from http://www.jmu.edu/madison/center/main_pages/madison_archives/era/native/franklin.htm

Jencks, Christopher. 1972. *Inequality: A Reassessment of the Effect of Family and Schooling in America.* New York: Basic Books.

———. 1979. *Who Gets Ahead? The Determinants of Economic Success in America.* New York: Basic Books.

———. 1994. *The Homeless.* Cambridge, MA: Harvard University Press.

Kanter, Rosabeth Moss. 1977. *Men and Women of the Corporation.* New York: Basic Books.

Kaplan, Robert D. 1998. "Travels into America's Future: Mexico and the Southwest." *Atlantic Monthly,* July.

Kennickell, A. B. (2003). *A Rolling Tide: Changes in the Distribution of Wealth in the U.S., 1989-2001,* Annandale-on-Hudson, NY: Levy Economics Institute, Bard College.

Kilbourne, Jean. 2000. *Killing Us Softly 3: Advertising's Image of Women* [Video]. Directed and produced by Sut Jhally. Northampton, MA: Media Education Foundation.

King, Martin Luther, Jr. [1967] 1992. "Black Power Defined." In *I Have a Dream: Writings and Speeches That Changed the World,* edited by James W. Washington. New York: HarperCollins.

Klinker, Philip. 2004. "Money Matters." *New Republic,* November 10. Retrieved from http://www.tnr.com/doc.mhtml

Kohn, Melvin L. 1977. *Class and Conformity.* 2d ed. Chicago: University of Chicago Press.

Korten, David C. 1995. *When Corporations Rule the World.* West Hartford, CT: Kumarian.

Kotlowitz, Alex. 1998. *The Other Side of the River: A Story of Two Towns, a Death, and America's Dilemma.* New York: Doubleday.

Kozol, Jonathan. 1991. *Savage Inequalities: Children in America's Schools.* New York: Crown.

———. 1995. *Amazing Grace: The Lives of Children and the Conscience of a Nation.* New York: Crown.

Kuznets, Simon. 1955. "Economic Growth and Income Inequality." *American Economic Review* 45:1–28.

Lachman, S. P., and Kosmin, B. A. 1993. *One Nation Under God.* New York: Harmony Books, a division of Random House, Inc.

Lamont, Michèle. 1992. *Money, Morals, and Manners: The Culture of the French and the American Upper-Middle Class.* Chicago: University of Chicago Press.

LeDuff, Charlie. 2001. "At a Slaughterhouse, Some Things Never Die." In *How Race Is Lived in America: Pulling Together, Pulling Apart,* by correspondents of the *New York Times.* New York: New York Times Books.

———. 2004. "Mexican Americans Struggle for Jobs." *New York Times,* October 13. Retrieved from http://www.nytimes.com/2004/10/13/national/13jobs.html

Lenin, Vladimir I. [1917] 1948. *Imperialism: The Highest Stage of Capitalism.* London: Lawrence & Wishart.

Lenski, Gerhard. 1966. *Power and Privilege: A Theory of Stratification.* New York: McGraw-Hill.

Lenski, Gerhard and Patrick Nolan. 1984. "Trajectories of Development: A Test of Ecological-Evolutionary Theory." *Social Forces* 63:1–23.

Lewis, Oscar 1961. *Children of Sanchez.* New York: Random House.

———. 1968. "The Culture of Poverty." In *On Understanding Poverty,* edited by Daniel Patrick Moynihan. New York: Basic Books.

Lieberson, Stanley. 1980. *A Piece of the Pie: Blacks and White Immigrants since 1880.* Berkeley: University of California Press.

Lipman, Mark and Leah Mahan. 1996. *Holding Ground: The Rebirth of Dudley Street* [Film]. Distributed by New Day Films, Harriman, NY.

Lipset, Seymour Martin. 1996. *American Exceptionalism: A Double-Edged Sword.* New York: W. W. Norton.

Loewen, James W. 1988. *The Mississippi Chinese: Between Black and White.* Prospect Heights, IL: Waveland.

Lomnitz, Larissa Adler. 1977. *Networks and Marginality: Life in a Mexican Shantytown.* New York: Academic Press.

Lubiano, Wahneema, ed. 1997. *The House That Race Built.* New York: Pantheon.

Lucal, Betsy. 1996. "Oppression and Privilege: Toward a Relational Concep-tualization of Race." *Teaching Sociology* 24:245–55.

Lurie, Alison. [1981] 2000. *The Language of Clothes*. New York: Owl.

Lynd, Robert S. and Helen Merrell Lynd. 1929. *Middletown: A Study in American Culture*. New York: Harcourt, Brace.

Lyson, Thomas A. 1989. *Two Sides to the Sunbelt: The Growing Divergence between the Rural and Urban South*. New York: Praeger.

MacLeod, Jay. 1995. *Ain't No Makin' It: Aspirations and Attainment in a Low-Income Neighborhood*. Boulder, CO: Westview.

Massey, Douglas S. and Nancy A. Denton. 1993. *American Apartheid: Segregation and the Making of the Underclass*. Cambridge, MA: Harvard University Press.

McGuire, Gail. 2002. "Gender, Race, and the Shadow Structure: A Study of Infor-mal Networks and Inequality in a Work Organization." *Gender & Society* 16: 303–322.

McIntosh, Peggy. 1995. "White Privilege and Male Privilege." In *Race, Class, and Gender: An Anthology*, 2d ed., edited by Margaret L. Andersen and Patricia Hill Collins. Belmont, CA: Wadsworth.

McLellan, David. 1977. *Karl Marx: Selected Writings*. Oxford: Oxford University Press.

———. 1988. *Marxism: Essential Writings*. Oxford: Oxford University Press.

McLuhan, Marshall. 1964. *Understanding Media: The Extensions of Man*. New York: Mentor.

McMichael, Philip. 2000. *Development and Social Change: A Global Perspective*. 2d ed. Thousand Oaks, CA: Pine Forge.

Medoff, Peter and Holly Sklar. 1994. *Streets of Hope: The Fall and Rise of an Urban Neighborhood*. Boston: South End.

Miller, D. T. and Michael Nowak. 1977. *The Fifties: The Way We Really Were*. Garden City, NY: Doubleday.

Mills, C. Wright. 1951. *White Collar*. New York: Oxford University Press.

———. 1956. *The Power Elite*. New York: Oxford University Press.

———. 1959. *The Sociological Imagination*. New York: Oxford University Press.

Mishel, Lawrence, Jared Bernstein, and John Schmitt. 1997. *The State of Working America 1996–97*. Armonk, NY: M. E. Sharpe.

Murray, Charles A. 1984. *Losing Ground: American Social Policy*. New York: Basic Books.

Myrdal, Gunnar. 1944. *An American Dilemma*. New York: Harper.

———. 1970. *The Challenge of World Poverty*. New York: Vintage.

Nakao, Keiko, and Treas, Judith. 1994. "Updating Occupational Prestige and Socioeconomic Scales: How the New Measures Measure Up." *Sociological Methodology* 24:1–72.

National Coalition for the Homeless. 2004. "Who Is Homeless?" (Fact Sheet 3). Retrieved from http://www.nationalhomeless.org/who.html

National Education Association. 2004. "No Child Left Behind Act/ESEA." Retrieved from http://home.nea.org/www/htmlmail.cfm

Nee, Victor, Jimy Sanders, and Scott Sernau. 1994. "Job Transitions in an Immigrant Metropolis: Ethnic Boundaries and Mixed Economy." *American Sociological Review* 59:849–72.

Newman, Katherine S. 1988. *Falling from Grace: The Experience of Downward Mobility in the American Middle Class*. New York: Free Press. *New York Times*. 2000. "Thaw Affects Pole." August 18.

"No Ice at Pole." *New York Times* (2000, August 18), p. 5.

Oakes, Jeannie. 1994. "More than Misapplied Technology: A Normative and Political Response to Hallinan on Tracking." *Sociology of Education* 67:84–89.

Oliver, Melvin L. and Thomas M. Shapiro. 1995. *Black Wealth/White Wealth: A New Perspective on Racial Inequality.* New York: Routledge.

Onishi, Norimitsu. 2000. "In the Oil Rich Nigeria Delta, Deep Poverty and Grim Fires." *New York Times,* August 11.

Parkin, Frank. 1979. *Marxism and Class Theory: A Bourgeois Critique.* New York: Columbia University Press.

Peña, Devon G. 1997. *The Terror of the Machine: Technology, Work, Gender, and Ecology on the U.S.-Mexico Border.* Austin: University of Texas Press.

Perlmann, Joel. 1988. *Ethnic Differences: Schooling and Social Structure among the Irish, Italians, Jews, and Blacks in an American City, 1880–1935.* Cambridge: Cambridge University Press.

Physician Task Force. 1985. *Hunger in America: The Growing Epidemic.* Middletown, CT: Wesleyan University Press.

Pitts, Leonard, Jr. 2004. "Where Is the Morality in Bush's Policy?" *Miami Herald,* November 8.

Porrit, Jonathan. 1991. *Save the Earth.* London: Dorling Kindersley.

Portes, Alejandro, Manuel Castells, and Lauren A. Benton, eds. 1989. *The Informal Economy: Studies in Advanced and Less Developed Countries.* Baltimore: Johns Hopkins University Press.

Portes, Alejandro and Ruben Rumbaut. 1990. *Immigrant America: A Portrait.* Berkeley: University of California Press.

Prebisch, Raul. 1950. *The Economic Development of Latin America and Its Principle Problems.* New York: United Nations.

Rebora, Anthony. 2004. "No Child Left Behind." *Education Week,* July 28. Retrieved from http://www.edweek.org/context/topics/issuespage.cfm

Reiman, Jeffrey. 1998. *The Rich Get Richer and the Poor Get Prison: Ideology, Class, and Criminal Justice.* 5th ed. Boston: Allyn & Bacon.

Riesman, David (with Nathan Glazer and Reuel Denney). 1953. *The Lonely Crowd: A Study of the Changing American Character.* New Haven, CT: Yale University Press.

Robinson, Eugene. 1999. *Coal to Cream: A Black Man's Journey beyond Color to an Affirmation of Race.* New York: Free Press. (Excerpted in *New York Times,* September 17, 1999; retrieved from http://www.nytimes.com/books/first/r/robinson-coal.html)

Rose, Stephen J. 2000. *Social Stratification in the United States: The New American Profile Poster.* Rev. ed. New York: New Press.

Rosen, Bernard. 1982. *The Industrial Connection.* New York: Aldine.

Rosenthal, Elisabeth. 2001. "Without 'Barefoot Doctors,' China's Rural Families Suffer." *New York Times,* March 14.

Rostow, W. W. 1960. *The Stages of Economic Growth: A Non-Communist Manifesto.* Cambridge: Cambridge University Press.

Rowe, Claudia. 1999. "Saving Children from Sweatshops: One Teen's Crusade." *Biography Magazine,* November.

Rubin, Beth A. 1996. *Shifts in the Social Contract: Understanding Change in American Society.* Thousand Oaks, CA: Pine Forge.

Rubin, Lillian B. 1994. *Families on the Fault Line*. New York: HarperCollins.

Ryan, William. 1971. *Blaming the Victim*. New York: Vintage.

Sahlins, Marshall. 1972. *Stone Age Economics*. Chicago: Aldine.

Salzman, Harold, and Domhoff, G. William. 1983. "Nonprofit Organizations and the Corporate Community, *Social Science History* 7: 205–216.

Sanders, Jimy and Victor Nee. 1987. "Limits of Ethnic Solidarity in the Enclave Economy." *American Sociological Review*. 52:745–73.

Scarr, Sandra, Deborah Phillips, and Kathleen McCartney. 1989. "Working Mothers and Their Families." *American Psychologist* 44:1402–9.

Schanberg, Sydney H. 1996. "Six Cents an Hour." *Life,* June.

Schor, Juliet B. 1998. *The Overspent American: Why We Want What We Don't Need*. New York: HarperPerennial.

Schumpeter, Joseph. 1949. *Change and the Entrepreneur*. Cambridge, MA: Harvard University Press.

Schwartz, Felice N. 1989. "Management, Women, and the New Facts of Life." *Harvard Business Review* 67(3):65–76.

Sen, Amartya. 1999. *Development as Freedom*. Oxford: Oxford University Press.

Sernau, Scott. 1993. "School Choices: Rational and Otherwise: A Comment on Coleman." *Sociology of Education* 66:88–90.

———. 1994. *Economies of Exclusion: Underclass Poverty and Labor Market Change in Mexico*. Westport, CT: Praeger.

———. 1996. "Economies of Exclusion: Economic Change and the Global Underclass." *Journal of Developing Societies* 12:38–51.

———. 2000. *Bound: Living in the Globalized World*. West Hartford, CT: Kumarian.

Sidel, Ruth. 1996. *Keeping Women and Children Last: America's War on the Poor*. New York: Penguin.

Simmel, Georg. [1904] 1957. "Fashion." *American Journal of Sociology* 62(May).

Skolnick, Arlene S. 1996. *The Intimate Environment*. 6th ed. New York: HarperCollins.

Snipp, Matthew. 1986. "The Changing Political and Economic Status of the American Indians: From Captive Nations to Internal Colonies." *American Journal of Economics and Sociology* 45:145–57.

Sorensen, Jesper B. and David Grusky. 1999. "Can Class Analysis Be Salvaged?" *American Journal of Sociology* 103:1187–1234.

South, Scott and Glenna Spitze. 1994. "Housework in Marital and Non-Marital Households." *American Sociological Review* 59:327–47.

Stack, Carol. 1974. *All Our Kin: Strategies for Survival in a Black Community*. New York: Harper & Row.

Stanley, Thomas J. 1996. *The Millionaire Next Door: The Surprising Secrets of America's Wealthy*. Atlanta, GA: Longstreet.

Stapinski, Helene. 1998. "Let's Talk Dirty." *American Demographics* 20(11):50–56.

Stark, Rodney. 1996. *Sociology*. 6th ed. Belmont, CA: Wadsworth.

Steinbeck, John. 1939. *The Grapes of Wrath*. New York: Modern Library.

Steinberg, Stephan. 1981. *The Ethnic Myth: Race, Ethnicity, and Class in America*. Boston: Beacon.

Stephens, W. N. 1963. *The Family in Cross-Cultural Perspective*. New York: Holt, Rinehart & Winston.

Sweezy, Paul. 1968. "Power Elite or Ruling Class?" In *C. Wright Mills and the Power Elite*, edited by G. William Domhoff and Hoyt B. Ballard. Boston: Beacon.

Takaki, Ronald. 1993. *A Different Mirror: A History of Multicultural America.* Boston: Little, Brown.

Takaki, Ronald. 1994. *From Different Shores: Perspectives on Race and Ethnicity in America.* 2d ed. New York: Oxford University Press.

Terkel, Studs. 1992. *Race.* New York: New Press.

Trebay, Guy. 2000. "Shopping the Madison Avenue of Manhasset." *New York Times,* July 25.

Tuan, Mia. 1998. *Forever Foreigners or Honorary Whites? The Asian Ethnic Experience Today.* New Brunswick, NJ: Rutgers University Press.

Tuma, Nancy Brandon and Michael T. Hannan. 1984. *Social Dynamics: Methods and Models.* Orlando, FL: Academic Press.

Tumin, Melvin M. 1953. "Some Principles of Stratification: A Critical Analysis." *American Sociological Review* 18:387–94.

UNICEF. *End of Decade Database Report on Basic Education.* Retrieved from www.childinfo.org/eddb.edu/index.htm.

UNICEF. The Progress of Nations. 1996. Report on the Industrial World. "Safety nets for children are weakest in US." Retrieved from www.unicef.org/pop96/indust4.htm

UNICEF. 2004, January. "Basic Education" at www.childinfo.org.

United Nations Development Program. 1996. *Human Development Report 1996: Economic Growth and Human Development.* New York: Oxford University Press.

———. 1998. *Human Development Report 1998: Consumption for Human Development.* New York: Oxford University Press.

———. 2000. *Human Development Report 2000: Human Rights and Human Development.* New York: Oxford University Press.

———. 2003. *Human Development Report 2003: Millennium Development Goals: A Compact among Nations to End Human Poverty.* New York: Oxford University Press.

———. 2004. *Human Development Report 2004: Cultural Liberty in Today's Diverse World.* New York: Oxford University Press.

Urrea, Luis Alberto. 1996. *By the Lake of Sleeping Children.* New York: Doubleday.

U.S. Bureau of Labor Statistics. 2004a. *American Time-Use Survey Summary.* Retrieved from http://www.bls.gov/news.release/atus.nr0.htm

U.S. Bureau of Labor Statistics, Employment Statistics Program. 2004b. "Occupational Employment and Wages." News release USDL 04-752.

U. S. Census Bureau Bicentennial Statistics (1976).

U.S. Census Bureau. *Statistical Abstract of the United States* 1999.

U.S. Census Bureau. 2002. American Community Survey.

U.S. Census Bureau, *Statistical Abstract of the United States,* 2003.

U.S. Census Bureau. 2004, November. Current Population Survey, "Voting and Registration in the Election of November 2000."

U.S. Census Bureau. 2005. Historical Income Data, Current Population Survey Tables, F-1, F-3. www.census.gov/hhes/income.

U.S. Department of Agriculture, Economic Research Service. 2004, July. "Rural Poverty at a Glance." Rural Development Research Report Number 100.

U.S. Department of Education, National Center for Education Statistics. 2002. *Digest of Education Statistics.*

U.S. Department of Labor. (1994, June). *Bureau of Labor Statistics News,* Table A-6, Selected employment indicators.

U.S. Department of Labor. 2001. *Employment and Earnings.*

U.S. Department of Labor, Bureau of Labor Statistics. (1988). *Occupational Outlook Handbook.* Washington, DC: Government Printing Office. Also available at www.bls.gov.

U.S. Department of Labor, Bureau of Labor Statistics. (2004, June). *Occupational Outlook Handbook.* Washington, DC: Government Printing Office. Also available at www.bls.gov.

U.S. Department of Labor Report 972 (2003, September). "Highlights of Women's Earnings in 2002," Chart 1.

U.S. Immigration and Naturalization Service. (1997). Statistical Yearbook.

U.S. Immigration and Naturalization Service. (2000). Statistical Yearbook.

U.S. Immigration and Naturalization Service. (2004). Statistical Yearbook.

U.S. National Center for Educational Statistics. 1999. *Digest of Educational Statistics.* Washington, DC: Government Printing Office.

Valentine, Charles A. 1968. *Culture and Poverty: Critique and Counterproposals.* Chicago: University of Chicago Press.

Van Natta, Don, Jr. and John M. Broder. 2000. "The Few, the Rich, the Rewarded Donate the Bulk of G.O.P. Gifts." *New York Times,* August 2.

Veblen, Thorstein. [1899] 1953. *The Theory of the Leisure Class.* New York: New American Library.

Waldinger, Roger. 1986. *Through the Eye of the Needle.* New York: New York University Press.

Waldinger, Roger and Michael I. Lichter. 2003. *How the Other Half Works: Immigration and the Social Organization of Labor.* Berkeley: University of California Press.

Wallerstein, Immanuel. 1974. *The Modern World System.* New York: Academic Press.

Walton, John. 1970. "A Systematic Survey of Community Power Research." In *The Structure of Community Power,* edited by Michael Aiken and Paul E. Mott. New York: Random House.

Washington Post. 1990. "Sentencing Challenged." May 27.

———. 1997. "They're Not Loud, They're Not Proud: Second National Conference on Whiteness." November 30.

———. 2004. "Food Stamp Use Common." August 24.

Weaver, James H., Michael T. Rock, and Kenneth Kusterer. 1997. *Achieving Broad-Based Sustainable Development.* West Hartford, CT: Kumarian.

Weber, Max. [1920] 1964. *The Theory of Social and Economic Organization.* Translated by A. M. Henderson and Talcott Parsons. Glencoe, IL: Free Press.

———. [1922] 1979. *Economy and Society.* 2 vols. Berkeley: University of California Press.

———. [1905] 1997. *The Protestant Ethic and the Spirit of Capitalism.* Los Angeles: Roxbury.

Weiner, Myron. 1966. *Modernization: The Dynamics of Growth.* New York: Basic Books.

Whyte, William H., Jr. 1956. *The Organization Man.* New York: Doubleday.

Wilson, Kenneth L. and Alejandro Portes. 1980. "Immigrant Enclaves: An Analysis of the Labor Market Experiences of Cubans in Miami." *American Journal of Sociology* 86:295–315.

Wilson, William Julius. 1978. *The Declining Significance of Race: Blacks and Changing American Institutions.* Chicago: University of Chicago Press.

Wilson, William Julius. 1987. *The Truly Disadvantaged: The Inner City, the Underclass, and Public Policy.* Chicago: University of Chicago Press.

———. 1996. *When Work Disappears: The World of the New Urban Poor.* New York: Alfred A. Knopf.

World Bank. 2000. *World Development Report 2000/2001: Attacking Poverty.* Washington, DC: World Bank.

———. 2002. *World Development Report 2003: Sustainable Development in a Dynamic World: Transforming Institutions, Growth, and Quality of Life.* Washington, DC: World Bank.

Wright, Erik Olin. 1985. *Classes.* New York: Schocken.

———. 1997. *Class Counts: Comparative Studies in Class Analysis.* New York: Cambridge University Press.

Wright, Erik Olin and Luca Perrone. 1977. "Marxist Class Categories and Income Inequality." *American Sociological Review* 42:32–55.

Yates, Diana. 1992. *Chief Joseph: Thunder Rolling Down from the Mountains.* New York: Ward Hill.

Zinn, Maxine Baca. 1989. "Family, Race, and Poverty in the Eighties." *Signs* 14:856–74.

Glossary/Index

About the Author

Scott Sernau (Ph.D. Cornell University) is Professor of Sociology at Indiana University South Bend, where he regularly teaches courses on social inequality as well as international inequalities, sociology of family, urban society, and race and ethnic relations. He is the author of *Economies of Exclusion: Underclass Poverty and Labor Market Change*, *Critical Choices: Applying Sociological Insight*, and *Bound: Living in the Globalized World*. For the past decade, he has edited the American Sociological Association Teaching Resource Center instructors' guide for social stratification courses. He has won numerous campus and statewide teaching awards, including the Sylvia Bowman Award for distinguished teaching on American society, and he serves on the IU Faculty Colloquium on Excellence in Teaching.